NATO ASI Series

Advanced Science Institutes Series

A series presenting the results of activities sponsored by the NATO Science Committee, which aims at the dissemination of advanced scientific and technological knowledge, with a view to strengthening links between scientific communities.

The Series is published by an international board of publishers in conjunction with the NATO Scientific Affairs Division

A Life Sciences	Plenum Publishing Corporation
B Physics	London and New York
C Mathematical and Physical Sciences	Kluwer Academic Publishers
D Behavioural and Social Sciences	Dordrecht, Boston and London
E Applied Sciences	
F Computer and Systems Sciences	Springer-Verlag
G Ecological Sciences	Berlin Heidelberg New York
H Cell Biology	London Paris Tokyo Hong Kong
I Global Environmental Change	Barcelona Budapest

PARTNERSHIP SUB-SERIES

1. Disarmament Technologies	Kluwer Academic Publishers
2. Environment	Springer-Verlag/Kluwer Academic Publishers
3. High Technology	Kluwer Academic Publishers
4. Science and Technology Policy	Kluwer Academic Publishers
5. Computer Networking	Kluwer Academic Publishers

The Partnership Sub-Series incorporates activities undertaken in collaboration with NATO's Cooperation Partners, the countries of the CIS and Central and Eastern Europe, in Priority Areas of concern to those countries.

NATO-PCO DATABASE

The electronic index to the NATO ASI Series provides full bibliographical references (with keywords and/or abstracts) to about 50000 contributions from international scientists published in all sections of the NATO ASI Series. Access to the NATO-PCO DATABASE compiled by the NATO Publication Coordination Office is possible in two ways:

- via online FILE 128 (NATO-PCO DATABASE) hosted by ESRIN, Via Galileo Galilei, I-00044 Frascati, Italy.

- via CD-ROM "NATO Science & Technology Disk" with user-friendly retrieval software in English, French and German (© WTV GmbH and DATAWARE Technologies Inc. 1992).

The CD-ROM can be ordered through any member of the Board of Publishers or through NATO-PCO, Overijse, Belgium.

Series H: Cell Biology, Vol. 101

Springer
Berlin
Heidelberg
New York
Barcelona
Budapest
Hong Kong
London
Milan
Paris
Santa Clara
Singapore
Tokyo

Molecular Mechanisms of Signalling and Membrane Transport

Edited by

Karel W. A. Wirtz

Centre for Biomembranes and Lipid Enzymology
Utrecht University, Padualaan 8
3584 TB Utrecht, The Netherlands

With 103 Figures

Springer
Published in cooperation with NATO Scientific Affairs Division

Proceedings of the NATO Study Institute on Molecular Mechanisms of Signalling and Targeting, held on the Island of Spetsai, Greece, August 18–30, 1996

Library of Congress Cataloging-in-Publication Data applied for

Die Deutsche Bibliothek - CIP-Einheitsaufnahme

Molecular mechanisms of signalling and membrane transport : [proceedings of the NATO Advanced Study Institute on Molecular Mechanisms of Signalling and Targeting, held on the Island of Spetsai, Greece, August 18 - 30, 1996] / ed. by Karel W. A. Wirtz. Publ. in cooperation with NATO Scientific Affairs Division. - Berlin ; Heidelberg ; New York ; Barcelona ; Budapest ; Hong Kong ; London ; Milan ; Paris ; Santa Clara ; Singapore ; Tokyo : Springer, 1997
(NATO ASI series : Ser. H, Cell biology ; Vol. 101)

ISBN-13: 978-3-642-64559-4 e-ISBN-13: 978-3-642-60799-8
DOI: 10.1007/978-3-642-60799-8

Softcover reprint of the hardcover 1st edition 1997

Typesetting: Camera ready by authors/editors
Printed on acid-free paper
SPIN 10525507 31/3137 - 5 4 3 2 1 0

PREFACE

A NATO Advanced Study Institute on "Molecular Mechanisms of Signalling and Targeting" was held on the Island of Spetsai, Greece, from August 18–30, 1996. As a continuation of a successful tradition that began in 1978, the aim of this tenth institute was to focus on some main principles of signal transduction mechanisms, with an emphasis on GTP-binding proteins, lipid signalling mechanisms, transcription factors, and membrane transporters. By bringing experts on these diverse topics together and into contact with an outstanding group of young and eager researchers, this Institute succeeded in generating an atmosphere of learning and of intellectual creativity. Students and lecturers returned for home with a deeper understanding of the challenging problems we are facing in this field.

Presentations and discussion focussed on the structure and function of heterotrimeric G-proteins and their receptors as potential targets for therapeutic drugs. This field was put into perspective by in-depth discussions on the function of the small GTP-binding proteins, Ras, Rho, Rab, Arf in signal transduction and morphogenesis. Growth factor receptor, Ca^{2+} and protein kinase C signal transduction pathways were discussed as leading principles in intracellular and intercellular communication. At the core of the Institute were the wide implications of the regulation of phosphoinositide 3OH-kinases (*e.g.* PI-3-kinase) and phospholipases C and D for cell function and the role of ceramides as modulators of stress response (apoptosis). These important principles of signalling mechanisms were extended to transcription factors and gene activation, to the relationship between antioxidants and NF-κB-activation, and to multidrug resistance transporters.

This book presents the content of the major lectures and a selection of the most relevant posters. These proceedings offer a comprehensive account of the most important topics discussed at the Institute.

February 1997 The Editor

CONTENTS

HETEROTRIMERIC AND SMALL G-PROTEINS

INTRA- AND INTERCELLULAR COMMUNICATION

ENDOCYTOSIS AND VESICLE FLOW

LIPIDS AND INTRACELLULAR SIGNALLING

REGULATION OF NUCLEAR TRANSCRIPTION

MEMBRANE TRANSPORT PROTEINS

Heterotrimeric Guanine Nucleotide Binding Proteins: Structure and Function

Thomas Wieland, Rüdiger Schulze and Karl H. Jakobs
Institut für Pharmakologie
Universität GH Essen
D-45122 Essen
Germany

The ability to sense and appropriately respond to extracellular signaling molecules is one of the most important properties of an individual cell within a multicellular organism. As many signaling molecules are hydrophilic and unable to penetrate the cell membrane, cells require efficient mechanisms to recognize mediators at the outer surface of the cell and to transduce the signal into the cell. Some systems are relatively self-contained in structure like ligand-operated ion channels and receptors with an intrinsic tyrosine kinase or other enzymatic activity. Other sytems are built from multiple protein components. The vast majority of the multi-component systems are those regulated by heterotrimeric guanine nucleotide binding proteins (G proteins). The G protein-regulated signal transduction machinary is generally composed of three distinct molecular entities.

1. *Receptor:* a seven times the plasma membrane-spanning protein that builds up the recognition site for specific ligands and the interaction site with G proteins at its extracellular and intracellular domains, respectively
2. *Transducer:* at least one subtype of heterotrimeric G proteins capable of translating receptor activation into regulation of specific effectors
3. *Effector:* a moiety that is involved in generating the intracellular signal, *e.g.*, altered second messenger concentrations or ion fluxes

The following brief review will focus on recent findings on the structure and function of heterotrimeric G proteins.

NATO ASI Series, Vol. H 101
Molecular Mechanisms of Signalling
and Membrane Transport
Edited by Karel W. A. Wirtz

Heterotrimeric G Protein Structure

Heterotrimeric G proteins are composed of three subunits (αβγ), of which, under non-denaturating conditions, the β and γ subunit form a permanent dimer. To date, 19 distinct α subunits (M_r 39,000 - 52,000) are known which can be divided into four subfamilies based on the degree of their sequence homology (Table 1) (Simon *et al.*, 1991). Five β subunits (M_r 35,000 - 44,000) have been identified so far (Table 2) (Watson *et al.*, 1994), from which 4 (β_1- β_4) are highly homologous. The recently identified β_5 shows less homology and exists in two forms due to alternative splicing (Watson *et al.*, 1994; 1996). In contrast to the β subunits, the so far known 12 γ subunits (Table 2) show a rather high degree of diversity (Ray *et al.*, 1995; Morishita *et al.*, 1995). Recently, the crystal structure of α_{i1}, α_t, $\beta_1\gamma_1$, $\alpha_{i1}\beta_1\gamma_2$ and $\alpha_t\beta_1\gamma_1$ have been resolved, which now allows insights into the molecular organization of these signal transducing proteins (Coleman *et al.*, 1994; Lambright *et al.*, 1994; 1996; Wall *et al.*, 1995; Sondek *et al.*, 1996).

Gα Subunit Structure

G protein α subunits belong to the ancient protein superfamily of GTPases. The primary structure of α subunits contains 5 regions that are common to all GTPases, so-called G regions, that are crucial for binding of GTP and Mg^{2+} as well as for GTP hydrolysis (Conklin & Bourne, 1993). G protein α subunits are distinguished from other GTPases by their N- and C-terminal extensions as well as 4 inserts. Comparisons of the crystal structures of G protein α subunits with other GTPases have shown that the homologies in the primary structure of the α subunit to other GTPases, *e.g.*, p21Ras, are also reflected in the spatial structure. The crystal structures of two different α subunits, α_t (Noel *et al.*, 1993; Lambright *et al.*, 1994) and α_{i1} (Coleman *et al.*, 1994) have been determined in their active and inactive states. Since the three dimensional structures of α subunits appear to be largely identical, the tertiary structure will be explained for $G\alpha_t$, the best characterized G protein. Fig. 1 shows the three major domains that can be discerned within α, the GTPase domain, the helical domain, and the N-terminal helix that projects away from the rest of α (Lambright *et al.*, 1996). Helical domain and GTPase domain are connected by 2 flexible linkers, L1 and L2. While the GTPase domain displays sequence homology to other GTPases like p21Ras and procaryotic GTPases, *e.g.*, EF-Tu, the helical domain is a unique feature of heterotrimeric G proteins. It represents the longest (120 amino acids) of the 4 inserts that do not display homology to other GTPases. The GTPase domain consists of a six stranded β sheet (β2-β7) which is surrounded by five helices (α2-α6). The β sheets are running parallel with the exception of β2 which runs antiparallel to the others. α2 is a 3_{10} helix while the others are α-helices. The helical domain has an α-helical secondary structure, containing five short helices (αB-αF) which surround a longer central helix (αA). The helical domain has been proposed to act as an intrinsic GTPase activator (Markby *et al.*, 1993). This would explain why α subunits, in contrast to Ras, hydrolyze GTP in the absence of an external activator. The guanine nucleotide is buried in a cleft between the GTPase domain and the helical domain, and make contacts to amino acids from both domains, as indicated in Fig. 1. Nucleotide exchange and GTP hydrolysis induce structural changes only to three distinct regions in the α subunit, termed switch I, II and III (Coleman *et al.*, 1994). Switch I (S173-T183) comprises linker 2 and a part of β2. Switch II (F195-T215) mainly consists of α2 and the α2-β4-loop and is involved in effector regulation and βγ binding (Coleman & Sprang, 1996).

In contrast to switch I and II, switch III (D227-R238) is unique to heterotrimeric G proteins since it encompasses insert 2 between β4 and α3. Arg174 (part of switch I) and Gln200 (part of switch II) are indispensable for GTPase activity, probably by stabilizing the transition state of the enzyme (Hamm & Gilchrist, 1996). Crystallographic studies of α_{i1} have shown that the switch regions II and III adopt a disordered conformation upon GTP hydrolysis (Mixon *et al.*, 1995). While N- and C-termini are disordered in the active state, they form a microdomain (so-called N-C-domain) when GTP is hydrolyzed. This microdomain interacts with a neighboring α subunit, forming head-to-tail polymers (Bourne, 1995). Release of the GDP molecule is probably facilitated by weakened contacts between the GTPase domain and the helical domain.

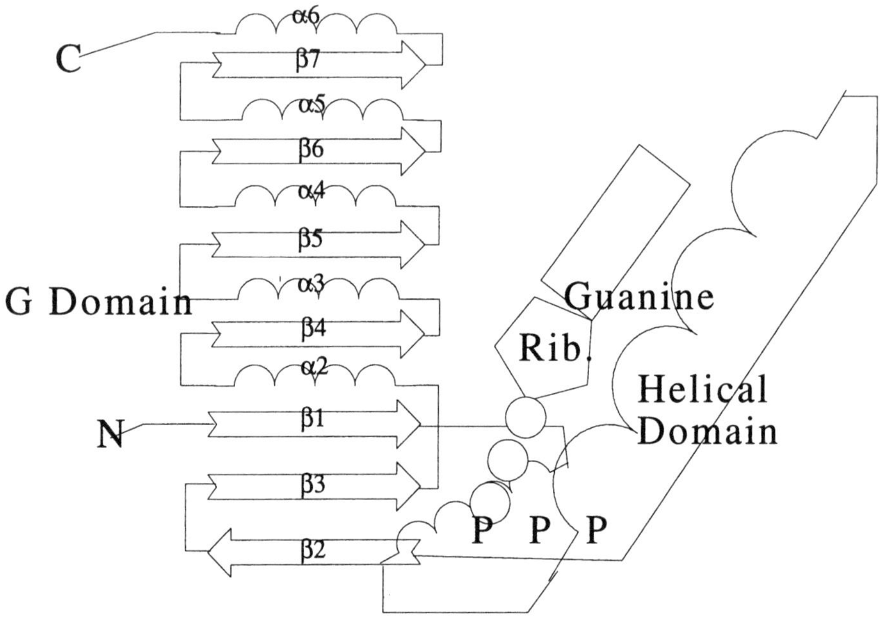

Fig. 1: Schematical representation of the secondary structure of α_t (after Wittinghofer, 1994).

Gβγ Dimer Structure

The primary structure of β subunits is composed of an N-terminal extension followed by seven so-called WD-motifs. These are typically about 30 amino acids long, commencing by a GH pair and terminating by a WD pair, hence the name. The WD domains are separated by sequences with variable length. The WD motif is an old evolutionary theme that can be traced back to the most primitive eucaryotes. It was first discovered in G protein β subunits but later was found in a variety of other regulatory proteins that apparently have no common function (Neer *et al.*, 1994). It has been proposed that they act as building modules for multimeric complexes. Since no WD protein has less than 4 WD domains, it was expected that they represent a highly cooperative structure. This has been confirmed by the recent X-ray diffraction analysis of G protein heterotrimers (Wall *et al.*, 1995; Lambright *et al.*, 1996). The three-dimensional structure of β contains an N-terminal helix that forms a coiled coil with the

N-terminal helix of γ, followed by seven β-sheets that are made up of 4 antiparallel strands (termed strands a, b, c and d).

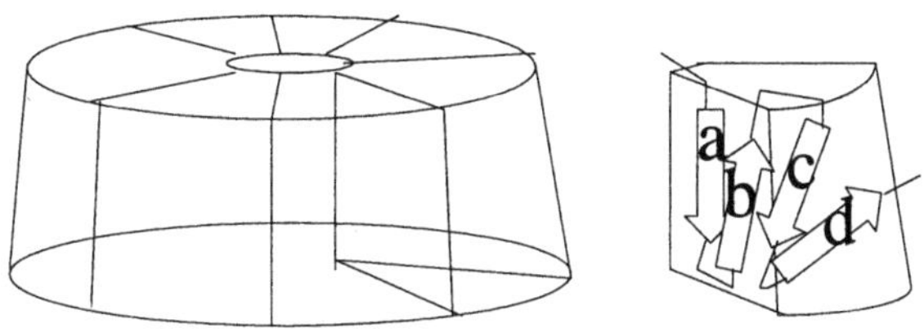

Fig. 2: Schematical representation of the Gβ propeller structure (after Neer & Smith, 1996).

Fig. 2 schematically shows that the β-sheets fold into a β-propeller structure that radiates outward with sevenfold symmetry, leaving a central channel with an average diameter of 12Å (Wall *et al.*, 1995). β-Propellers have previously been observed in a few other proteins, *e.g.*, methanol dehydrogenase and galactose oxidase, that are functionally unrelated to β. However, they bear some remote sequence similarity to WD proteins which suggests an evolutionary relationship between β-propeller and WD proteins (Sondek *et al.*, 1996). The seven β-sheets are not formed by one WD domain each. Instead, a WD domain begins between the third and the fourth strand and ends at the corresponding position in the following sheet (Sondek *et al.*, 1996). The N-terminal part of the first WD repeat forms the outermost strand of the seventh sheet, thereby establishing a link between the beginning and the end of the propeller. Conserved residues establish a hydrophobic environment between the β-sheets. Stabilization of the propeller structure is achieved by inter- and intrablade hydrogen bond triads that are formed by a conserved aspartate between strands b and c, the histidine of the GH motif, and a serine or threonine in strand b (Neer & Smith, 1996). The interaction between β and γ has long been known to be extremely tight. This can now be explained by the spatial structure of the heterotrimer: Beside the coiled coil that is formed by the N-termini of β and γ, the latter subunit makes other extensive contacts with β, binding to 4 propeller blades, but not to itself (Neer & Smith, 1996).

Intersubunit Contacts

Different regions of α subunit and βγ dimer contribute to intersubunit contact surfaces: The N-terminal helix of α binds to the d strands of blades 1 and 7 of the β-propeller, parallel to the propeller's channel, while the switch II region within α's GTPase domain binds to the top of the β-propeller where it is held in place by multiple salt bridges and binding of a conserved tryptophan residue in β to a hydrophobic pocket in α (Neer & Smith, 1996). The heterotrimer is further stabilized by simultaneous insertion of the prenyl moiety at the C-terminus of γ and the acyl moiety at the N-terminus of α into the plasma membrane (Lambright *et al.*, 1996). In contrast, direct interaction of α and γ was not observed in the crystal structure (Wall *et al.*, 1995). The intersubunit contacts are disrupted in the GTP-bound state due to structural changes in the α subunit. Most importantly, the switch II region is twisted and thereby withdrawn from β (Hamm & Gilchrist, 1996). Different surfaces on both α and βγ are exposed that do not interact with each other but with effectors and regulators.

Interactions with Other Proteins

Since the effector binding regions within α encompass the switch II region and the N-terminus, *i.e.*, regions that directly contact β in the heterotrimer, βγ may directly compete for binding sites (Wall *et al.*, 1995). However, it is clear that multiple sites in α are involved in receptor interactions and that βγ also contributes by making direct contacts to the receptor. The interacting sites include the N- and C-termini of α and the C-terminus of β (Conklin & Bourne, 1993). The crucial role of α's C-terminus has been demonstrated by stabilization of the active conformation of rhodopsin using a peptide that corresponds to the last 10 amino acids of α_t (Martin *et al.*, 1996). Cross-linking studies have revealed that the third intracellular loop of the α_2-adrenergic receptor interacts with the C-terminal helix of α and WD domain 7, close to the C-terminus of β (Taylor *et al.*, 1994). A part of the G3 region in α is also involved in receptor interaction. Since G3 is essential for nucleotide binding, it has been suggested that receptor-catalyzed GDP release is transmitted via G3 (Wittinghofer, 1994). The helical domain of α does not seem to participate in receptor binding.

G proteins have so far not been crystallized in complex with effector molecules. However, mutational analyses have revealed regions within both α and the βγ dimer that contribute to these interactions. Interestingly, regions in the α subunit known to contact effectors exclusively reside in the G domain. For α_o and α_{i2}, helices α2, α3, α4 and the loop regions connecting them to the following β sheets have been found to be important for effector binding (Hamm & Gilchrist, 1996). In α_t, helix α4 and the following loop are sufficient for specific interaction with the γ subunit of the retinal cGMP phosphodiesterase (PDE) (Rarick *et al.*, 1992). The βγ dimer binds to a variety of proteins that do not have an obvious common βγ binding domain. Interactions with βγ have not only been reported for various effectors, *e.g.*, adenylyl cyclase subtypes, phospholipase Cβ (PLCβ) isoforms, G protein-gated K^+ channels (C_K), but also to many regulatory proteins, including β-adrenergic receptor kinase (βARK), phosducin and small G proteins like ADP-ribosylation factor (Neer, 1995). However, the regions in βγ that constitute the binding site have been identified in very few cases only. For the yeast β subunit, Ste4, mutational analysis indicated that the βγ coiled coil is important for interaction with the downstream effector, Ste5 (Leberer *et al.*, 1992). Apparently, many effectors bind to the same regions of β: A peptide derived from adenylyl cyclase II that prevents βγ-mediated activation also inhibits binding of βγ to βARK and C_K (Chen *et al.*, 1995). In contrast to α, the spatial structure of the βγ dimer does not change when the heterotrimer dissociates. This suggests that the regions in βγ contacting effectors and regulators are identical or close to the regions involved in binding of α or that these regions are inaccessible to other proteins when α is bound. To further elucidate the interaction mechanisms of G proteins with receptors and effectors, crystal structures of such complexes are eagerly awaited.

Covalent Modifications of Heterotrimeric G Proteins

Gα Subunits - Heterotrimeric G proteins are subject to multiple cotranslational and posttranslational modifications. These include phosphorylation, acylation and isoprenylation. In addition, certain subtypes of α subunits can be ADP-ribosylated by bacterial toxins. Modification by the toxins produced by *Vibrio cholerae* and *Bordetella pertussis* is highly specific and can be used as a tool for distinguishing G protein subtypes: Pertussis toxin catalyzes the ADP-ribosylation of a cysteine residue located four amino acids from the C-terminus. All α subunits that have a cysteine residue in this position, *i.e.*, α_{i1}, α_{i2}, α_{i3}, α_{o1}, α_{o2}, α_t and α_{gust} (Table 1), are subject to modification by pertussis toxin. This modification, which requires interaction with the $\beta\gamma$ dimer, results in uncoupling from the receptor. α_t can also be ADP-ribosylated by cholera toxin, whereas α_s and α_{olf} can be modified by cholera but not pertussis toxin. Cholera toxin ADP-ribosylates an arginine residue, leading to diminished GTPase activity and, thus, constitutive activation of the α subunit (Hepler and Gilman, 1992). Some α subtypes (α_q, α_z, α_{11-16}) are not subject to modification by either pertussis or cholera toxin.

α_i, α_o, and α_z are myristoylated at their N-terminus (Mumby *et al.*, 1990). The fatty acid moiety is transferred in a cotranslational process from myristoyl-CoA by a protein N-myristoyl transferase. In contrast, α_t is modified by different saturated and non-saturated 12- and 14-carbon fatty acids (Kokame *et al.*, 1992). An N-terminal glycine and a serine residue at position 6 are components of the myristoylation signal. Subunits which lack these features are not myristoylated, like α_s, α_q and α_{11-16}. The fatty acylation facilitates membrane attachment of α subunits and increases their affinity for $\beta\gamma$ dimers (Linder *et al.*, 1991).

In addition to this irreversible lipid modification, some α subunits, *e.g.*, α_s, are reversibly palmitoylated at cysteine 3 (reviewed by Casey, 1994). Palmitoylation probably has a regulatory function since activation of the β-adrenergic receptor leads to depalmitoylation of α_s which is then incapable of stimulating adenylyl cyclase (Wedegaertner & Bourne, 1994).

Several G protein α subunits can be phosphorylated on serine, threonine or tyrosine residues. α_z is phosphorylated by protein kinase C (PKC, Ser/Thr-protein kinase) on serine 27, close to the N-terminus of the protein which is supposed to be important for binding of $\beta\gamma$ (Fields & Casey, 1995). Similar data have been reported for α_{12} (Kosaza *et al.*, 1996). α_{i2} is subject to phosphorylation on different serine residues by protein kinases A and C (Houslay, 1991). α_s can also be phosphorylated by p60^{src} *in vitro* (Moyers *et al.*, 1995). The reports on α phosphorylation available so far suggest that these modifications play a role in assembly of macromolecular complexes by altered affinities for other proteins. In contrast, the activity of the G protein hardly changes upon phosphorylation (Neer, 1995).

Table 1: Mammalian G Protein α Subunits.

Family	Subtype	Expression	Effectors
	α_{sS} (2 forms)*	Ubiquitous	Adenylyl cyclase ↑ (all types)
α_s	α_{sL} (2 forms)*	Ubiquitous	Ca^{2+} channel ↑ (L-type)
	α_{olf}	Olfact. Epithelium	Adenylyl cyclase ↑ (type V)
	α_{gust}	Taste buds, gut	?
	$\alpha_{t\text{-}r}$	Retinal rods, taste buds	cGMP phosphodiesterase ↑
	$\alpha_{t\text{-}c}$	Retinal cones	
	α_{i1}	Widely	Adenylyl cyclase ↓
α_i	α_{i2}	Ubiquitous	(types I, III, V, VI)
	α_{i3}	Nearly ubiquitous	K^+ channel ↑
	α_{o1}*	Neuronal and	Ca^{2+} channels ↓
	α_{o2}*	Neuroendocrine	(L- and N-type)
	α_z	Neuronal, platelets	Adenylyl cyclase ↓ ?
	α_q	Ubiquitous	Phospholipase C β ↑
	α_{11}	Ubiquitous	($\beta_4 \geq \beta_1 \geq \beta_3 > \beta_2$)
α_q	α_{14}	Kidney, lung, spleen	
	α_{15} (mouse)	Hematopoetic cells	
	α_{16} (human)		
α_{12}	α_{12}	Ubiquitous	?
	α_{13}	Ubiquitous	?

*splice variants

Gβ Subunits - Unlike α subunits, β subunits can be transiently phosphorylated specifically by GTP at a histidine residue (Wieland *et al.*, 1993). Stimulation of phosphorylation by receptor activation suggests that this modification of β subunits represents an alternative pathway of G protein activation (Kaldenberg-Stasch *et al.*, 1994). For the yeast G protein β subunit, STE4, phosphorylation at multiple serine residues has been found to occur in response to the mating pheromone. The phosphorylation sites reside in an extension that is unique to STE4 (Cole & Reed, 1991). Finally, β1 has been reported to be acetylated at its N-terminus (Matsuda *et al.*, 1994).

Gγ Subunits - G protein γ subunits are either farnesylated or geranylgeranylated in a multi-step posttranslational process. The modification occurs via a thioester bond at a cysteine residue which is three amino acids from the C-terminus. If the C-terminal amino acid is a leucine residue like in γ2,3,5,7 a geranylgeranyltransferase modifies the cysteine. In contrast, a serine, glutamine or methionine residue in this position is the signal for a farnesyltransferase. Thus, γ1, γ8 and γ11 are farnesylated. After the attachment of the isoprenoid moiety to the cysteine residue, the three last amino acids are proteolytically removed, and the now C-terminal prenylcysteine is methylated. Mutation of the cysteine residue prevents isoprenylation. Such mutant γ subunits still form βγ dimers but fail to associate with the plasma membrane (Higgins & Casey, 1994). The most recently found γ subunit γ12 can be phosphorlyated by PKC when the βγ dimer is not complexed with an α subunit (Morishita *et al.*, 1995).

Table 2: Mammalian G Protein β and γ Subunits.

Subtype	Expression	Effectors interacting with βγ
β_1	Ubiquitous	Adenylyl cyclase ↓ (type I)
β_2	Ubiquitous	Adenylyl cyclase ↑ (types II, IV)
β_3	Ubiquitous	Phospholipase Cβ ↑
β_4	Ubiquitous	$(\beta_3 \geq \beta_2 \geq \beta_1 >> \beta_4)$
β_{5S}*	Mainly brain	K^+ channel ↑
β_{5L}*	Retina	Ca^{2+} channel ↓
		Receptor kinases (type 2, 3) ↑
γ_1+	Retinal rods	Phospholipase A_2 ↑ ?
γ_2	Mainly brain	Phosphoinositide 3-kinase ↑ ?
γ_3	Mainly brain	
γ_4	Mainly brain	βγ combinations apparently not formed:
γ_5	Ubiquitous	$\beta_2\gamma_1$, $\beta_2\gamma_{11}$
γ_7	Widely	Tissue specific combinations:
γ_8+	Retinal cones	$\beta_1\gamma_1$ Retinal rods
γ_{10}	Widely	$\beta_3\gamma_8$ Retinal cones
γ_{11}+	Widely	
γ_{12}	Ubiquitous	

*splice variants

+these γ subunits are farnesylated, all others are geranyl-geranylated

Heterotrimeric G Protein Function

Regulation of G Protein Activation

A functional interaction between receptors and G proteins requires all components of the heterotrimer, the α subunit and the βγ dimer. An activated receptor induces the release of GDP bound to the α subunit of the inactive heterotrimer and thus, GTP, which is present in higher concentrations in the cell than GDP, can bind to the α subunit nucleotide-binding site (for review see Birnbaumer *et al.*, 1990; Birnbaumer, 1992; Simon *et al.*, 1991) (Fig. 3). As a single receptor can activate as many as 500 G protein molecules, *e.g.*, rhodposin-transducin interaction, this GDP/GTP exchange is assumed to be the first step of signal amplification (Chabre & Deterre, 1989). Binding of GTP induces conformational changes leading to i) dissociation of the G protein from the receptor and ii) dissociation of the heterotrimer into a GTP-liganded α subunit and a βγ dimer. Termination of G protein activation is achieved by two processes, i) the hydrolysis of GTP to GDP by the GTPase activity inherent to the α subunit and ii) the reassociation of the α subunit with the βγ dimer.

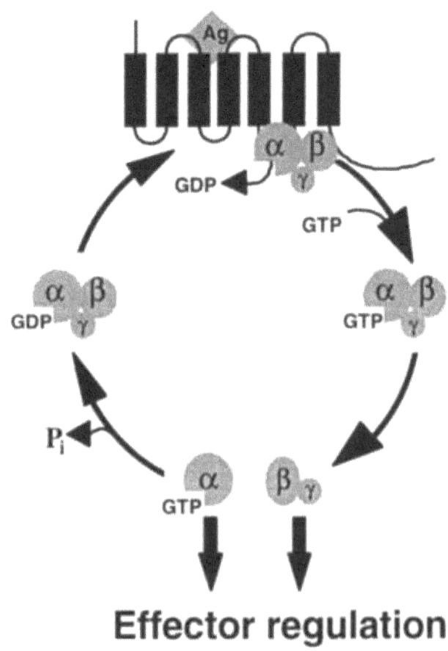

Fig. 3: Activation cycle of heterotrimeric G proteins.

However, it has become evident that signal transduction by heterotrimeric G proteins is more complicated as described above. First, the complexity arises by the diversity of the G protein components themselves (see above). The currently 19 subtypes of α subunits, 5 of β subunits and 12 of γ subunits are raising the possibility of hundreds of specific combinations. Indeed, it has been found that in the case of receptor-G protein interaction signaling can converge or diverge (for review see Offermanns & Schultz, 1994). Many receptors are capable of activating different G protein subtypes leading to regulation of different signaling pathways. In contrast, receptors which mediate functionally similar responses in a specialized cell often activate the same subtypes of G proteins. At low concentrations of different agonists this convergent activation may serve to integrate the information, and on the other hand, when a given pool of G proteins is fully activated by one type of receptors it may prevent overstimulation of the cell by activation of a second receptor.

In addition to the receptor-induced GDP/GTP exchange, two transphosphorylation reactions may also be involved in G protein activation. The first can be attributed to membranous isoforms of nucleoside diphosphate kinase (NDPK), an enzyme capable of transferring high energetic phosphate from a nucleoside triphosphate to another nucleoside diphoshate. The participation of this enzyme to G protein activation has recently be reviewed (Piacentini & Niroomand, 1996). Membranous NDPK seems to be able to form complexes with G proteins and ,thus, to catalyze the formation of GTP from ATP and GDP nearby the G protein. In comparison to exogenous GTP, the GTP formed by NDPK has been reported to have an advantage in G protein activation (Wieland & Jakobs, 1992). The second reaction involves G protein β subunits intermediately phosphorylated by GTP. The histidine-bound high energetic phosphate can be transferred to GDP at G protein α subunits, thereby leading to formation of GTP and α subunit activation (Wieland *et al.*, 1991; 1992; 1993). GTP-specific phosphorylation of β subunits was observed in a variety of mammalian tissues and apparently requires a so far unknown cofactor (Hohenegger *et al.*, 1996; Nürnberg *et al.*, 1996). It is assumed that this reaction is involved in signal amplification and diversification (Kaldenberg-Stasch *et al.*, 1994)

There are several processes known to attenuate receptor-G protein interaction or G protein activation. The first of these processes involves phosphorylation of the activated receptor by specific receptor kinases (GRK's), *e.g.*, βARK (for reviews see Lohse, 1993; Premont *et al.*, 1995). The phosphorylation of the receptor by GRK's does not interfere with the

G protein activation itself, instead it favours the binding of a second protein, *i.e.*, arrestin or arrestin-like proteins to the receptor. Binding of these proteins to the receptor apparently is at the interaction site with the G protein and thereby competitively inhibits G protein activation. Some receptors are also substrates for protein kinases activated downstream of G proteins, *e.g.*, PKA and PKC. In this case, phosporylation of the receptor occurs within the domain interacting with the G protein and thereby blocks G protein activation (Lohse, 1993).

Besides these processes modifying the function of receptors, there is evidence for a regulation at the level of the G protein itself. Recently, a knew family of proteins, called regulator of G protein signaling (RGS), was described (Druey *et al.*, 1996; Koelle & Horvitz, 1996). When expressed, members of the RGS family have been shown to decrease signaling through interleukin-8 and platelet-activating factor receptors most likely by direct interaction with G protein α subunits (De Vries *et al.*, 1995; Druey *et al.*, 1996). The molecular mechanism by which RGS proteins interfere with α subunit activation is currently not known. However, three possibilities can be envisaged: i) RGS proteins may either simply sequester activated α subunits, ii) inhibit the receptor-driven GDP/GTP exchange or iii) stimulate the GTPase activity of α subunits (Berman *et al.*, 1996). The data so far obtained suggest that expression of RGS proteins may be induced by signals through G protein-coupled receptors to trigger a desensitization mechanism.

Several other cytosolic proteins have been shown to interact with βγ dimers. Phosducin is a 33 kDa protein which is expressed most abundantly in the retina and the pineal gland (Lee *et al.*, 1987). At its N-terminus phosducin contains a structural motif called PH domain which is also present in some effectors regulated by βγ dimers (Touhara *et al.*, 1987; Koch *et al.*, 1993). Through this motif the N-terminus of phosducin binds βγ dimers very tightly and thus inhibits interaction of free βγ dimers with α subunits as well as effectors (Bauer *et al.*, 1992; Lee *et al.*, 1992b; Hawes *et al.*, 1994; Xu *et al.*, 1995). A phosducin-like protein has been found in rat brain (Miles *et al.*, 1993) which regulates βγ function in a manner similar to phosducin (Schröder & Lohse, 1996). In addition, calmodulin, a widely expressed Ca^{2+}-binding protein that modulates numerous Ca^{2+}-dependent processes, has been shown to tightly interact with βγ dimers in the presence of Ca^{2+} (Asano *et al.*, 1986; Katada *et al.*, 1987). Similar to phosducin, Ca^{2+}-calmodulin can prevent interaction of α subunits with $\beta\gamma$ by binding the dimer. Therefore, at least the above mentioned proteins should be capable to interact with $\beta\gamma$ dimers and thereby limit their availability *in vivo*.

Both α subunits and $\beta\gamma$ dimers have been shown to be substrates of Ser/Thr-protein kinases, *e.g.*, PKC, which gains in altered functional properties. Recently, it has been shown that free α_z and $\alpha 12$ subunits can be phosphorylated with PKC on a homologous N-terminal Ser residue (Fields & Casey, 1995; Kozasa & Gilman, 1996). After phosphorylation, the interaction of the α subunits with free $\beta\gamma$ dimers was greatly reduced. As the phosphorylation of these particular α subunits can also be observed *in vivo* after activation of PKC (Lounsbury *et al.*, 1991; Kozasa & Gilman, 1996), it is likely that this phosphorylation is involved in either a sensitization or desensitization process in living cells. Indeed, phosphorylation of members of the α_i subfamily, *e.g.*, α_{i2}, by PKC has been shown to attenuate the ability to inhibit adenylyl cyclase (Katada *et al.*, 1987; Bushfield *et al.*, 1990; Strassheim & Malbon, 1994). Members of these particular subfamily can also be phosphorylated by cGMP-dependent protein kinase *in vivo* and *in vitro*. Moreover, it has been shown that the phosphorylation somehow interferes with the coupling of receptors to effectors, *e.g.*, phospholipase C (PLC). Similar to the above mentioned α subunits, $\beta\gamma$ dimers containing the newly identifed γ subunit $\gamma 12$ are substrate to phosphorylation by PKC (Morishito *et al.*, 1995). In free $\beta\gamma 12$ dimers but not the heterotrimeric form a N-terminal Ser residue was found to be the most likely phosphorylation site. Compared to unphosphorylated $\beta\gamma 12$, the phosphorylated protein was more resistent against digestion by calpain and seemed to interact with α subunits more tightly.

G Protein-Effector Interaction

Gα Subunits - Direct interaction with activated G protein α subunits has so far been demonstrated for a variety of effectors, *e.g.*, isoforms of adenylyl cyclase, PLCβ and the retinal cGMP PDE (Table 1).

(α_s family) All four forms of the α_s protein stimulate the so far known isoforms of adenylyl cyclase (Olate *et al.*, 1988; Tang & Gilman, 1992). In addition, α_s can apparently stimulate Ca^{2+} channels (Yatani *et al.*, 1987; 1988a; Hamilton *et al.*, 1991; Birnbaumer, 1992). A related G protein α subunit which is found in the olfactory system (α_{olf}) similarly stimulates adenylyl cyclase.

(α_i family) The three subtypes of α_i proteins are also interacting with a variety of effectors. They inhibit specific members of the adenylyl cyclase family, *e.g.*, type I, III, V and VI (Simonds *et al.*, 1989; Wong *et al.*, 1991; Taussig *et al.*, 1993b; 1994) and stimulate K^+ channels (Yatani *et al.*, 1988b; Kirsch *et al.*, 1990). In addition, a receptor-induced appearance of activated α_i subunits at microfilaments has recently been reported in HL-60 cells (Wieland *et al.*, 1996). The appearance of activated α_i coincidently occurred with receptor-regulated actin polymerization, suggesting an involvement of α_i in regulation of F-actin formation. α_z, an α subunit highly homologous to α_i but not pertussis toxin-sensitive, has also the potential to inhibit adenylyl cyclase (Kozasa & Gilman, 1995). The two subtypes of α_o have been shown to inhibit N- and L-type Ca^{2+} channels (Kleuss *et al.*, 1991; Degtiar *et al.*, 1996; Kalkbrenner *et al.*, 1995). The two isoforms of transducin α subunits, $\alpha_{t\text{-}r}$ and $\alpha_{t\text{-}c}$, specifically expressed in retinal rods and cones, respectively, stimulate the retinal cGMP PDE by releasing the inhibitory PDEγ subunit from the enzyme (Chabre & Deterre, 1989). The effector for α_{gust}, an α subunit specifically expressed in taste buds and gut and involved in the signal transduction induced by bitter and sweet substances (Hoefer *et al.*, 1996; Wong *et al.*, 1996), has so far not been identified.

(α_q family) All members of the the α_q family, *i.e.*, α_q, α_{11}, α_{14}, α_{15} and α_{16}, although showing different expression patterns (Table 1), stimulate isoforms of PLCβ with a similar preference pattern (PLC$\beta_4 \geq \beta_1 \geq \beta_3 > \beta_2$) (Smrcka *et al.*, 1991; Taylor *et al.*, 1991; Lee *et al.*, 1992a; Wu *et al.*, 1992, 1993). Most interestingly, α_{15} and α_{16} have the potential to serve as promiscuitive G protein α subunits. They apparently couple any activated receptor to PLCβ (Offermans & Simon, 1995; Liu & Wess, 1996). However, the physiological relevance of these findings remains to be determined.

(α_{12} family) Only limited information about the signal transduction by the two members of the α_{12} family, α_{12} and α_{13}, is currently available. No effector has so far been identified with certainty. However , both α subunits are apparently involved in cytoskeleton rearrangement and mitogenesis (Buhl *et al.*, 1995; Dhanasekaran & Dermott, 1996). In addition, α_{13} is engaged in the stimulation of a Ca^{2+} channel (Will-Blasczak *et al.*, 1994) and the ubiquitous Na^+/H^+ antiporter, the activation of which however most likely represents the final event in a multi-step cascade (Voyno-Yasenetskaya *et al.*, 1993).

Gβγ Dimers - During the last few years it has become obvious that not only activated α subunits but also free βγ dimers are directly interacting with and thereby regulating effector activity (Birnbaumer, 1992; Sternweis, 1994). The adenylyl cyclase type I can be inhibited by free βγ dimers, whereas types II and IV are strongly stimulated by βγ in the additional presence of activated α_S (Taussig *et al.*, 1993a). Equally well documented is the stimulation of PLCβ isoforms (PLCβ3 > β2 > β1) by βγ (Camps *et al.* 1992a; b; Blank *et al.*, 1992; Boyer *et al.*, 1992; 1994; Katz *et al.*, 1992; Carozzi *et al.*, 1993; Smrcka & Sternweis, 1993; Park *et al.*, 1993; Watson *et al.*, 1994; 1996). The direct interaction with K^+ and Ca^{2+} channels leading to stimulation and inhibition of these channels, respectively, has been described recently (Reuveny *et al.*, 1994; Ikeda, 1996; Herlitze *et al.*, 1996). A peptide derived from adenylyl cyclase type II was identified which may map the interaction site of some effectors with βγ (Chen *et al.*, 1995). In addition to these apparently direct interaction with effectors, βγ dimers are involved in activation of PLA_2, phosphoinositide 3-kinase and Ras-dependent pathways (Mayer & Marshall, 1993; Stephens *et al.*, 1994; Crespo *et al.*, 1994).

Despite some evidence which argues for specific β and γ combinations acting as transducer in a distinct signaling pathway (Kleuss *et al.*, 1992; 1993; Degtiar *et al.*, 1996; Kalkbrenner *et al.*, 1995), no particular specificity of different βγ combinations was observed in reconstitution experiments with effectors (Iñiguez-Lluhi *et al.*, 1992; Sternweis, 1994). Nevertheless, a certain degree in specificity of βγ combinations was observed in targeting β-adrenergic receptor kinase to its substrate (Müller *et al.*, 1993). Moreover, some specific βγ combinations are more likely to be formed than others (Yan *et al.*, 1996).

Thus, it has to be determined in the future whether the specificity of G proteins involved in distinct signal transduction pathways (Kleuss *et al.*, 1991; 1992; 1993; Degtiar *et al.*, 1996; Kalkbrenner *et al.*, 1995) is due to the selective interaction of a certain receptor with (a) defined heterotrimer(s) and/or a selective interaction of the activated G protein subunits with specific effectors or is not thus restricted at all.

Acknowledgements

The authors' studies reported herein were supported by the Deutsche Forschungsgemeinschaft and the Fonds der Chemischen Industrie.

References

Asano T, Ogasawara N, Kitajima S, Sano M (1986) Interaction of GTP binding proteins with calmodulin. FEBS Lett 203:135-138

Bauer PH, Müller S, Puzicha M, Pippig S, Obermaier B, Helmreich, EJM, Lohse, MJ (1992) Phosducin is a protein kinase A-regulated G protein regulator. Nature 358:73-76

Berman DM, Wilkie TM, Gilman, AG (1996) GAIP and RGS4 are GTPase-activating proteins for the G_i subfamily of G protein α subunits. Cell 86:445-452

Birnbaumer L (1992) Receptor-to-effector signaling through G proteins: roles for βγ dimers as well as α subunits. Cell 71:1069-1072

Birnbaumer L, Abramowitz J, Brown AM (1990) Receptor-effector coupling by G proteins. Biochim Biophys Acta 1031:163-224

Blank JL, Brattain KA, Exton JH (1992) Activation of cytosolic phosphoinositide phospholipase C by G protein βγ-subunits. J Biol Chem 267:23069-23075

Bourne HR (1995) Trimeric G proteins: Surprise witness tells a tale. Science 270:933-934

Boyer JL, Waldo GL, Harden TK (1992) βγ subunit activation of G protein regulated phospholipase C. J Biol Chem 267: 5451-25456

Buhl AM, Lassignal-Johnson N, Dhanasekaran N, Johnson GL (1995) $G\alpha_{12}$ and $G\alpha_{13}$ stimulate Rho-dependent stress fiber formation and focal adhesion assembly. J Biol Chem 270:24631-24643

Camps M, Carozzi A, Schnabel P, Scheer A, Parker PJ, Gierschik P (1992a) Isozyme selective stimulation of phospholipase C-β2 by G protein βγ-subunits. Nature 360:684-686

Camps M, Hou C, Sidiropoulos D, Stock JB, Jakobs KH, Gierschik P (1992b) Stimulation of phospholipase C by G protein βγ-subunits. Eur J Biochem 206:821-831

Carozzi A, Camps M, Gierschik P, Parker PJ (1993) Activation of phosphatidylinositol lipid-specific phospholipase C-β3 by G protein βγ-subunits. FEBS Lett 315:340-342

Casey, PJ (1994) Lipid modifications of G proteins. Curr Opin Cell Biol 6:219-225

Chabre M, Deterre P (1989) Molecular mechanisms of visual transduction. Eur J Biochem 179:255-266

Chen J, DeVivo M, Dingus J, Harry A, Li J, Sui J, Carty DJ, Blank JL,Exton, JH, Stoffel RH (1995) A region of adenylylcyclase II critical for regulation by G protein βγ subunits. Science 268:1166-1169

Clapham DE (1996) The G protein nanomachine. Nature 379:297-299

Cole GM, Reed SI (1991) Pheromone-induced phosphorylation of a G protein β subunit in *S. cerevisiae* is associated with an adaptive response to mating pheromone. Cell 64:703-716

Coleman .DE, Sprang SR (1996) How G proteins work: A continuing story. Trends Biochem Sci 21:41-44

Coleman DE, Berghuis AM, Lee E, Linder ME, Gilman AG, Sprang SR (1994) Structures of active conformations of $G_{i\alpha 1}$and the mechanism of GTP hydrolysis. Science 265:1405-1412

Conklin BR, Bourne HR (1993) Structural elements of G_α subunits that interact with Gβγ, receptors, and effectors. Cell 73:631-641

Crespo P, Xu N, Simonds WF, Gutkind JS (1994) Ras-dependent activation of MAP kinase pathway mediated by G protein βγ subunits. Nature 369:418-420

Degtiar VE, Wittig B, Schultz G, Kalkbrenner F (1996) A specific G_o heterotrimer couples somatostatin receptors to voltage-gated calcium channels in RinM5f cells. FEBS Lett 380:137-141

DeVries L, Mousli M, Wurmser A, Farquahr MG (1995) GAIP, a protein that interacts with the trimeric G protein $G_{\alpha i3}$, is a member of a protein family with a highly conserved core domain. Proc Natl Acad Sci USA 92:11916-11920

Dhanasekaran N, Dermott JM (1996) Signaling by G_{12} class of G protein. Cell. Signalling:in press

Druey KM, Blumer KJ, Kang VH, Kehrl JH (1996) Inhibition of G protein mediated MAP kinase activation by a new mammalian gene family. Nature 379:742-746

Fields TA, Casey PJ (1995) Phosphorylation of $G_{z\alpha}$ by protein kinase C blocks interaction with the $\beta\gamma$ complex. J Biol Chem 270:23119-23125

Hamilton SL, Codina J, Hawkes MJ, Yatani A, Sawada T, Frickland FM, Froehner SC, Spiegel AM, Toro L, Stefani E, Birnbaumer L, Brown AM (1991) Evidence for direct interaction of $G_{s\alpha}$ with the Ca^{2+}channel of skeletal muscle. J Biol Chem 266:19528-19535

Hamm HE, Gilchrist A (1996) Heterotrimeric G proteins. Curr Opinion Cell Biol 8:189-196

Hawes BE, Touhara K, Kurose H, Lefkowitz RJ, Inglese J (1994) Determination of the $G\beta\gamma$ binding domain of phosducin. J Biol Chem 269:29825-29830

Hepler JR, Gilman AG (1992) G-proteins. Trends Biochem Sci 17:383-387

Herlitze S, Garcia DE, Mackie K, Hille B, Scheuer T, Catteral WA (1996) Modulation of Ca^{2+}channels by G protein $\beta\gamma$ subunits. Nature 380:258-262

Higgins JB, Casey PJ (1994) In vitro processing of recombinant G protein γ subunits. J Biol Chem 269:9067-9073

Höfer D, Püschel B, Drenckhahn D (1996) Taste receptor-like cells in the rat gut identified by expression of α-gustducin. Proc Natl Acad Sci USA 93:6631-6634

Hohenegger M, Mitterauer T, Voss T, Nanoff C, Freissmuth M (1996) Thiophosphorylation of the G protein β subunit in human platelet membranes: Evidence against a direct phosphate transfer reaction to $G\alpha$ subunits. Mol Pharmacol 49:73-80

Houslay MD (1991) „Cross-talk“: A pivotal role for protein kinase C in modulating relationships between signal transduction pathways. Eur J Biochem 195:9-27

Ikeda SR (1996) Voltage-dependent modulation of N-type calcium channels by G protein $\beta\gamma$ subunits. Nature 380:255-258

Iñiguez-Lluhi JA, Simon MI, Robishaw JD, Gilman AG (1992) G protein $\beta\gamma$-subunits synthesized in Sf9 cells. J Biol Chem 267:23409-23417

Kokame K, Fukada Y, Yoshizawa T, Takao T, Shimonishi Y (1992) Lipid modification at the N terminus of photoreceptor G-protein α-subunit. Nature 359:749-752

Kaldenberg-Stasch S, Baden M, Fesseler B, Jakobs KH, Wieland T (1994) Receptor-stimulated guanine-nucleotide-triphosphate binding to guanine-nucleotide-binding regulatory proteins. Eur J Biochem 221:25-33

Kalkbrenner F, Degtiar VE, Schenker M, Brendel S, Zobel A, Schultz G, Wittig B (1995) Subunit composition of G_o proteins functionally coupling galanin receptors to voltage-gated calcium channels. EMBO J 14:4278-4737

Katada T, Kusakabe K, Oinuma M, Ui M (1987) A novel mechanism for the inhibition of adenylate cyclase via inhibitory GTP binding proteins. J Biol Chem 262:11897-11900

Katz A, Wu DQ, Simon MI (1992) Subunits $\beta\gamma$ of heterotrimeric G protein activate $\beta 2$ isoform of phospholipase C. Nature 360:686-689

Kirsch G, Codina J, Birnbaumer L, Brown AM (1990) Coupling of ATP-sensitive K+ channels to purinergic receptors by G proteins in rat ventricular myocytes. Am J Physiol 259:H280-H286

Kleuss C, Hescheler J, Ewel C, Rosenthal W, Schultz G, Wittig B (1991) Assignement of G protein subtypes to specific receptors inducing inhibition of calcium currents. Nature 353:43-48

Kleuss C, Scherübl H, Hescheler J, Schultz G, Wittig B (1992) Different β-subunits determine G protein interaction with transmembrane receptors. Nature 358:424-426

Kleuss C, Scherübl H, Hescheler J, Schultz G, Wittig B (1993) Selectivity in signal transduction determined by γ-subunits of heterotrimeric G proteins. Science 359:832-834

Koch WJ, Inglese J, Stone WC, Lefkowitz RJ (1993) The binding site for the βγ subunits of heterotrimeric G proteins on the β-adrenergic receptor kinase. J Biol Chem 268:8256-8260

Koelle MR, Horvitz HR (1996) Egl-10 regulates G protein signaling in the C. elegans nervous system and shares a conserved domain with many mammalian proteins. Cell 84:115-125

Kozasa T, Gilman AG (1995) Purification of recombinant G proteins from Sf9 cells by hexahistidine tagging of associated subunits. J Biol Chem 270:1734-1741

Kozasa T, Gilman AG (1996) Protein kinase C phosphorylates $G_{12\alpha}$ and inhibits its interaction with Gβγ. J Biol Chem 271:12562-12567

Lambright DG, Noel JP, Hamm HE, Sigler PB (1994) The 1.8Å crystal structure of transducin α-GDP: Structural determinants for activation of a heterotrimeric G-protein α subunit. Nature 369:621-628

Lambright DG, Sondek J, Bohm A, Skiba NP, Hamm HE, Sigler PB (1996) The 2.0Å crystal structure of a heterotrimeric G protein. Nature 379:311-319

Leberer E, Dignard D, Hougan L, Thomas DY, Whiteway M (1992) Dominant-negative mutants of a yeast G protein β subunit identify two functional regions involved in pheromone signaling. EMBO J 11:4805-4813

Lee RH, Lieberman BS, Lolley RN (1987) A novel complex from bovine visual cells of a 33000 Dalton phosphoprotein with β and γ transducin. Biochemistry 26:3983-3990

Lee CH, Park D, Wu D, Rhee SG, Simon MI (1992a) Members of the G_o alpha subunit gene family activate phospholipase C-β isozymes. J Biol Chem 267:16044-16047

Lee RH, Ting TD, Lieberman BS, Tobias BE, Lolley RN, Ho YK (1992b) Regulation of retinal cGMP cascade by phosducin in bovine rod photoreceptor cells. J Biol Chem 267:25104 -25112

Linder ME, Pang IH, Duronio RJ, Gordon JI, Sternweis PC, Gilman AG (1991) Lipid modifications of G protein subunits: Myristoylation of $G_{o\alpha}$ increases its affinity for βγ. J Biol Chem 266:4654-4659

Liu J, Wess J (1996) Different single receptor domains determine the distinct G protein coupling profiles of members of the vasopressin receptor family. J Biol Chem 271:8772-8778

Lohse MJ (1993) Molecular mechanisms of membrane receptor desensitization. Biochim Biophys Acta 1179:171-188

Lounsbury KM, Casey PJ, Brass LF, Manning DR (1991) Phosphorylation of G_z in human platelets. J Biol Chem 266:22051-22056

Markby DW, Onrust R, Bourne HR (1993) Separate GTP binding and GTPase activating domains of Gα subunit. Science 262:1895-1901

Martin EL, Rens-Domiano S, Schatz PJ, Hamm HE (1996) Potent peptide analogues of a G protein receptor binding region obtained with a combinatorial library. J Biol Chem 271:361-366

Matsuda T, Takao T, Shimonishi Y, Murata M, Asano T, Yoshizawa T, Fukada Y (1994) Characterization of interactions between transducin α/βγ-subunits and lipid membranes. J Biol Chem 269:30358-30363

Mayer RJ, Marshall LA (1993) New insight on mammalian phospholipase A_2 (s); comparison of arachidonyl-selective and non-selective enzymes. FASEB J 7:339-348

Miles MF, Barhite S, Sganga M, Elliott M (1993) Phosducin-like protein: An ethanol-responsive potential modulator of guanine nucleotide-binding protein function. Proc Natl Acad Sci USA 90:10831-10835

Mixon MB, Lee E, Coleman DE, Berghuis AM, Gilman AG, Sprang SR (1995) Tertiary and quarternary structural changes in $G_{i\alpha 1}$ induced by GTP hydrolysis. Science 270:954-960

Morishita R, Nakayama H, Isobe T, Matsuda T, Hashimoto Y, Okano T, Fukada Y, Mizuno K, Ohno S, Kozawa O, Kato K, Asano T (1995) Primary structure of a γ subunit of G protein, γ_{12}, and its phosphorylation by protein kinase C. J. Biol. Chem. 270:29469-29475

Moyers JS, Linder ME. Shannon JD , Parsons SJ (1995) Identification of the *in vitro* phosphorylation sites on Gsα mediated by $pp60^{c\text{-}src}$. Biochem. J. 305:411-417

Müller S, Hekman M, Lohse MJ (1993) Specific enhancement of β-adrenergic receptor kinase activity by defined G protein β and γ subunits. Proc Natl Acad Sci USA 90:10439-10443

Mumby SM, Heukeroth RO, Gordon JI, Gilman AG (1990) G protein α subunit expression, myristoylation, and membrane association in COS cells. Proc Natl Acad Sci USA 87:728-732

Neer EJ, Smith TF (1996) G protein heterodimers: New structures propel new questions. Cell 84:175-178

Neer EJ (1995) Heterotrimeric G proteins: Organizers of transmembrane signals. Cell 80:249-257

Neer EJ, Schmidt CJ, Nambudripad R, Smith TF (1994) The ancient regulatory-protein family of WD-repeat proteins. Nature 371:297-300

Noel JP, Hamm HE, Sigler PA (1993) The crystal structure of the GTPγS-bound α subunit of the rod G-protein. Nature 366:654-663

Nürnberg B, Harhammer R, Exner T, Schulze RA, Wieland T (1996) Species- and tissue-dependent diversity of G protein β-subunit phosphorylation. Biochem J 318:717-722

Offermanns S, Schultz G (1994) Complex information processing by the transmembrane signaling system involving G proteins. Naunyn-Schmiedeberg's Arch Pharmacol 350:329-338

Offermanns S, Simon MI (1995) $G\alpha_{15}$ and $G\alpha_{16}$ couple a wide variety of receptors to phospholipase C. J Biol Chem 270:15175-15180

Olate J, Mattera R, Codina J, Birnbaumer L (1988) Reticulocyte lysates synthesize an active α-subunit of the stimulatory G protein G_s. J Biol Chem 263:10394-10400

Park D, Jhon DY, Lee CW, Lee KH, Rhee SG (1993) Activation of phospholipase C isozymes by G protein βγ-subunits. J Biol Chem 268:4573-4576

Piacentini L, Niroomand F. (1996) Phosphotransfer reactions as means of G protein activation. Mol Cell Biochem 157:59-63

Premont RT, Inglese J, Lefkowitz RJ (1995) Protein kinases. 3. Protein kinases that phosphorylate activated G protein coupled receptors. FASEB J 9:175-183

Rarick HM, Artemyev NO, Hamm HE (1992) A site on rod G protein α subunit that mediates effector activation. Science 256:1031-1033

Ray K, Kunsch L, Bonner LM, Robishaw JD (1995) Isolation of cDNA clones encoding eight different human G protein γ subunits, including three novel forms designated the γ_4, γ_{10} and γ_{11} subunits. J Biol Chem 270:21765-21771

Rens-Domiano S, Hamm HE (1995) Structural and functional relationship of heterotrimeric G-proteins. FASEB J 9:1059-1066

Reuveny E, Slesinger PA, Inglese J, Morales JM, Iñiguez-Lluhi JA, Lefkowitz RJ, Bourne H, Jan YN, Jan LY (1994) Activation of a cloned muscarinic potassium channel by G protein βγ subunits. Nature 370:143-146

Schröder S, Lohse MJ (1996) Inhibition of G protein βγ subunit functions by phosducin-like protein. Proc Natl Acad Sci USA 93:2100-2104

Simon MI, Strathmann MP, Gautam N (1991) Diversity of G proteins in signal transduction. Science 252:802-808

Simonds WF, Goldsmith PK, Codina J, Unson CG, Spiegel AG (1989) G_{i2} mediates α_2-adrenergic inhibition of adenylyl cyclase in platelet membranes: in situ identification with Gα C-terminal antibodies. Proc Natl Acad Sci USA 86:7809-7813

Smrcka AV, Sternweis PC (1993) Regulation of purified subtypes of phosphatidylinositol lipid-specific phospholipase C-β by G protein βγ-subunits. J Biol Chem 268:9667-9674

Smrcka AV, Hepler JR, Brown KO, Sternweis PC (1991) Regulation of polyphosphoinositide-specific phospholipase C activity by purified G_q.Science 251:804-807

Sondek J, Bohm A, Lambright DG, Hamm HE, Sigler PB (1996) Crystal structure of a G protein βγ dimer at 2.1Å resolution. Nature 379:369-374

Stephens L, Smrcka A, Cooke FT, Jackson TR, Sternweis PC, Hawkins PT (1994) A novel phosphoinositide 3 kinase activity in myeloid-derived cells is activated by G protein βγ subunits. Cell 77:83-93

Sternweis PC (1994) The active role of βγ in signal transduction. Curr Opin Cell Biol 6:198-203

Tang WJ, Gilman AG (1992) Adenylyl cyclases. Cell 70:869-872

Taussig R, Quarmby LM, Gilman AG (1993a) Regulation of purified type I and type II adenylylcyclases by G protein βγ subunits. J Biol Chem 268:9-12

Taussig R, Iñiguez-Lluhi JA, Gilman AG (1993b) Inhibition of adenylyl cyclase by $G_{i\alpha}$. Science 261:218-221

Taussig R, Tang WJ, Hepler JR, Gilman AG (1994) Distinct patterns of bidirectional regulation of mammalian adenylyl cyclases. J Biol Chem 269:6093-6100

Taylor J.M, Jacob-Moisier GG, Lawton RG, Remmers AE, Neubig RR (1994) Binding of an α_1-adrenergic receptor third intracellular loop to Gβ and the amino terminus of Gα. J Biol Chem 269:27618-27624

Taylor SJ, Chae HZ, Rhee SG, Exton JH (1991) Activation of the β1 isozyme of phospholipase C by α subunits of the G_q class of G proteins. Nature 350:516-518

Touhara K, Inglese J, Pitcher JA, Shaw G, Lefkowitz RJ (1994) Binding of G protein βγ subunits to pleckstrin homology domains. J Biol Chem 269:10217-10220

Voyono-Yasenetskaya T, Conklin BR, Gilbert RL, Hooley R, Bourne HR, Barber DL (1994) $G\alpha_{13}$ stimulates Na-H exchange. J Biol Chem 269:4721-4724

Wall MA, Coleman DE, Lee E, Iñiguez-Lluhi JA, Posner BA, Gilman AG, Sprang SR (1995) The structure of the G protein heterotrimer $G_{i\alpha 1}\beta_1\gamma_2$. Cell 83:1047-1058

Watson JA, Katz A, Simon MI (1994) A fifth member of the mammalian G-protein β–subunit family: Expression in brain and activation of the β2 isotype of phospholipase C. J Biol Chem 269:22150-22156

Watson JA, Aragay AM, Slepak VZ, Simon MI (1996) A novel form of the G protein β subunit Gβ5 is specifically expressed in the vertebrate retina. J Biol Chem:in press

Wedegaertner PB, Bourne HR (1994) Activation and depalmitoylation of $G_{s\alpha}$.Cell 77:1063-1070

Wieland T, Jakobs KH (1992) Evidence for nucleoside diphosphokinase-dependent channeling of guanosine 5'-[γ-thio]triphosphate to guanine nucleotide-binding proteins. Mol Pharmacol 42:731-735

Wieland T, Meyer zu Heringdorf D, Schulze RA, Kaldenberg-Stasch S, Jakobs KH (1996) Receptor-induced translocation of activated guanine nucleotide-binding protein α_i subunits to the cytoskeleton in myeloid differentiated human leukemia (HL-60) cells. Eur J Biochem:in press

Wieland T, Nürnberg B, Ulibarri I, Kaldenberg-Stasch S, Schultz G, Jakobs KH (1993) Guanine nucleotide-specific phosphate transfer by guanine nucleotide-binding regulatory protein β-subunits. J Biol Chem 268:18111-18118

Wieland T, Ronzani M, Jakobs KH (1992) Stimulation and inhibition of human platelet adenylyl cyclase by thiophosphorylated transducin βγ-subunits. J Biol Chem 267:20791-20797

Wieland T, Ulibarri I, Gierschik P, Jakobs KH (1991) Activation of signal-transducing guanine-nucleotide-binding regulatory proteins by guanosine 5'-[γ-thio]triphosphate: information transfer by intermediately thiophosphorylated βγ-subunits. Eur J Biochem 196:707-716

Will-Blasczak MA, Singer WD, Gutowski S, Sternweis PC, Belardetti F (1994) The G protein G_{13} mediates inhibition of voltage-dependent calcium current by bradykinin. Neuron 13:1215-1224

Wittinghofer A (1994) The structure of transducin Gαt: More to view than just Ras. Cell 76:201-204

Wong GT, Gannon KS, Margolskee RF (1996) Transduction of bitter and sweet taste by gustducin. Nature 381:796-800

Wong UH, Federman A, Pace AM, Zachary I, Evans T, Pouysségur J, Bourne HR (1991) Mutant α subunits of G_{i2} inhibit cyclic AMP accumulation. Nature 351:63-65

Wu DQ, Katz A, Simon MI (1993) Activation of phospholipase C β2 by the α subunit and βγ subunit of trimeric GTP binding protein. Proc Natl Acad Sci USA 90:5297-5301

Wu DQ, Lee CH, Rhee SG, Simon MI (1992) Activation of phospholipase C by the α-subunit of the G_q and G_{11} protein in transfected Cos-7 cells. J Biol Chem 267:1811-1817

Xu J, Wu DQ, Slepak VZ, Simon MI (1995) The N-terminusof phosducin is involved in binding of βγ subunits of G protein. Proc Natl Acad Sci USA 92:2086-2090

Yan K, Kalyanaraman V, Gautam N (1996) Differential ability to form G protein βγ complexes among members of the β and γ subunit families. J Biol Chem. 271:7141-7146

Yatani A, Codina J, Imoto Y, Reeves JP, Birnbaumer L, Brown AM (1987) A G protein directly regulates mammalian cardiac calcium channels. Science 238:1288-1292

Yatani A, Imoto Y, Codina J, Hamilton SL, Brown AM, Birnbaumer L (1988a) The stimulatory G protein of adenylyl cyclase, G_s, directly stimulates dihydropyridine sensitive

skeletal muscle Ca^{2+} channels: evidence for direct regulation independent of phosphorylation by cAMP-dependent protein kinase. J Biol Chem 263:9887-9895
Yatani A, Mattera R, Codina J, Graf R, Okabe K, Padrell E, Iyengar R, Brown AM, Birnbaumer L (1988b) The G protein-gated atrial K^+ channel is stimulated by three distinct $G_i\alpha$-subunits. Nature 336:680-682

G protein coupled receptors: structure, functions and mutation

Joël Bockaert
CNRS - UPR 9023
CCIPE
141, rue de la Cardonille
34094 Montpellier Cedex 5
France

The evolution of multicellular organisms has been conditioned by the capacity of their cells to communicate in order to coordinate their cell division, their cellular differentiation, their embryonic development as well as their different physiological functions. The messages used for cellular communication are hormones, neurotransmitters, growth and differentiation factors, cytokines, chemokines. The receptors for these messages as well as receptors for messages of the environment such as odors, light, tasteful molecules are, in most cases, membrane-bound molecules with steroid and thyroid hormone receptors as the important exception. During the last 10 years, purification and cloning of most of these receptors have revealed that they belong to a small number of protein families characterized by their structure and their functions. These families include: 1) receptor which are channels, such as the nicotinic receptor; 2) receptors which are enzymes such as the tyrosine kinases (insulin, EGF, PDGF receptors), the guanylate cyclase receptors (atrial natriuretic factor receptors), or serine/threonine kinases (TGF-β receptors); 3) receptors for cytokines (Il2, Il3,GH, prolactin, interferon) which activate separate tyrosine kinases (Jak and fyn kinases); 4) GTP binding protein (G proteins) coupled receptors (GPCRs).

The use of GPCRs to communicate:

The GPCR family is certainly the most important, with probably more than one thousand members (0.1% of the genome). Among them, several hundred are involved in the recognition of odors and tasteful molecules (sweet and bitter) (Bockaert 1991; Buck and Axel 1991; Dohlman, Thorner *et al.* 1991; Lancet and Ben-Ari 1993; Bockaert 1995; Kolesnikov and Margolskee 1995). It is this family which has been the more successful during evolution,

NATO ASI Series, Vol. H 101
Molecular Mechanisms of Signalling
and Membrane Transport
Edited by Karel W. A. Wirtz

being able to recognize very different chemical structures as messages, such as photons, ions (Ca^{2+}), organic odorant molecules, amino-acids and derivatives, nucleotides, nucleosides, lipids, peptides, proteins (Figure 1).

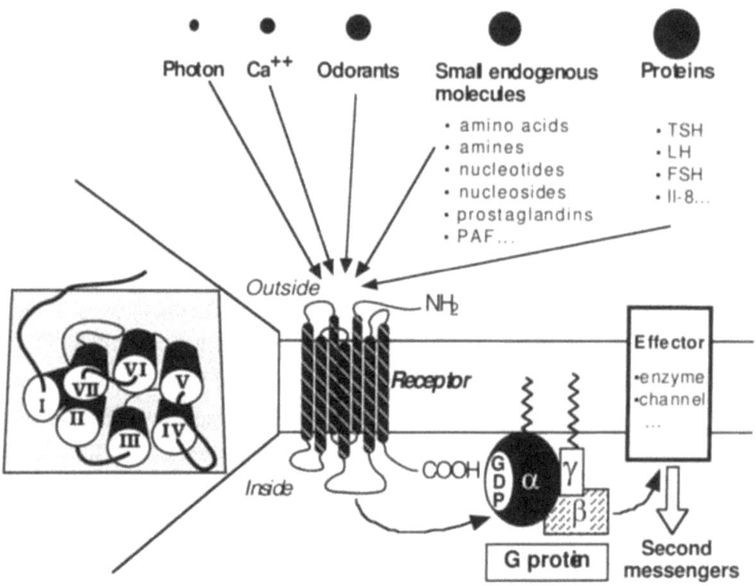

Figure 1: General characteristics of GPCRs.
Left part: a view from the top of GPCR showing the seven transmembrane domains (TM). Note the triedric structure made by TM-I, TM-II, and TM-VII.
Right part: A lateral view of the membrane showing: 1) a GPCR, 2) a trimeric G protein composed of the an α subunit which bind GTP an a tight complex formed by the β and γ subunits. The α and βγ subunits are associated to the membrane by lipids , myristate and palmitate for the α subunits and farnesyl or geranyl-geranyl for γ subunits (Neer 1995; Wedegaertner, Wilson *et al.* 1995). 3) effectors which can be enzymes such as adenylyl cyclases and phosphodiesterases or channels such as Ca^{2+} or K^{+} channels (Neer 1995).

GPCRs are monomeric proteins which have a hydropathy profile (according to methods developed by Kyte and Doolittle, 1982), that suggests the presence of seven transmembrane domains (TM) with an α helix structure joined by three intracellular (i1, i2, i3) and three extracellular loops (e1, e2, e3). The TMs are tightly associated to form a narrow dihedral pocket in which agonists and antagonists are buried, as is the case in Group Ia receptors such as adrenergic receptors (see figure 8). It has been established for gonadotropin receptors, that TM-II and TM-VII are in contact. In particular, two amino acid residues well conserved in Group I receptors, *i.e.* an aspartic acid (D) in TM-II and an asparagine (N) in TM-VII, are interacting. One can exchange those amino acids without modifying the receptor structure having two aspartic acid or two asparagine residues in those TMs disorganizes the receptor (Zhou, Flanagan *et al.* 1994). Other experiments indicate that TM-I and TM-VII are also close together (Suryanarayana, von Zastrow *et al.* 1992) suggesting that TM-I, TM-II and TM-VII make a triedric structure (Figure 1). Chimeric receptors of α2 adrenergic receptors (TM-I to TM-V) and muscarinic receptors (TM-VI to TM-VII) or of α2 adrenergic receptors (TM-VI to TM-VII) and muscarinic receptors (TM-I to TM-V) are not functional. However, when these receptors are co-expressed both α2 adrenergic and muscarinic specific binding can be detected (Maggio, Vogel *et al.* 1993). This suggests: 1) that GPCR is formed by two functional domains consisting of TM-I to TM-V and TM-VI–TM-VII joined by the i3 loop and 2) that receptor dimers are formed in which domains TM-I to TM-V from one chimera associate with homologous receptor domains TM-VI and TM-VII from the other chimera. While this does not prove that dimerization occurs with natural receptors, it indicates that functional receptor dimers could form.

It should be emphasized that the membrane arrangement of most GPCRs is largely speculative, primarily based on the hydropathicity analyses of the sequence, on mutagenesis experiments as well as on immunolocalization of sequence-directed antibodies. However, none of these data are in contradiction with the seven transmembrane model. In addition, we have more information about rhodopsin, the light sensitive GPCR (Chabre and Deterre 1989). This receptor belongs to the Group Ia (Figure 8) and a two-dimentional crystal has been obtained (Schertler, Villa *et al.* 1993). An electron microscopic analysis with a 9Å resolution has been compared with a similar analysis of the bacteriorhodopsin (a seven TM protein which is not a GPCR but a H^+ pump) (Figure 2). For rhodopsin, the distinct 4 electronic density regions represent the four α helices perpendicularly oriented within the membrane, whereas the electronic density region forming a bow-like structure, represents the three other α helices. The morphology of bacteriorhodopsin is less wide and more spread suggesting a different organization of the α helices (Figure 2).

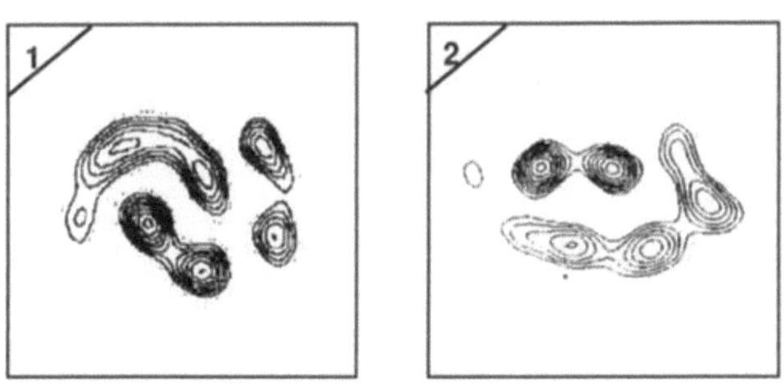

Figure 2: Morphology of rhodopsin (1) and bacteriorhodopsin (2) with a 9Å resolution. (Modified from Schertler, Villa *et al.* 1993).

The essential function of GPCRs is, on the one hand, to recognize the message and, on the other hand, to catalyse the GDP/GTP exchange at the level of G proteins. GPCRs are just exchange factors similar, in term of function, to the exchange factors of small G proteins such as Sos or Ras-GRF (Boguski and McCormick 1993). When an agonist interacts with a GPCR, this receptor associates with a (or several) specific heterotrimeric G proteins in the GDP form (Figure 3). The α subunit loses its affinity for GDP and a short-life intermediary complex is formed (RCPG-αβγ) without any nucleotide bound. (Chabre and Deterre 1989). *In vivo*, GTP rapidly occupies the nucleotide site. GTP binding to the α subunit triggers the dissociation of the three partners, the receptor, the α subunit and the βγ dimers (Figure 3). *In vitro*, the RCPG-αβγ complex without nucleotide bound is very stable. It represents the high-affinity state of the receptor for the ligand.

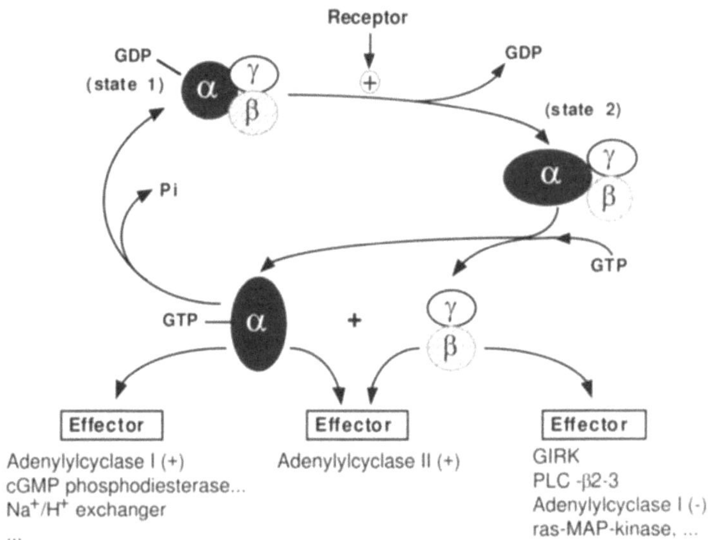

Figure 3: Activation of heterotrimeric G protein by GPCRs.
For explanation see text.

For a long time, it has been thought that the α subunit was the only functional subunit regulating effectors such as adenylyl cyclase, cGMP phosphodiesterase, channels, phospholipase C β1, Na^+/H^+ exchanger (see Figure 4). We now realize that the βγ dimers modulate alone or in combination with a great number of effectors (listed in Figure 4).

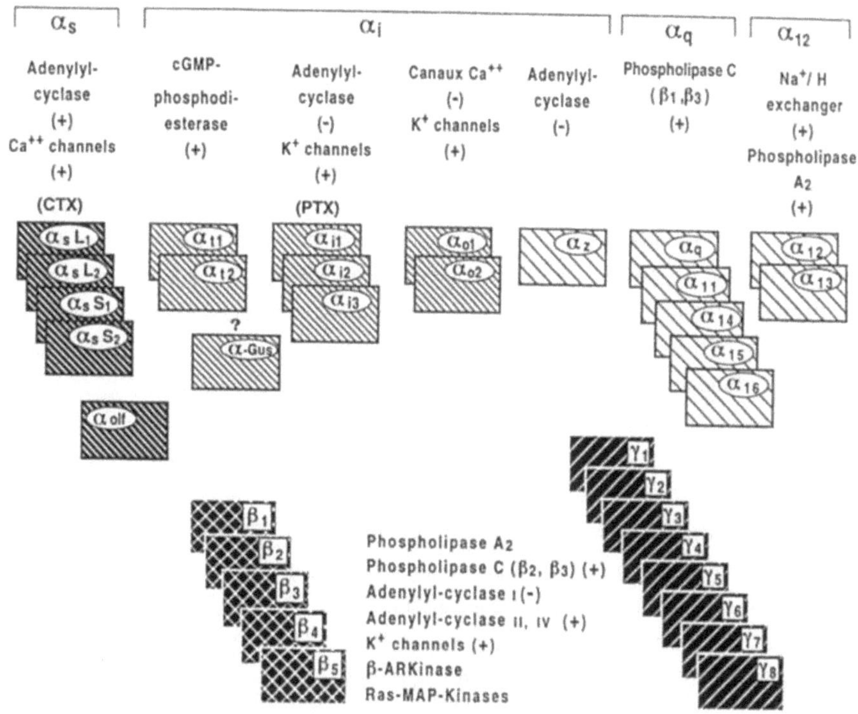

Figure 4: Diversity and functions of the α and the βγ dimers of G proteins. Based on primary sequence homologies, 4 classes of a subunits have been recognized: αs, αi, αq, and α12. The main biological effectors of these subunits are indicated. (-) indicates an inhibition whereas (+) indicates a stimulation. Some α subunits are ADPribosylated by the cholera-toxin (CTX), others by the Bortetella pertussis toxin (PTX), some by both. Finally some are not affected by these toxins. At least 8 different γ and 5 different β subunits have been cloned. The main effectors of βγ dimers are indicated.

GPCRs stimulate pleiotropic transduction cascades:

With the exception of the photon-retinal complex which activates only one receptor, the rhodopsin, each message molecule activates several GPCRs. Serotonin is certainly the leader with regard to the number of receptors (13) that a neurotransmitter is able to stimulate. Each GPCR can stimulate several G proteins and, therefore, several effectors. To take an example, the D2 dopaminergic receptors which control prolactin secretion are able to inhibit the adenylyl cyclase via Gi2, voltage sensitive Ca^{2+} channels via Go, as well as stimulate K^{+} channels via Gi3 (Llédo, Homburger *et al.* 1992). The complexity is even higher since each α subunit activates or inhibits several effectors. Generally those effectors exist under several entities (Figure 4). There are, for example, eight different adenylyl cyclase molecules, each of them having a different pattern of regulations. Adenylyl cyclase of type I is activated by αs and the complex Ca^{2+}/calmodulin acting in a coincident manner. It will be activated in neurons in which αs coupled receptors are activated together with an increase in Ca^{2+}. Such an adenylyl cyclase molecule is called a coincident detector and adenylyl cyclase I has been implicated in pre-synaptic plasticity events. Adenylyl cyclase II and IV are activated by αs in the presence of relatively high amounts of βγ released by GPCRs activating Gi or Go. Therefore, the co-stimulation of a receptor which activates Gs, and of a receptor which activates Gi (which inhibit the adenylyl cyclase when stimulated alone) induces a higher stimulation of the type II and IV adenylyl cyclases than that produced by the stimulation of Gs alone (Tang and A.G. 1991). βγ dimers have also numerous effectors (Figure 4). One of them is particularly interesting because it illustrates the cross-talk between the transduction mechanisms of GPCR and the tyrosine kinase receptors (TKRs). Indeed, many GPCRs coupled to Gi, such as serotonin 5-HT1A receptors (Figure 5) stimulate DNA synthesis and mitogenic processes when associated with the stimulation of TKRs such as EGF receptors (Varrault, Bockaert *et al.* 1992). Whereas the weak stimulation of DNA synthesis by EGF receptors is not inhibited by PTX, the potent stimulation of DNA synthesis obtained with the co-stimulation of 5-HT1A plus EGF receptors is inhibited by PTX (Figure 6) indicating that a Gi protein is involved in the action of 5-HT1A receptors.

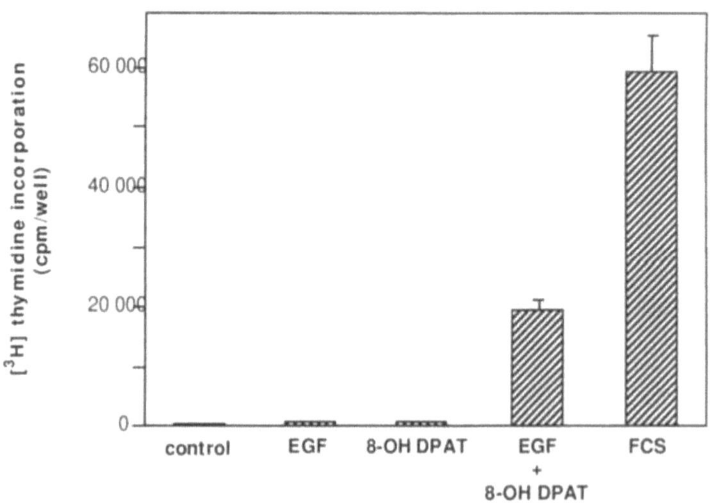

Figure 5: Effects of different mitogens on DNA synthesis reinitiation.
The NIH-3T3 cells used were transfected with 5-HT1A receptors. The concentration of different drugs were 10ng/ml EGF, 0.1μM 8-OH DPAT (a 5-HT1A agonist). Modified from (Varrault, Bockaert *et al.* 1992).

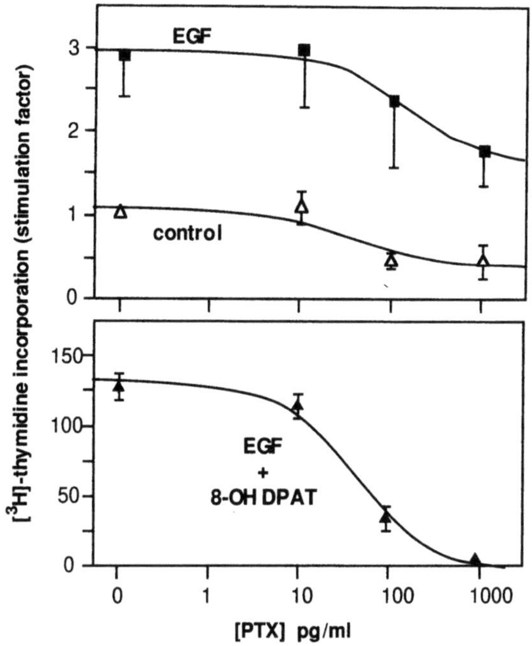

Figure 6: Bordetella pertussis (PTX) does not inhibits the weak EGF induced DNA synthesis but does inhibit the potent DNA synthesis induced by the coincident stimulation of EGF and the GPCR, 5-HT1A receptor.
Drugs concentrations are as in figure 5. Modified from (Varrault, Bockaert *et al.* 1992)

We know that these Gi coupled GPCRs activate the proto-oncogene ras via the βγ released from Gi (Crespo, Xu *et al.* 1994; Faure, Voyno-Yasenetskaya *et al.* 1994). The exact transduction cascade is not completely understood. One possibility is that βγ phosphorylates Shc, an adaptor protein which then activates Sos and ras. Ras activates the Raf/MEK/ERK (=MAP kinase) mitogenic cascade. Therefore, the synergistic effect between GPCRs and TRKs could be at the level of Sos (Figure 7). How βγ induces phosphorylation of Sch is not known.

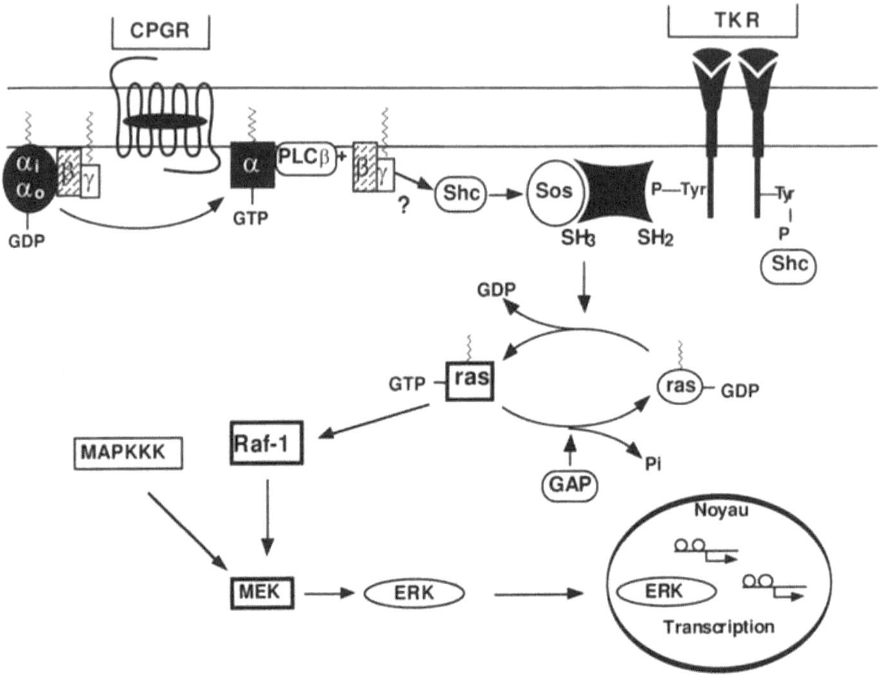

Figure 7: Control of mitogenic signals by GPCRs and Tyrosine kinase receptors(TKR) Autophosphorylated TKR are recognized by SH2 domains of Grb2, an adaptor protein. The two SH3 domains of Grb2 interact with the exchange factor Sos (or other exchange factors). The complex Grb2-Sos catalyse the GDP/GTP exchange on ras. ras-GTP interacts and activates the kinase Raf-1. kaf-1 activate the MEK kinase which then activate the ERK kinases (=MAP-Kinases). The ERK kinases are translocated in the nucleus and phosphorylate transcription factors leading to cell division. βγ released from heterotrimeric G proteins of Gi (or Go) subtypes phosphorylates Sch which activates Sos and then the ERK kinase pathway.

Different genes but only one topology:

It is possible to classify GPCRs into three main groups (I, II, III) using two main criteria: their primary sequence and the localisation of the agonist binding site. The main difference between these three groups is the absence of primary sequence homologies. Therefore, if they have evolved from a common ancestral gene, the divergence probably occurred early during evolution. If these three groups represent an example of convergent evolution between genes derived from three different ancestral genes, one can wonder about the nature of the selection which finally selected proteins having seven TM to be coupled to G proteins. Perhaps it is the more adequate topology for an efficient coupling to G proteins.

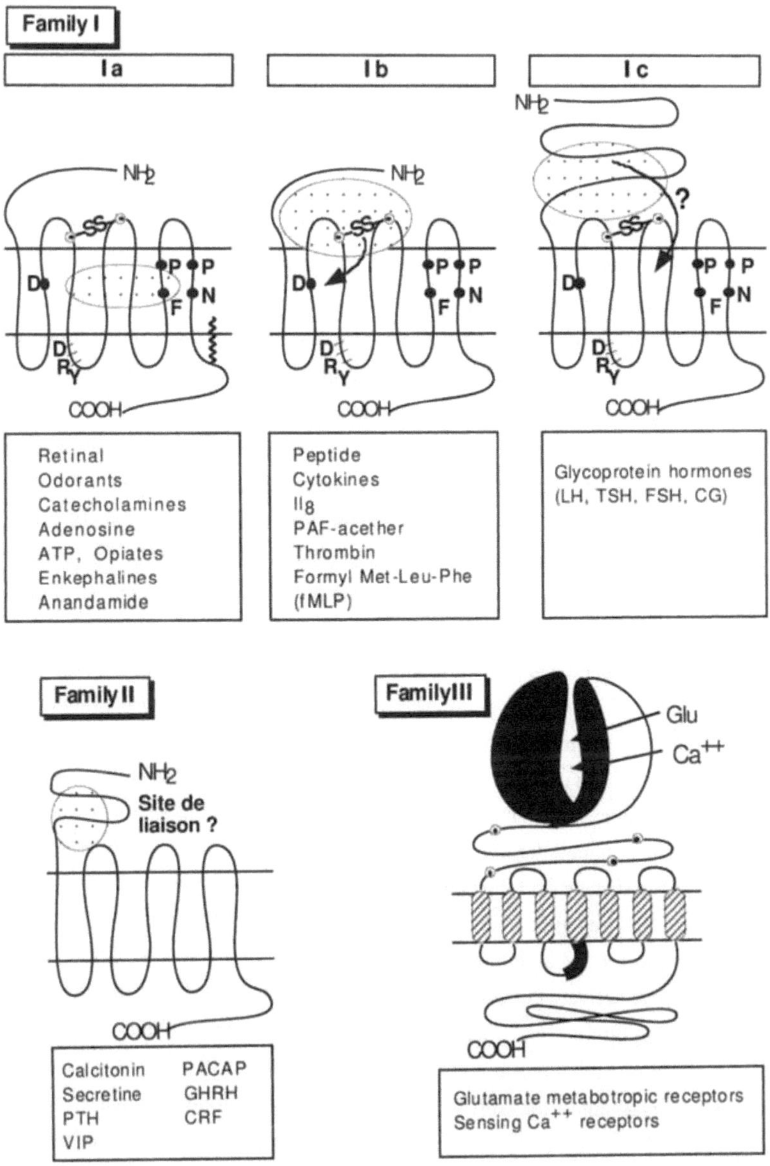

Figure 8: GPCR classification
For explanation see text

Group I:

It comprises most of the G protein-coupled receptors including olfactory receptors (Lancet and Ben-Arie 1993). The prototypes of the group are the β-adrenergic receptors and rhodopsin (Bockaert 1991; Dohlman, Thorner *et al.* 1991; Hibert, Hoflack *et al.* 1993; Bockaert 1995) (Figure 8). The group I can be divided into three subgroups depending upon the possible location of the agonist-binding site. In group Ia (Figure 8), the binding site is localized within the TMs (the distance from the membrane surface to the antagonist-binding site has been estimated to be equal to 12Å in β-adrenergic receptors (Tota and Strader 1990). Group Ia includes receptors for small ligands as indicated in Figure 8. In group Ib, the binding site for small peptides involves the N-terminal domain, the external loops as well as the external part of the pocket within the TMs (Fong, Huang *et al.* 1992; Gerszten, Chen *et al.* 1994; Trumpp-Kallmeyer, Chini *et al.* 1995). It is proposed that the addressing part of the peptide molecule (conferring a high affinity to the peptide) binds to the external domains of the receptors including the external loops whereas the message part of the molecule (triggering activity) binds within the TMs (Trumpp-Kallmeyer, Chini *et al.* 1995). A special commentary can be given on thrombin receptors in which the agonist is a peptidic sequence localized just before the TM-I within the N-terminal domain. This sequence becomes a new N-terminal domain, following proteolytic cleavage of the native N-terminal domain by thrombin. This sequence becomes at the same time the thrombin receptor agonist (Vu, Hung *et al.* 1991). Recently another proteinase-activated receptor has been cloned (Nystedt, Emilsson *et al.* 1994). Finally, group Ic includes receptors for glycoprotein hormones (TSH, LH, FSHhCG) (Figure 8). The N-terminal domain is very long (350-400 amino-acids) and contains repeated leucine enriched domains similar to those found in ribonuclease inhibitor. The folding of the extracellular domain of the TSH receptor would be a doughnut sector with the inside concave surface of the sector composed of β sheets which provides the recognition surface of the TSH receptor.(Van Sande, Parma *et al.* 1995). It is not known how the consequence of hormone binding on the N-terminal domain is transduced to an activation of the receptor able to activate the G proteins (at the intracellular surface). In FSH receptors, the role of e1 both in binding and transduction has been described, although the amino-acids involved in these two functions are different (Ji and Ji 1995).

The sequence alignment of the GPCRs of group I reveals the following consensus sequences: a) an aspartate (D) in TM-II which is important for the allosteric regulation of the receptor by monovalent cations, protons and for the coupling to G proteins (discussed in Zhou, Flanagan *et al.* 1994); b) an aspartate in TM-III is conserved in receptors that have an agonist containing a cationic amino group such as cathecholamines, serotonine and acetylcholine; c) the DRY sequence (aspartate, arginine, tyrosine) or ERW (glutamate, arginine, tryptophane) at the N-terminal part of i2 is the signature of group I. It has been proposed that the arginine play a role in the coupling to G proteins (Oliviera, Paiva *et al.* 1994); d) a cysteine, relatively well conserved at the N-terminal part of the C-terminal tail is often palmitoylated in adrenergic receptors (figure 8); e) two cysteines are linked by an S-S bridge between e1 and e2.

Group II:
Group II contains fewer members. They are receptors for peptides like secretine, calcitonin, PTH, VIP, PACAP (pituitary adenylyl cyclase activating peptide; see Figure 8). Few structure-function studies have been done with these receptors. The N-terminal domain, the first extracellular loop, but also TM-II are important for binding and selective recognition of the agonists (Houssami, Findlay *et al.* 1994; Turner, Bambino *et al.* 1996). We have recently found a splice variant in the N-terminal domain of the PACAP receptor which modifies the affinity and potency of agonists for the receptor (Pantaloni, Brabet *et al.* 1996).

Group III:
This group includes 8 genes coding for the metabotropic glutamate receptors (mGluRs) and one gene coding for a Ca^{2+} sensing receptor localized in parathyroid glands, kidneys and brain (Nakanishi 1992; Brown, Gamba *et al.* 1993; O'Hara, Sheppard *et al.* 1993; Pin and Bockaert 1995) (Figure 8). It is surprising that mGluRs and the Ca^{2+} sensing receptor share 30% identity, the position of 20 cysteines (17 in the N-terminal, 1 in e1 and e2, 1 in TMV). Their homology, in the N-terminal domain, with the periplasmic amino-acid binding proteins of bacteria (especially with the leucine, isoleucine, valine binding protein: LIVBP) has been the basis of a model proposed for the glutamate binding site in mGluRs but also in ionotropic glutamate receptors (O'Hara, Sheppard *et al.* 1993; Stern-Bach, Bettler *et al.* 1994; Pin and Bockaert 1995). In this model, similar to the amino-acid binding site of periplasmic proteins of bacteria, the glutamate binding site of glutamate receptors has two globular domains and a hinge region. The two globular domains bind and then trap the amino-acids or the glutamate (Figure 8). A direct confirmation that the binding site for glutamate is in the extracellular N-terminal domain was recently provided. Indeed, the production of this domain of the AMPA receptor subunit (GluR4), gives a soluble protein which binds glutamate and has the same pharmacology as the native subunit (Kuusinen, Arvola *et al.* 1995). It is noteworthy that loss-of-function mutations were identified in close proximity of the hinge region of the Ca^{2+} sensing receptor in patients suffering from Ca^{2+} homeostasis (Pollak, Brown *et al.* 1993).

Allosteric binding sites in GPCRs for pharmaceutical companies:

Pharmaceutical companies have found a great number of non-peptidic antagonists for peptidergic receptors. In general these antagonists have a high affinity and specificity for their receptors, although they have a chemical structure completely different from that of the natural agonist. In the case of the drug named CP-96345, substance P receptor NK1 is inhibited with great potency (Ki=14nM) (Figure 9). CP-96345 binds to the NK1 receptor at a site which is distinct from the site which binds substance P. Indeed, mutations localized in the external part of TM-V and TM-VI suppress CP-96345 binding but not substance P binding (Gether, Johansen *et al.* 1993). This suggests that CP-96345 binding sites and substance P binding sites are partially distinct. It is supposed that CP-96345 is freezing the modifications of the receptor conformation induced by substance P. This hypothesis is supported by the elegant experiment of the laboratory of T.W. Schwartz, which has replaced two residues at the top of TM-V and one at the top of TM-VI, in the CP-96345 binding site, with His residues. Under these conditions, Zn^{2+} is able to inhibit the activation of the NK1 receptor (Elling, Moller Nielsen *et al.* 1995). Similar experiments have been done with k-opioid receptors (Thirstrup, Elling *et al.* 1996). Even more surprising is the observation that septide, a shorter form of substance P (Figure 9) activates NK1 receptor but bind to a site partially different from the site to which substance P binds (Pradier, Ménager *et al.* 1994). All these experiments indicate that it is possible to find drugs which are not structurally related to the

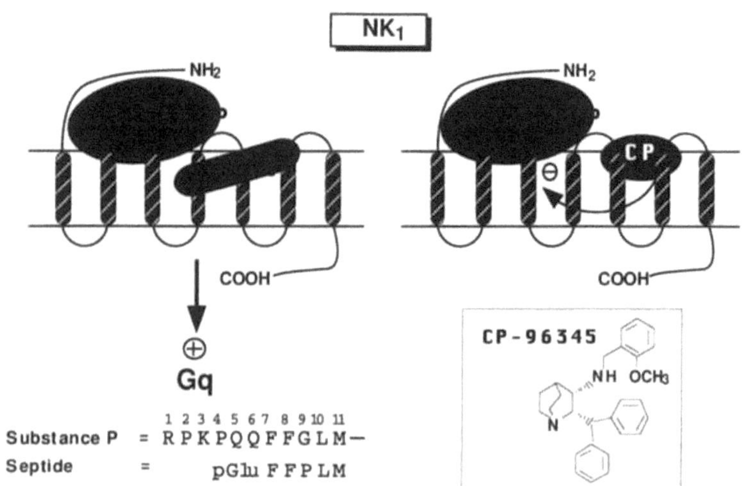

Figure 9: Putative allosteric binding sites in substance P receptors.
Two agonists of NK1 receptor, substance P, the natural agonist and septide a synthetic one bind to partially different sites in the receptor but are both agonists. Similarly, CP-96345 is an antagonist of NK1 receptors which bind to an "allosteric" site different from the substance P binding site. pGlu is pyro Glu.

agonists but which freeze the receptor under an inactive conformation. The existence of such allosteric sites within the pocket made by TMs has certainly no physiological significance, but is certainly very useful for the pharmaceutical companies.

Domains of GPCRs involved in coupling to G proteins:

A great number of experiments indicate that i3 loop is the most important domain in GPCRs for their coupling to G proteins (Bockaert 1991). This is indeed the case in group I and II of GPCRs but not in group III in which we have demonstrated that it is the i2 loop which is the most important domain for this function (Pin, Joly *et al.* 1994; Gomeza, Joly *et al.* 1995). More precisely, it is the N-terminal and the C-terminal part of i3 loop which are important since the internal region of i3 can be deleted without any effect (Lefkowitz and Caron 1988). This is also illustrated by the two following examples:

1) In muscarinic receptors, of which there are five types (m1-m5), the N-terminal part of i3 is highly conserved in m1, m3, and m5 (coupled to phospholipase C) but not in m2 or m4 (negatively coupled to adenylyl cyclase). Introduction of a short stretch of 16-17 residues from the N-terminal domain of i3 or m3 into the corresponding region of m2 produces a chimera as potent as m3 to stimulate phospholipase C. The reciprocal chimera allows the negative coupling of m3 to adenylyl cyclase.
2) We have cloned six different splice variants of the PACAP receptors (pituitary adenylyl cyclase activating peptide) which are characterized and compared to the basic receptor (PACAP-R) by additional insertion(s) of two cassettes named "hip" and "hop" in the loop i3, either alone or in combination (Figure 10). PACAP-R is coupled both to adenylyl cyclase and to phospholipase C. The characterization of transduction pathways of these splice variants clearly demonstrates, in naturally expressed GPCRs , the important role of i3. Indeed, the presence of the hop cassette alone has no influence on the coupling to adenylyl cyclase and phospholipase C (Figure 10). The hip cassette alone abolishes coupling to phospholipase C and alters potency of coupling to adenylyl cyclase. The combination of the two cassettes hip-hop gives an intermediate phenotype, displaying slightly altered efficiency for adenylyl-cyclase stimulation and modified pattern of phospholipase C activation (Spengler, Waeber *et al.* 1993; Journot, Spengler *et al.* 1994)

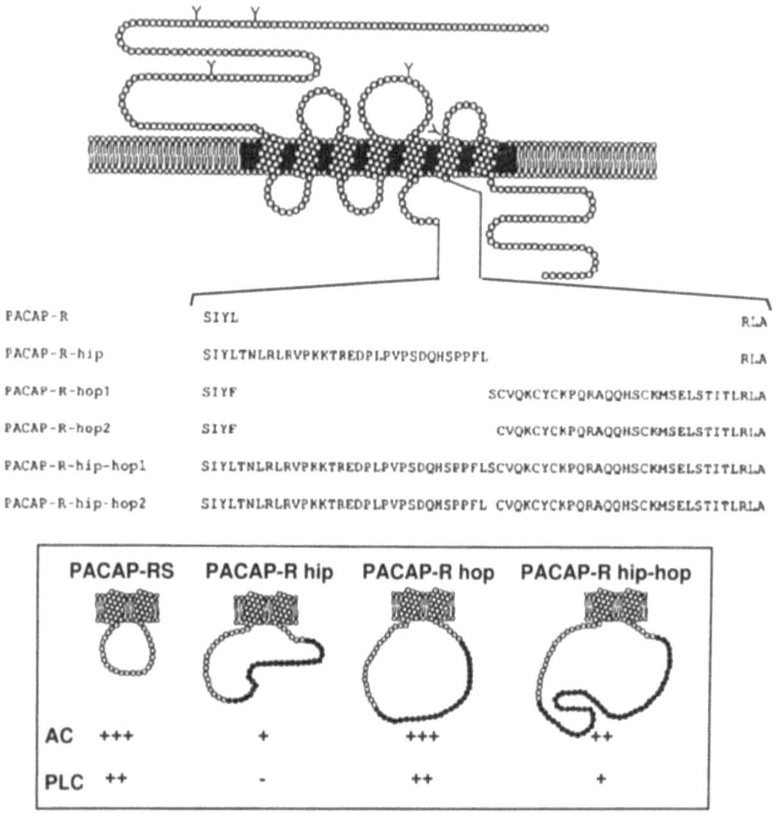

Figure 10: Architecture and functional properties of the PACAP receptors.
For explanations see text and (Spengler, Waeber *et al.* 1993; Journot, Spengler *et al.* 1994).

In metabotropic glutamate receptors (mGluR), a receptor of group III, we have demonstrated, by constructing chimeras between the phospholipase C-coupled mGluR1 and the negatively adenylyl cyclase coupled mGluR3, that i2 plays a critical role in G-protein coupling (Pin, Joly *et al.* 1994; Gomeza, Joly *et al.* 1995). However, as for other GPCRs, we found that all intracellular segments are indeed involved in the coupling and activation of the G-protein (Gomeza, Joly *et al.* 1995) Without i2 there is no coupling, but i2 alone is not sufficient, either another loop (i1 or i3) or part of the C-terminal domain is needed (Figure 11).

Interestingly, both i3 of group I and i2 of group III possesses amphipathic α−helices extending TM-3 and TM4 towards the cytoplasm. This feature is believed to be important for the coupling of GPCRs to G proteins and is found in mastoparan, an amphipathic peptide from the wasp venom, known to activate directly Gi and Go proteins. This stimulation is implicated in the histamine release from mast cells induced by the peptide (Higashijima, Burner *et al.* 1990). Peptides from the end of i3 have been found to mimic receptor action, further indicating a primordial role of this domain (Varrault, Le Nguyen *et al.* 1994).

In mGluRs, but also in prostaglandin EP3 receptors, analysis of the splice variants of the C-terminal domain clearly indicates an important modulatory role of this domain in the regulation of coupling to G proteins (Pin, Waeber *et al.* 1992; Journot, Spengler *et al.* 1994; Prezeau, Gomeza *et al.* 1996). GPCRs interact with different domains of Gα proteins: 1) a domain localized in the β6-L10-α5 region (see Conklin and Bourne 1993) and 2) the N and the C-terminal domain of G α. This latter region appears to be crucial for interaction but also for the specificity of interaction. Mutations in the C terminus of Gα, its covalent modification by PTX in Gi or Go, peptide-specific antibodies directed against it and peptides mimicking C-terminal sequences, all inhibit receptor-mediated activation of G proteins (discussed in (Conklin, Farfel *et al.* 1993)). Few amino-acids of the C-terminus of Gα control the specificity of interaction. For example, Gi/Go coupled receptors (such as α2-adrenergic receptors or D2 dopaminergic receptors) are generally unable to activate Gαq when expressed

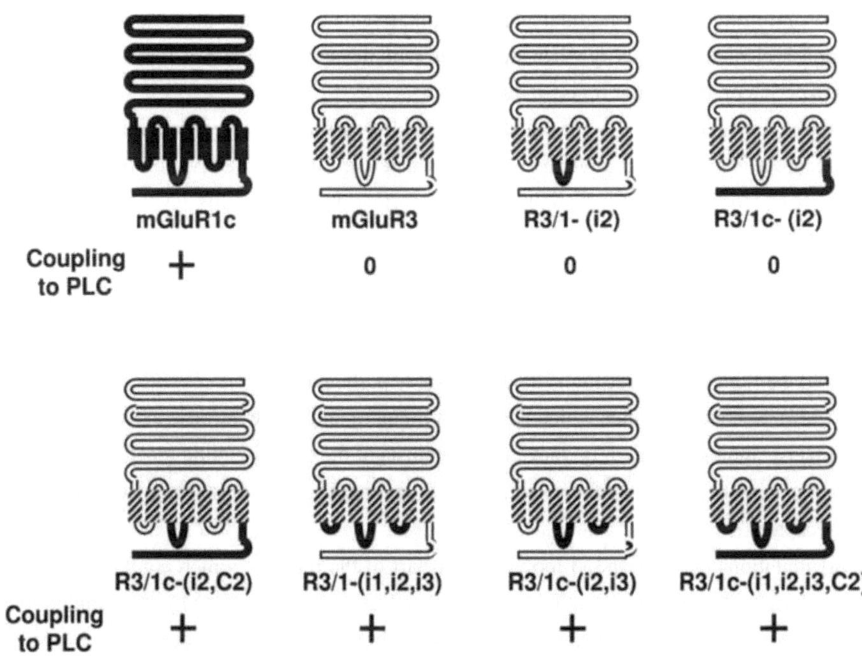

Figure 11: Chimeras between mGluR1c coupled to PLC (black) and mGluR3 negatively coupled to adenylyl cyclase (white) and their ability to activate the phospholipase C (PLC).
C: C-terminal domain

at a reasonable level. However, those receptors activate a chimera made with Gqα in which a minimum of 3 amino-acids of its C-terminal domain (N-L-V) is replaced by the corresponding residues of Gi2α (G-L-F) or Goα (G-L-Y) (Conklin, Farfel *et al.* 1993). This is also true for the mGluRs coupled to Gi/Go, despite the fact that their coupling domains are different (see above) from that of other GPCRs. Interestingly, the G15α and G16α subunits (which stimulate phospholipase C) can be activated by a great variety of receptors normally coupled to Gi /Go and even Gs (Offermanns and Simon 1995).

GPCRs and pathology:

We have discussed the role of the C-terminal part of i3 in the coupling to G proteins without referring to a series of experiments in which point mutations of key residues have been done. For example, the replacement of alanine 293 in the α1-adrenergic receptor by each of the 19 others amino-acids gives a receptor which has an intrinsic activity. This indicates that the receptor has now the ability to activate the G protein without agonist (Kjelsberg, Cotecchia *et al.* 1992) (Figure 12). When glutamate (E) replaces alanine (A), the intrinsic activity of the receptor is close to the activity when stimulated with the agonist (Figure 12).

This experiment and similar others indicate that:

1) the native receptor is the one which gives the lowest basal activity, probably because the receptor is maintained under a constrained inactive form. The activation of the receptor will consist in modifying slightly the conformation of i3 at this level to release this inhibition.
2) It is possible that some native receptors may have some tonic intrinsic activity. This is the case for splice variants of mGluR1 and mGluR5 having a long C-terminal domain (Prezeau, Gomeza *et al.* 1996), for the 5-HT2 receptors (Barker, Westphal *et al.* 1994) and the β-adrenergic receptors at least in some particular environment (Götze and Jakobs 1994). Some compounds which have been found to inhibit the intrinsic activity are called inverse agonists. Generally, they are antagonists of the corresponding receptor. However, only some of the receptor antagonists have inverse agonism on constitutively activated receptors.
3) If mutations occur in the i3 loop or other part of the receptor which confer an intrinsic activity to the receptor, a pathology is expected, corresponding to an hyper stimulation of the receptor. Indeed, somatic mutations were first found in TSH receptors expressed in toxic adenoma (9 over 11 adenoma studied) (Parma, Duprez *et al.* 1993; Van Sande, Parma *et al.* 1995) (Figure12). One of the first found was Ala623Ile which affects the exact TSH homolog of Ala293 of the α1-adrenergic receptor (Figure 12). Recent data

have demonstrated that mutations responsible for toxic adenoma are actually scattered over the e1, e2, TM-III to TM-VII, and i3. Germline mutations have also been found in TM-III and TM-VII (Van Sande, Parma *et al.* 1995). Several other activating mutations, responsible for hereditary pathologies, have been described; for example, in rhodopsin, leading to retinitis pigmentosa (autosomal dominant) and in LH receptors (autosomal dominant) leading to familial precocious puberty (Coughlin 1994). In addition to these activating mutations of GPCRs, there are numerous inactivating mutations (Spiegel, Weinstein *et al.* 1993; Coughlin 1994), some of which are indicated in figure 13. In addition to these mutations, polymorphism of genes coding for GPCRs may be responsible for significant differences in phenotypes. Two examples have been described: 1) in rhodopsin, 62% of the population have a serine in position 180, whereas the remaining population has an alanine. The former have a better sensitivity to red light than the latter (Mollon 1992) and 2) in β3-adrenergic receptors, the presence of an arginine at position 64 instead of a tryptophane is associated with a higher frequency of obesity (Clement, Vaisse *et al.* 1995).

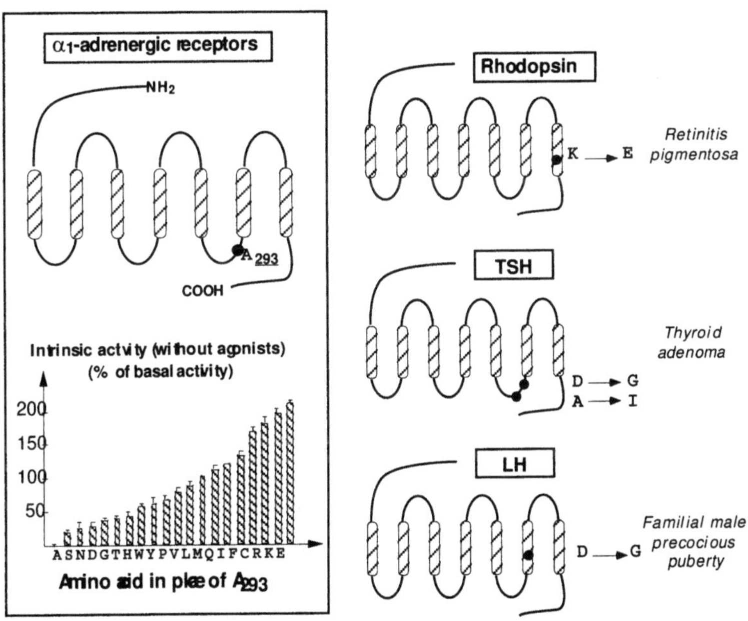

Figure 12: Activating mutations in GPCRs
Left part: experimental mutations of alanine (A) 293 in i3 confers an intrinsic activity to α1-adrenergic receptors, the intensity of which is depending on the nature of the replacing residue (adapted from (Kjelsberg, Cotecchia *et al.* 1992).
Right part: some activating mutations of GPCRs inducing pathologies.

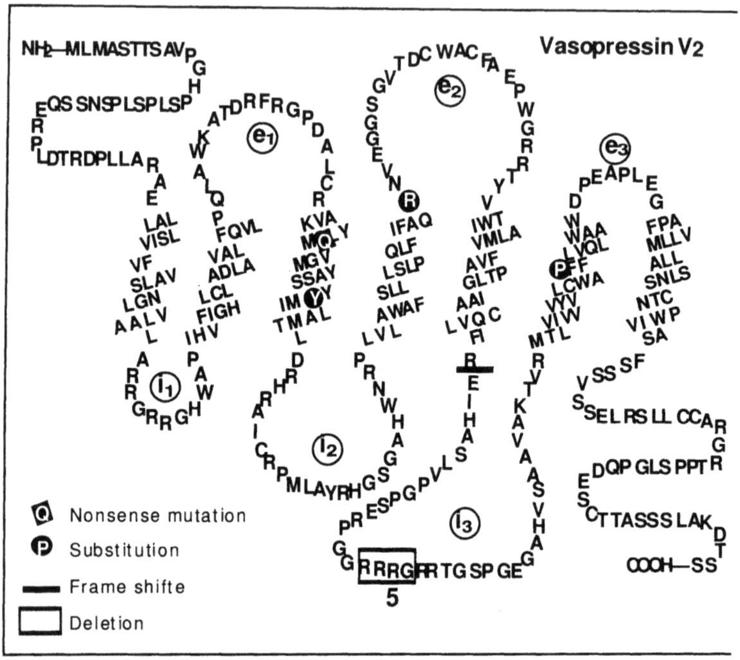

Rhodopsin	• Retinitis pigmentosa
Vasopressin (V_2)	• Diabetes insipidus
ACTH	• Familial glucocorticoid deficiency
Ca^{++} Receptor	• Familial hypocalciuric hypercalcemia
	• Neonatal hyperparathyroidism
GRF	• Little mouse, Dwarfism
MSH	• Mouse coat color (recessive yellow)
Endothelin B	• Multigenic Hirschsprung's disease
LH	• Male pseudohermaphrodism
Thromboxane A_2 receptor	• Inherited bleeding disorder

Figure 13: Loss of function of some GPCRs associated with pathologies.

Conclusion:

Started more than 25 years ago, with the studies of M. Rodbell 's laboratory, the studies of GPCRs have rapidly expanded during the last 5 years. Two difficulties remain. The first concerns the analysis of the three-dimensional structure of these proteins and analysis of their conformational changes during activation by their regulatory ligands. The second difficulty is related to the diversity of the proteins involved in the transduction pathways and their complex relationships. The necessity to have an integrated idea of the puzzling array of multiple transduction pathways generated by GPCRs is certainly a challenge for students of this summer school.

References

Barker, E. L., R. S. Westphal, *et al.* (1994) Constitutively active 5-hydroxytryptamine$_{2c}$ receptors reveal novel inverse agonist activity of receptor ligands. J. Biol. Chem. 269:11687-11690.

Bockaert, J. (1991) G proteins and G-protein-coupled receptors: structure, function and interactions. Curr. Opin. Neurobiol. 1:32-42.

Bockaert, J. (1995) Les récepteurs à sept domaines transmembranaires: physiologie et pathologie de la transduction. médecine/science 11:382-394.

Boguski, M. S. and F. McCormick (1993) Proteins regulating ras and its relatives. Nature 366:643-654.

Brown, E. M., G. Gamba, *et al.* (1993) Cloning and characterization of an extracellular Ca^{++}-sensing receptor from bovine parathyroid. Nature 366:575-580.

Buck, L. and R. Axel (1991) A novel multigene family may encode odorant receptors: a molecular basis for odor recognition. Cell 65:175-187.

Chabre, M. and P. Deterre (1989) Molecular mechanism of visual transduction. Eur. J. Biochem. 179:255-256.

Coughlin, S. R. (1994) Expanding horizons for receptors coupled to G proteins: diversity and disease. Curr. Opin. Biol. 6:191-197.

Crespo, P., N. Xu, *et al.* (1994) Ras-dependent activation of MAP kinase pathway mediated by G-protein βγ subunits. Nature 369:418-420.

Dohlman, H. G., J. Thorner, *et al.* (1991) Model systems for the study of seven-transmembrane-segment receptors. Annu. Rev. Biochem. 60:653-688.

Faure, M., T. A. Voyno-Yasenetskaya, *et al.* (1994) cAMP and bg subunits of heterotrimeric G proteins stimulate the mitogen-activated protein kinase pathways in COS-7 cells. J. Biol. Chem. 269:7851-7854.

Fong, T. M., R. R. C. Huang, *et al.* (1992) The extracellular domain of the neurokinin-1 receptor is required for high-affinity binding of peptides. Biochemistry 31:11806-11811.

Gerszten, R. E., J. I. Chen, *et al.* (1994) Specificity of the thrombin receptor for agonist peptide is defined by its extracellular surface. Nature 368:648-651.

Gether, U., T. E. Johansen, *et al.* (1993) Different binding epitopes on the NK1 receptor for substance P and a non-peptide antagonist. Nature 362:345-348.

Gomeza, J., C. Joly, *et al.* (1995) The second intracellular loop cooperates with the other intracellular domains to control coupling to G-proteins. J. Biol. Chem. 271:2199-2205.

Götze, K. and K. H. Jakobs (1994) Unoccupied β-adrenoceptor-induced adenylyl cyclase stimulation in turkey erythrocyte membranes. Eur. J. Pharmacol. 268:151-158.

Hibert, M. F., J. Hoflack, *et al.* (1993) Modèles tridimentionnels des récepteurs couplés aux protéines G. médecine/sciences 9:31-40.

Houssami, S., D. M. Findlay, *et al.* (1994) Isoforms of the rat calcitonin receptor: consequences for ligand biding and signal transduction. Endocrinology 135:183-190.

Ji, I. and T. H. Ji (1995) Differential roles of exoloop 1 of the human follicle-stimulating hormone in hormone binding and receptor activation. J. Biol.Chem. 270:15970-15973

Kjelsberg, M. A., S. Cotecchia, *et al.* (1992) Constitutive activation of the α_{1B}-adrenergic receptor by all amino acid substitutions at a single site. J.Biol.Chem. 267:1430-1433.

Kolesnikov, S. S. and R. F. Margolskee (1995) A cyclic-nucleotide-suppressible conductance activated by transducin in taste cells. Nature 376:85-88.
Kuusinen, A., M. Arvola, *et al.* (1995) Molecular dissection of the agonist binding site of an AMPA receptor. EMBO J. 14:6327-6332.
Kyte, J. and R. F. Doolittle (1982) A simple method for displaying the hydropathic character of a protein. J. Mol. Biol. 157:105-132.
Lancet, D. and N. Ben-Arie (1993) Olfactory receptors. Current Biology 3:668-674.
Lefkowitz, R. J. and M. G. Caron (1988) Adrenergic receptors. Models for the study of receptors coupled to guanine nucleotide regulatory proteins. J. Biol. Chem. 263:4993-4996.
Llédo, P., V. Homburger, *et al.* (1992) Differential G protein-mediated coupling of D2 dopamine receptors to K^+ and Ca^{++} channels in rat anterior pituitary cells. Neuron 8:455-463.
Maggio, R., Z. Vogel, *et al.* (1993) Coexpression studies with mutant muscarinic/adrenergic receptors provide evidence for intermolecular cross-talk between G-protein-linked receptors. Proc. Natl. Acad. Sci., USA 90:3103-3107.
Mollon, J. (1992) Colour vision. Worlds of difference. Nature 356:378-379.
Nakanishi, S. (1992) Molecular diversity of glutamate receptors and implications for brain function. Science 258:597-603.
Neer, E. J. (1995) Heterotrimeric G proteins: organizers of transmembrane signals. Cell 80:249-257.
Nystedt, S., K. Emilsson, *et al.* (1994) Molecular cloning of a potential proteinase activated receptor. Proc. Natl. Acad. Sci. USA 91:9208-9212.
O'Hara, P. J., P. O. Sheppard, *et al.* (1993) The ligand-binding domain in metabotropic glutamate receptors is related to bacterial periplasmic binding proteins. Neuron 11:41-52.
Offermanns, S. and M. I. Simon (1995) $G\alpha_{15}$ and $G\alpha_{16}$ couple a wide variety of receptors to phospholipase C. J. Biol. Chem. 270:15175-15180.
Oliviera, L., A. C. M. Paiva, *et al.* (1994) A common step for signal transduction in G protein-coupled receptors. Trends Pharmacol. Sci. 15:170-172.
Pantaloni, C., P. Brabet, *et al.* (1996) Alternative splicing in the N-terminal extracellular domain of the PACAP receptor modulates receptor selectivity and relative potency of PACAP-27 and-38 in phospholipase C activation. J. Biol. Chem. in press:
Parma, J., L. Duprez, *et al.* (1993) Somatic mutations in the thyrotropin receptor gene cause hyperfunctioning thyroid adenoma. Nature 365:649-651.
Pin, J. P. and J. Bockaert (1995) Get receptive to metabotropic glutamate receptors. Current opinion in Neurobiology 5:342-349.
Pin, J. P., C. Joly, *et al.* (1994) Domains involved in the specificity of G-protein activation in phospholipase C coupled metabotropic glutamate receptors. EMBO J. IN PRESS
Pollak, M. R., E. M. Brown, *et al.* (1993) Mutations in the human Ca^{++}-sensing receptor gene cause familial hypocalciuric hypercalcemia and neonatal severe hyperparathyroidism. Cell 75:1297-1303.
Pradier, L., J. Ménager, *et al.* (1994) Septide: an agonist for the NK1 receptor acting at a site distinct from substance P. Mol. Pharmacol. 45:287-293.
Schertler, G. F. X., C. Villa, *et al.* (1993) Projection structure of rhodopsin. Nature 362:770-772.
Spiegel, A. M., L. S. Weinstein, *et al.* (1993) Abnormalities in G protein-coupled signal transduction pathways in human disease. J. Clin. Invest. 92:1119-1125.
Stern-Bach, Y., B. Bettler, *et al.* (1994) Agonist selectivity of glutamate receptors is specified by two domains structurally related to bacterial amino acid-binding proteins. Neuron 13:1345-1357.
Stroop, S. D., R. E. Kuestner, *et al.* (1995) Biochemistry 34:183-190.
Suryanarayana, S., M. von Zastrow, *et al.* (1992) Identification of intramolecular interactions in adrenergic receptors. J. Biol. Chem. 267:21991-21994.
Tang, W. and G. A.G. (1991) Type-specific regulation of adenylyl cyclase by G protein $\beta\gamma$ subunits. Science 254:1500-1503.

Thirstrup, K., C. E. Elling, *et al.* (1996) Construction of a high affinity zinc switch in k-opioid receptor. J. Biol. Chem. 271:7875-7878.
Tota, M. R. and C. D. Strader (1990) Characterization of the binding domain of β-adrenergic receptor with the fluorescent antagonist carazolol. Evidence for a buried ligand binding site. J. Biol. Chem. 26:16891-16897.
Trumpp-Kallmeyer, S., B. Chini, *et al.* (1995) Towards understanding the role of the first extracellular loop for the binding of peptide hormones to G-protein coupled receptors. Pharmaceutica Acta Helvetiae 70:255-262.
Turner, P. R., T. Bambino, *et al.* (1996) A putative selectivity filter in the G-protein-coupled receptors for parathyroid hormone and secretin. J. Biol. Chem. 271:9205-9208.
Van Sande, J., J. Parma, *et al.* (1995) Genetic basis of endocrine disease. Somatic and germline mutations of the TSH receptor gene in thyroid diseases. J. Clin. Endocr. and Metabolism. 80:2577-2585.
Varrault, A., J. Bockaert, *et al.* (1992) Activation of $5\text{-}HT_{1A}$ receptors expressed in NIH-3T3 cells induces focus formation and potentiates EGF effect on DNA synthesis. Mol. Biol. Cell 3:961-969.
Varrault, A., D. Le Nguyen, *et al.* (1994) $5\text{-}Hydroxytryptamine_{1A}$ receptor synthetic peptides. Mechanisms of adenylyl cyclase. J.Biol.Chem. 269:16720-16725.
Vu, T.-K., H, D. T. Hung, *et al.* (1991) Molecular cloning of a functional thrombin receptor reveals a novel proteolytic mechanism of receptor activation. Cell 64:1057-1068.
Wedegaertner, P. B., P. T. Wilson, *et al.* (1995) Lipid modifications of trimeric G proteins. J. Biol. Chem. 270:503-506.
Zhou, W., C. Flanagan, *et al.* (1994) A reciprocal mutation supports helix 2 and helix 7 proximity in gonodotropin-releasing hormone receptor. Mol. Pharmacol. 45:165-170.

Heterologous Expression of G Protein-Coupled Receptors in Yeasts: an Approach to Facilitate Drug Discovery

Bart J.B. Francken, Walter H.M.L. Luyten and Josée E. Leysen

Department of Biochemical Pharmacology
Janssen Research Foundation
Turnhoutseweg 30
B-2340 Beerse
Belgium

Abstract

Mammalian G protein-coupled receptors can be functionally expressed in yeast cells. Co-expression of a heterologous receptor with an appropriate G protein generates a yeast expression system that can be used to study structural and functional properties of receptor/ligand and receptor/G protein interactions. In addition, heterologously expressed receptors can be functionally linked to the endogenous *Saccharomyces cerevisiae* pheromone response pathway. Yeast strains can be developed that respond to heterologous receptor activation by an easily measurable growth response or by the specific induction of a reporter gene. Since G protein-coupled receptors are the target for many therapeutic drugs, heterologous expression in yeast may provide an appropriate tool for the pharmaceutical industry to be used for the discovery of new drugs.

Introduction

G protein-coupled receptors (GPCRs) are protein monomers with seven transmembrane domains (reviewed by Dohlman *et al.*, 1991) (Fig. 1). They form a very large receptor superfamily with over thousand members. Signals that can interact with GPCRs are very diverse: alkaloids, ions, neurotransmitters (including amino acids), nucleotides, lipids, peptides, glycoprotein hormones, light and olfactants. A major characteristic of GPCRs is

NATO ASI Series, Vol. H 101
Molecular Mechanisms of Signalling
and Membrane Transport
Edited by Karel W. A. Wirtz

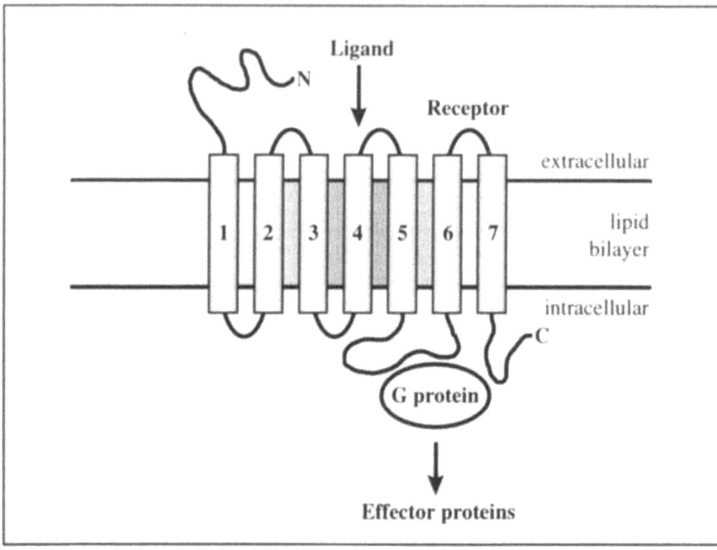

Figure 1: Generalised schematic diagram of a G protein-coupled receptor.

their intracellular association with GTP-binding proteins (G proteins; reviewed by Neer, 1995). These trimeric proteins are composed of the subunits: G_{α}, G_{β} and G_{γ}. Binding of an agonist to its cognate receptor triggers the exchange of bound GDP for GTP on the G_{α} subunit, which then dissociates from the receptor and the $G_{\beta\gamma}$ heterodimer. Both G_{α} and $G_{\beta\gamma}$ are able to activate or inhibit specific intracellular effector proteins. Effectors include enzymes that catalyse the generation of second messengers (e.g. adenylyl cyclase, phospholipase A, phospholipase C), ion channels and protein kinases.

G protein-coupled receptors are the target of many therapeutic drugs. For several decades the discovery of new biologically active ligands was based on the testing of chemicals on animal tissue containing receptors presumed to be similar to human receptors (Williams, 1991). Such an animal receptor system has shortcomings, however, in that slight differences between human and animal receptors can have profound effects on drug activity and that cells may express structurally similar, but pharmacologically different receptors.

Advances in molecular biology, especially in cloning and expression techniques, have greatly reduced the need for reliance on animal models (Lester, 1988; Strange, 1991; Luyten and Leysen, 1993). A wide variety of heterologous expression systems has already been used for

expression of GPCRs: *Xenopus laevis* oocytes, stably or transiently transfected mammalian cells, insect cells (baculovirus expression system), *Escherichia coli* and yeasts (Strosberg and Marullo, 1992; Grisshammer and Tate, 1995; Tate and Grisshammer, 1996).
The pharmaceutical industry can apply heterologous expression of cloned human GPCRs to discover new drugs targeting GPCRs or to find ligands that are more selective for one particular receptor subtype. For the screening of potential drugs, the host system should provide the ability to test the ligand binding characteristics of the recombinant receptor and its functional properties by measuring a coupled biological response. Since receptor activation by agonist binding is affected by the presence and state of coupling to a G protein, the host cells should express the appropriate G protein. An ideal cell-based high-throughput screening procedure for the identification of new ligands would consist of a cell system which contains a single human GPCR and a relevant G protein that can couple to an automatically measurable effector system (Broach and Thorner, 1996).

Heterologous expression of GPCRs in yeast cells provide several advantages over commonly used mammalian expression systems. Mammalian cells are relatively expensive to culture and difficult to propagate for an high-throughput screening procedure, while growing yeasts is cheap and easy. Yeast-based screening assays can be developed that are based on either positive growth or reporter gene expression and that yield an easily quantifiable (e.g. spectrophotometrically measurable) signal. Such yeast-based bioassays can be readily adapted for high-throughput screening, e.g. in 96-well microtitre plates. Moreover, the presence of multiple endogenous receptor subtypes in mammalian cells often confuses ligand binding results. Yeast cells, on the other hand, contain only a limited set of endogenous GPCRs, the mating pheromone receptors. The genetic background of yeasts expressing mammalian GPCRs can be modified relatively easily by homologous recombination: elimination of the mating pheromone receptor creates a system that is free of any endogenous GPCR.

In this paper, we attempted to review studies of heterologous expression of mammalian GPCRs in yeasts. Yeast cells possess GPCRs, the pheromone-binding receptors, homologous to those found in mammalian cells. Binding of mating pheromone to its cognate pheromone receptor activates an intracellular G protein-linked signal transduction pathway, the yeast pheromone response pathway, and eventually leads to cell mating. This pathway is summarised briefly. Further, an overview is given of reports that describe functional coupling of a heterologously expressed GPCR to the yeast pheromone response pathway. Finally, we pointed out some possible applications of heterologous expression of GPCRs in yeasts for drug discovery.

Heterologous expression of G protein-coupled receptors in yeasts

Different yeast species have already been used for the expression of GPCRs: *Saccharomyces (S.) cerevisiae*, *Schizosaccharomyces (Sch.) pombe* and *Pichia (P.) pastoris*. Table 1 gives an overview of the GPCRs which have been expressed in yeasts to date.

Expression level and pharmacological profile of G protein-coupled receptors heterologously expressed in yeasts

The reported expression levels of GPCRs expressed in yeasts are highly variable, ranging from 0.02 pmol/mg protein (Payette *et al.*, 1990) to 115 pmol/mg protein (King *et al.*, 1990). Many tricks have been applied to increase the amount of receptor expressed.

(1) The use of protease-deficient yeast strains can decrease proteolytic receptor degradation and thereby improve the receptor expression level (Sander *et al.*, 1994a; Weiß *et al.*, 1995; Bach *et al.*, 1996), although this effect is not always seen (Talmont *et al.*, 1996).

(2) Most known GPCRs lack an N-terminal signal peptide for membrane insertion of the receptor protein or for trafficking of the receptor to the cell surface. To facilitate membrane translocation and to direct the heterologous receptor to the plasma membrane, N-terminal fusions of the receptor protein to various yeast leader or signal sequences have been performed. Replacement of the 5' untranslated region and the first 63 base pairs (bp) of coding sequence of the human β_2-adrenergic gene with 11 bp of non-coding and 42 bp of coding region from the *S. cerevisiae STE2* gene (encoding the α-factor pheromone receptor) resulted in a very high expression level (115 pmol/mg protein) in *S. cerevisiae* (King *et al.*, 1990) and in a moderate level (7.5 pmol/mg protein) in *Sch. pombe* (Ficca *et al.*, 1995). An N-terminal fusion of the human dopamine D_{2S} receptor cDNA to the first 75 bp of the *STE2* coding sequence, however, did not improve expression (Sander *et al.*, 1994a, b). For the expression of the rat muscarinic M_5 receptor on the cell surface of *S. cerevisiae*, an N-terminal fusion with the *S. cerevisiae* α-factor signal sequence was required (Huang *et al.*, 1992). A

Table 1: Heterologous expression of G protein-coupled receptors in yeasts.

Receptor	Subtype	Species	Amount	Comments	Reference
SACCHAROMYCES CEREVISIAE					
Adenosine	A_{2a}	rat	0.45 pmol/mg	Deletion of *STE2* increased sensitivity to the ligand	Price *et al.*, 1996
				Deletion of *SST2*	
				Co-expression with yeast or yeast/rat chimeric G_α protein	
				Functional coupling to pheromone response pathway (bioassay: deletion of *FAR1*, *FUS1* replaced by reporter gene, deletion of *GPA1*)	
				Native receptor binding profile	
Adrenergic	α_{2C2}	human	3 pmol/mg	Non-native receptor binding profile	Marjamäki *et al.*, 1994
	β_2	human	115 pmol/mg	N-terminal fusion with the amino-terminus of the *S. cerevisiae* α-factor receptor	King. *et al.*, 1990
				Co-expression with transactivator protein (LAC9)	
				Co-expression with rat G_α protein	
				Inclusion of β_2-adrenergic receptor antagonist in growth medium	
				Functional coupling to pheromone response pathway (assay: *FUS1*-lacZ fusion integrated into genome, deletion of *GPA1*)	
				Native receptor binding profile	
Dopamine	D_{2L}	rat	0.03 pmol/mg		Presland *et al.*, 1993
	D_{2S}	human	1-2 pmol/mg	N-terminal fusion with the amino-terminus of the *S. cerevisiae* α-factor receptor did not improve expression level	Sander *et al.*, 1994a
				Protease-deficient strains improved expression level	
				Localization restricted to vacuole	
				Non-native receptor binding profile	
	D_{2S}	human	1-2 pmol/mg	N-terminal fusion with the amino-terminus of the *S. cerevisiae* α-factor receptor	Sander *et al.*, 1994b
				Non-native receptor binding profile	
	D_{2S}	human	2.8 pmol/mg	Non-native receptor binding profile	Sander *et al.*, 1994c
Muscarinic	M_1	human	0.02 pmol/mg	Native receptor binding profile	Payette *et al.*, 1990
acetylcholine	M_5	rat	0.13 pmol/mg	N-terminal fusion with the *S. cerevisiae* α-factor signal sequence	Huang *et al.*, 1992
				Radioligand binding on intact cells	
Serotonin	5A	human	16 pmol/mg	N-terminal fusion with the *Bacillus macerans* β-glucanase signal sequence	Bach *et al.*, 1996
				Protease-deficient strain	
				Heat shock treatment increased expression level two-fold	
				Native receptor binding profile	

Table 1 (continued)

Receptor	Subtype	Species	Amount	Comments	Reference
SACCHAROMYCES CEREVISIAE					
Somatostatin	2	rat	0.19 pmol/mg	Deletion of *SST2* increased sensitivity to the ligand ten-fold Overexpression of a transactivator protein (GAL4) Co-expression with yeast or yeast/rat chimeric G_{α} protein Functional coupling to pheromone response pathway (bioassay: deletion of *FAR1*, *FUS1* replaced by reporter gene, deletion of *GPA1*) Native receptor binding profile	Price *et al.*, 1995
SCHIZOSACCHAROMYCES POMBE					
Adrenergic	β_2	human	7.5 pmol/mg	N-terminal fusion with the amino-terminus of the *S. cerevisiae* α-factor receptor required to detect receptor expression Native receptor binding profile	Ficca *et al.*, 1995
Dopamine	D_{2S}	human	14.8 pmol/mg	Non-native receptor binding profile	Sander *et al.*, 1994c
Neurokinin	NK2	human	1.16 pmol/mg	N-terminal fusion with *Sch. pombe PHO1* signal sequence improved receptor trafficking to the cell surface Co-expression with mammalian G protein subunits failed to modulate agonist binding Radioligand binding on intact cells	Arkinstall *et al.*, 1995
PICHIA PASTORIS					
Opioid	μ	human	0.42 pmol/mg	N-terminal fusion with the *S. cerevisiae* α-factor signal sequence N-terminal fusion with the amino-terminus of the *S. cerevisiae* α factor receptor Protease-deficient strain did not improve expression level Native receptor binding profile	Talmont *et al.*, 1996
Serotonin	5A	human	22 pmol/mg	N-terminal fusion with the *S. cerevisiae* α-factor signal sequence increased receptor expression level two-fold N-terminal fusion with the *P. pastoris PHO1* signal sequence did not improve receptor expression level Protease-deficient strain increased expression level three-fold Localization restricted to endoplasmic reticulum and vacuole Native receptor binding profile	Weiβ *et al.*, 1995

fusion of the human serotonin $5HT_{5A}$ receptor to the *S. cerevisiae* α-factor signal (Weiβ *et al.*, 1995), and a fusion of the human neurokinin NK2 receptor with the *Sch. pombe PHO1* signal sequence (Arkinstall *et al.*, 1995) both resulted in an increase in receptor expression level in *P. pastoris* and *Sch. pombe*, respectively.

(3) If expression of the receptor is placed under the control of a galactose-inducible promoter, overexpression of the transcriptional transactivator protein GAL4 (or LAC9, a homologue of GAL4) may improve expression levels (King *et al.*, 1990; Price *et al.*, 1995).

(4) King *et al.* (1990) reported the inclusion of a β_2-adrenergic receptor antagonist in the growth medium during induction of human β_2-adrenergic receptor expression in *S. cerevisiae*.

(5) Heat shock treatment of *S. cerevisiae* cells expressing the human serotonin $5HT_{5A}$ receptor significantly increased the receptor level and also improved the affinity of the receptor as measured by radioligand binding (Bach *et al.*, 1996). This effect could be explained by a heat shock-induced production of so-called heat shock proteins, of which some function as molecular chaperones and help proteins to retain their correct conformation.

None of these tricks were necessary, however, for expressing the human muscarinic M_1 receptor (Payette *et al.*, 1990) or the rat adenosine A_{2a} receptor (Price *et al.*, 1996) in *S. cerevisiae*.

In many of the reported expression studies, the functional integrity of the receptor is established by determination of the pharmacological profile of the receptor in yeast membrane fractions. Mostly, the affinities for the ligands tested and the rank order of potencies for a series of receptor agonists or antagonists were comparable to the data observed for the same receptor in native tissues or in mammalian cells heterologously expressing the receptor. However, this was not the case for the human adrenergic α_{2C2} receptor expressed in *S. cerevisiae* (Marjamäki *et al.*, 1994), the human dopamine D_{2S} receptor expressed in *S. cerevisiae* or in *Sch. pombe* (Sander *et al.*, 1994a, b, c), the human serotonin $5HT_{5A}$ receptor and the human μ-opioid receptor expressed in *P. pastoris* (Weiβ *et al.*, 1995; Talmont *et al.*, 1996).

Receptors that are able to form a productive interaction with a G_α protein show high-affinity agonist binding. High- and low-affinity agonist-binding sites can be identified in radioligand binding assays in the presence of nonhydrolysable GTP analogues, which shift high-affinity binding to a low-affinity state. King *et al.* (1990) reported the co-expression of the β_2-

adrenergic receptor and the rat $G_{\alpha s}$ protein, but did not demonstrate high-affinity agonist binding. Co-expression of the human neurokinin NK2 receptor and mammalian G_{α}, G_{β} and G_{γ} subunits in *Sch. pombe* failed to modulate agonist binding, suggesting the absence of functional interaction between these components (Arkinstall *et al.*, 1995). Only two reports of functional mammalian GPCRs (the rat adenosine A_{2a} and the rat somatostatin subtype 2 receptor) expressed in *S. cerevisiae* cells demonstrated high-affinity agonist binding in membrane fractions (Price *et al.*, 1995; Price *et al.*, 1996).

The *Saccharomyces cerevisiae* pheromone response pathway

The yeast *S. cerevisiae* has two haploid cell types, **a** and α, that can mate and form **a**/α diploids. Cells of **a** mating type secrete **a**-factor pheromone and express the α-factor receptor (encoded by *STE2*); cells of α mating type secrete α-factor pheromone and express the **a**-factor receptor (encoded by *STE3*). Binding of the peptide mating factors to cognate receptors activates an intracellular signal transduction pathway that is common in **a** and α cells. Propagation of a signal along this pheromone response pathway leads to a number of physiological changes that prepare the cells for mating. The transcriptional induction of many genes, whose products are required for both transient arrest in G1 phase of the mitotic cell cycle and alterations in cell polarity and morphology, culminates in cell and nuclear fusion (reviewed by Marsh *et al.*, 1991; Kurjan, 1993).

Figure 2 gives an overview of the pheromone response pathway in *S. cerevisiae*. Binding of mating pheromone to its cognate receptor activates the heterotrimeric G protein, whose subunits G_{α}, G_{β} and G_{γ} are encoded by *GPA1*, *STE4* and *STE18*, respectively. This activation causes dissociation of G_{α} from $G_{\beta\gamma}$. $G_{\beta\gamma}$ then activates the downstream mitogen-activated protein kinase (MAPK) cascade in an as yet unknown manner involving STE20 and STE5 (reviewed by Herskowitz, 1995). The enzymes of the core module are members of families known as MAPKs (also called extracellular signal regulated kinases, ERKs), MAPK/ERK kinases (MEKs), and MEK kinases (MEKKs). STE20 is proposed to be directly controlled by $G_{\beta\gamma}$ (Leberer *et al.*, 1992). STE5 is proposed to act as a scaffold protein to organise the kinases, since it associates with all the components of the MAPK module: STE11, STE7 and the MAPKs FUS3 and KSS1 (Marcus *et al.*, 1994). STE20 phosphorylates and thereby activates STE11, a member of the MEKK family. STE11 activates the MEK STE7, which in

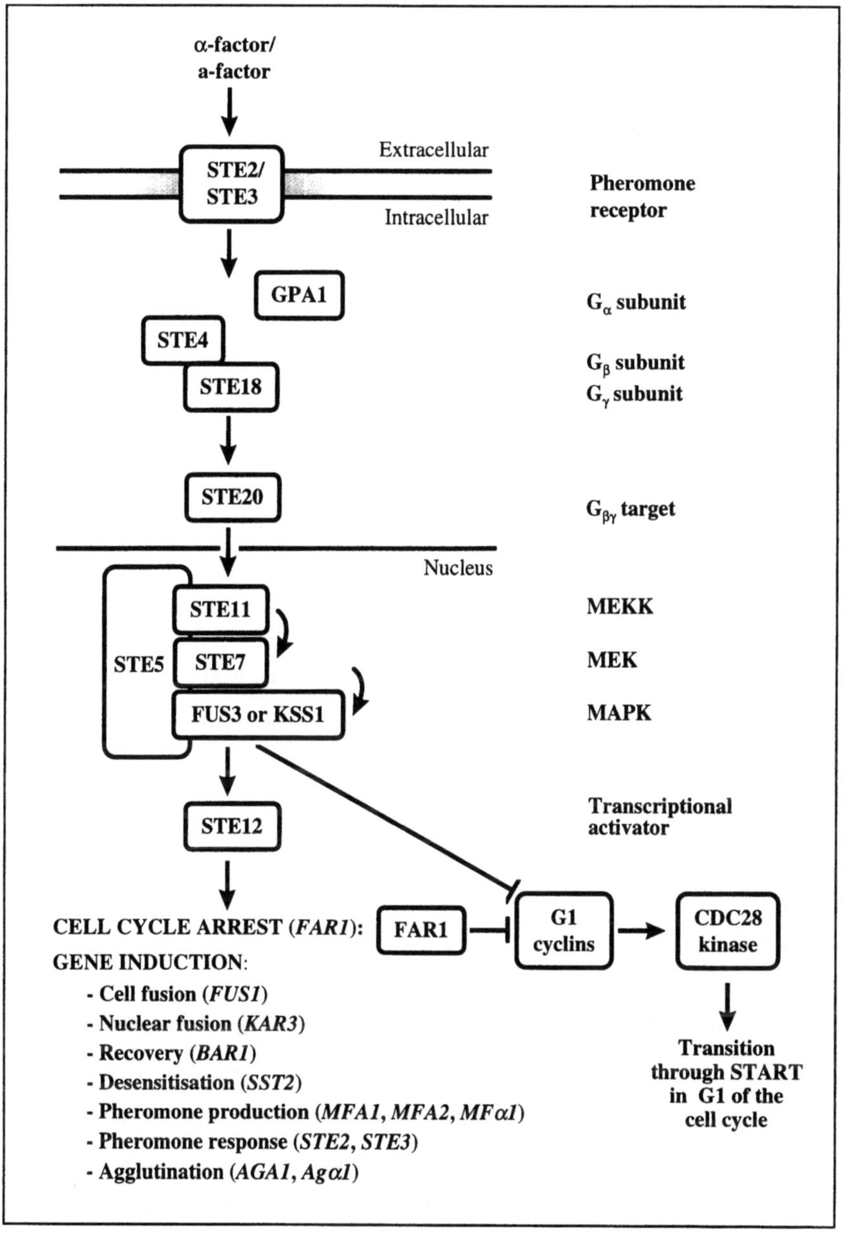

Figure 2: The pheromone response pathway in *Saccharomyces cerevisiae*.

turn phosphorylates FUS3 and KSS1. The activated MAPKs activate the transcription factor STE12. STE12 then mediates transcriptional induction of many genes whose products are involved in cell cycle arrest (*FAR1*, *FUS3*), cell fusion (*FUS1*), nuclear fusion (*KAR3*), recovery (*BAR1*), desensitisation (*SST2*), pheromone production (*MFA1*, *MFA2*, *MFαl*), pheromone response (*STE2*, *STE3*), agglutination (*AGA1*, *AGαl*) and changes in cell morphology.

Coupling of heterologously expressed G protein-coupled receptors to the *Saccharomyces cerevisiae* pheromone response pathway

The functional coupling of a heterologously expressed GPCR to the endogenous *S. cerevisiae* G protein-linked effector system, the pheromone response pathway, has been first demonstrated for the human β_2-adrenergic receptor (King *et al.*, 1990). Recently, the rat adenosine A_{2a} (Price *et al.*, 1995) and the rat somatostatin subtype 2 receptor were also shown to functionally couple to the *S. cerevisiae* pheromone response pathway (Price *et al.*, 1996).

In the study reported by King *et al.*, the human β_2-adrenergic receptor was co-expressed with the rat $G_{\alpha s}$ protein in mutant *S. cerevisiae* cells lacking the wild-type *GPA1* gene, which encodes the yeast G_α protein. In these cells, β_2-adrenergic receptor activation by agonist induced the characteristic morphological changes (so-called shmoo-formation) of yeast in response to activation of the mating signal transduction pathway by pheromone. To confirm that the pheromone response pathway was activated by the heterologous receptor and the rat $G_{\alpha s}$ protein, a yeast strain in which activation of the β_2-adrenergic receptor leads to induction of the β-galactosidase reporter gene was constructed by stable integration into the genome of a *FUS1-lacZ* gene fusion. *FUS1* encodes a kinase required for transition from mitosis into conjugation and *FUS1* gene expression is controlled by a pheromone-responsive promoter (Elion *et al.*, 1990). In cells co-expressing β_2-adrenergic receptor and rat $G_{\alpha s}$ protein, receptor agonist treatment induced β-galactosidase activity. Agonist responsiveness was dependent on expression of both the β_2-adrenergic receptor and the rat $G_{\alpha s}$ protein and required a yeast strain in which the endogenous G_α gene *GPA1* was disrupted. These results indicate that the rat $G_{\alpha s}$ protein can productively interact with both the human β_2-adrenergic receptor and the *S. cerevisiae* $G_{\beta\gamma}$ protein.

Price *et al.* reported the functional coupling of the rat somatostatin subtype 2 receptor to the pheromone response pathway in *S. cerevisiae* (Price *et al.*, 1995). To examine the ability of the rat receptor to couple to the pheromone response pathway, a bioassay was developed in which genetically modified *S. cerevisiae* strains respond to agonist activation of the rat receptor by exhibiting cell growth instead of cell cycle arrest. Several modifications were introduced into the yeast strain, as follows. (1) The *FAR1* gene was deleted. *FAR1* ençodes a negative regulator of G_1 cyclins (see Figure 2) and deletion of *FAR1* allows for continued cell growth and for transcriptional induction of pheromone-responsive genes in the presence of an activated mating signal transduction pathway (Chang and Herskowitz, 1990). (2) The *HIS3* reporter gene was placed under the control of the pheromone-responsive *FUS1* promoter. Therefore, the *FUS1* gene was replaced by *HIS3* and a yeast strain in which the *HIS3* gene was deleted was used. Receptor activation by an agonist permits growth of auxotrophic *his3* yeast cells on medium lacking histidine, since receptor stimulation consecutively leads to activation of the pheromone response pathway, induction of the *FUS1* promoter and increased HIS3 protein expression. (3) The *GPA1* gene was deleted and replaced with a plasmid-borne *GPA1* gene or a hybrid *GPA1*/rat $G_{\alpha i2}$ gene. As a result of these genetic modifications, *S. cerevisiae* cells expressing both the somatostatin receptor and a G_α protein exhibited a dose-dependent increase in the growth response upon agonist treatment.

In the 'agar plate bioassay', agonists were applied to inoculated agar medium lacking histidine. The responding *S. cerevisiae* cells formed an observable zone of growing yeast cells around the site of agonist administration. The heterologous receptor retained agonist selectivity in the bioassay: the diameter of growth zones was proportional to the reported affinity of the receptor ligands. Both GPA1 and the chimeric GPA1/rat $G_{\alpha i2}$ protein, composed of an amino-terminal $G_{\beta\gamma}$ interaction domain from GPA1 and a carboxy-terminal receptor interaction domain from rat $G_{\alpha i2}$, were able to form productive interactions with the rat receptor and the yeast $G_{\beta\gamma}$ protein. The sensitivity of the bioassay was increased by mutation of the *SST2* gene. SST2 is thought to be involved in adaptation and recovery from pheromone-dependent cell cycle arrest. Lack of functional SST2 greatly increases sensitivity to the mating pheromone and prolongs recovery (Chan and Otte, 1982a, b; Dietzel and Kurjan, 1987). This 'agar plate bioassay' has been used for the identification and characterisation of novel somatostatin antagonists (Bass *et al.*, 1996). To assay the effects of potential antagonist ligands, it was necessary to first stimulate cell growth within the agar plate by addition of agonist. Application of an antagonist then blocked the growth-promoting effect of the agonist.

Recently, the rat adenosine A_{2a} receptor has also been co-expressed with a plasmid-borne yeast *GPA1* gene or a hybrid *GPA1*/rat $G_{\alpha s}$ gene in similar *far1 his3 gpa1 S. cerevisiae* strains (Price *et al.*, 1996). As before, receptor agonist induced a dose-dependent increase in growth response. Deletion of *STE2*, the gene which encodes the α-factor pheromone receptor, increased the ligand sensitivity of the receptor in the bioassay. It has been proposed that STE2 may compete directly with the adenosine receptor for a limited amount of G protein or, alternatively, that STE2 may exert a desensitisation effect *in trans* in cells responding to agonist-bound adenosine receptor. In addition to the 'agar plate bioassay', a 'liquid bioassay' was developed in which the agonist-induced activation of yeast growth was monitored by measurement of increases in optical density (OD_{620}) of a liquid yeast culture.

Heterologous expression of G protein-coupled receptors in *Saccharomyces cerevisiae*: applications in drug discovery and in the study of G protein-coupled receptors

Heterologous expression in *S. cerevisiae* allows many applications in the study of GPCRs. (1) Heterologous GPCRs expressed alone in *S. cerevisiae* cells can be used in radioligand binding screening assays to identify novel ligands and subtype-selective drugs. Heterologous expression of cloned GPCRs in *S. cerevisiae* can be combined with site-directed mutagenesis and with the construction of chimeric receptors to obtain new insights in the ligand binding, functional and structural aspects of the receptors. Advantages of the yeast expression system over traditional mammalian expression systems include the low cost and ease of cultivation. Moreover, *S. cerevisiae* cells contain only a few endogenous GPCRs and genetic modification of yeast strains is relatively easy.

(2) *S. cerevisiae* cells that co-express heterologous receptors and G proteins may serve as a useful *in vivo* reconstitution system for the study of receptor/G protein interaction specificity. In addition, co-expression of receptor and G protein can be used for determining the agonist or antagonist properties of new ligands. Activation of the heterologously expressed receptor by

agonists can be distinguished from antagonist activation and monitored, for example by an increase in binding to G proteins of radiolabeled, non-hydrolysable GTP analogues.

(3) The effects of ligands on GPCRs can also be monitored in yeast-based bioassays that use functional coupling of heterologous receptors to the endogenous pheromone response pathway. Ligand binding to the heterologous receptor then activates a yeast kinase cascade and results in a detectable cell response, e.g. positive growth or reporter gene expression. For high-throughput screening, yeast-based bioassays in 96-well microtitre plate format can be readily developed.

Functional coupling of a mammalian receptor to the *S. cerevisiae* G protein-linked effector system can be achieved either by expression of the heterologous receptor alone or by co-expression with an appropriate mammalian or chimeric (yeast/mammalian) G_{α} protein. In the first case, a productive interaction of the endogenous G_{α} subunit (GPA1) with the heterologously expressed receptor is required, in addition to $G_{\alpha}/G_{\beta\gamma}$ association. In the second case, the heterologous G_{α} protein needs to interact functionally with both the receptor and the $G_{\beta\gamma}$ heterodimer.

In a wild-type *S. cerevisiae* strain expressing a mammalian receptor that is functionally coupled to the MAPK cascade, agonist stimulation results in cell cycle arrest (growth inhibition) and typical morphological changes. However, *S. cerevisiae* strains can be genetically modified to develop a receptor system that induces a more quantitative response upon agonist stimulation. By placing a reporter gene (encoding an enzyme) under the control of a pheromone-responsive promoter, activation of the heterologous receptor leads to enzyme activity that can be detected in a biochemical assay. Furthermore, *S. cerevisiae* strains can be genetically engineered that do not respond to receptor activation by growth inhibition, but by yeast cell growth, which can be measured quantitatively in an 'agar plate assay' (diameter of growth zones) or in a 'liquid assay' (optical density of the culture).

(4) Recently, an autocrine yeast-based system was developed that allowed the identification of agonists and antagonists of the *S. cerevisiae* α-factor pheromone receptor STE2 (Manfredi *et al.*, 1996). A *S. cerevisiae* strain that secretes peptides from a random peptide library and that responds to activation of its α-factor receptor by growth was constructed. Yeast transformants secreting STE2 agonist peptides were identified as growing colonies, due to autocrine activation of the endogenous α-factor receptor. Similarly, cells producing STE2 antagonists

prohibited the growth arrest response elicited by α-factor (incorporated in the growth medium) and exhibited cell growth. Since mammalian GPCRs can be functionally expressed in *S. cerevisiae* strains, this autocrine yeast expression system may be used to discover novel peptide agonists and antagonists for mammalian GPCRs.

References

Arkinstall S., Edgerton M., Payton M. and Maundrell K. (1995) Co-expression of the neurokinin NK2 receptor and G-protein components in the fission yeast *Schizosaccharomyces pombe*. FEBS Lett. 375, 183-187.

Bach M., Sander P., Haase W. and Reiländer H. (1996) Pharmacological and biochemical characterization of the mouse $5HT_{5A}$ serotonin receptor heterologously produced in the yeast *Saccharomyces cerevisiae*. Receptors and Channels 4, 129-139.

Bass R.T., Buckwalter B.L., Patel B.P., Pausch M.H., Price L.A., Strnad J. and Hadcock J.R. (1996) Identification and characterization of novel somatostatin antagonists. Mol. Pharmacol. 50, 709-715.

Broach J.R. and Thorner J. (1996) High-throughput screening for drug discovery. Nature 384 (Supp.), 14-16.

Chan R.K. and Otte C.A. (1986a) Isolation and genetic analysis of *Saccharomyces cerevisiae* mutants supersensitive to G1 arrest by a factor and α factor pheromones. Mol. Cell. Biol. 2, 11-20.

Chan R.K. and Otte C.A. (1986b) Physiological characterization of *Saccharomyces cerevisiae* mutants supersensitive to G1 arrest by a factor and α factor pheromones. Mol. Cell. Biol. 2, 21-29.

Chang F. and Herskowitz I. (1990) Identification of a gene necessary for cell cycle arrest by a negative growth factor of yeast: FAR1 is an inhibitor of a G1 cyclin, CLN2. Cell 63(5), 999-1011.

Dietzel C. and Kurjan J. (1987) Pheromonal regulation and sequence of the *Saccharomyces cerevisiae SST2* gene: a model for desensitization to pheromone. Mol. Cell. Biol. 15, 3635-3643.

Dohlman G.H., Thorner J., Caron M.C. and Lefkowitz R.J. (1991) Model studies for the studies of seven-transmembrane-segment receptors. Ann. Rev. Biochem. 60, 653-688.

Elion E.A., Grisafi P.L. and Fink G.R. (1990) *FUS3* encodes a $cdc2^{+}$/CDC28-related kinase required for transition from mitosis into conjugation. Cell 60, 649-664.

Ficca A.G., Testa L. and Tocchini Valentini G.P. (1995) The human β2-adrenergic receptor expressed in *Schizosaccharomyces pombe* retains its pharmacological properties. FEBS Lett. 377, 140-144.

Grisshammer R. and Tate C.G. (1995) Overexpression of integral membrane proteins for structural studies. Q. Rev. Biophys. 28, 315-422.

Herskowitz I. (1995) MAP kinase pathways in yeast: for mating and more. Cell 80, 187-197.

Huang H., Liao C., Yang B. and Kuo T. (1992) Functional expression of rat M5 muscarinic acetylcholine receptor in yeast. Biochem. Biophys. Res. Comm. 182 (3), 1180-1186.

King K., Dohlman H.G., Thorner J., Caron M.G. and Lefkowitz R.J. (1990) Control of yeast mating signal transduction by a mammalian β_2 adrenergic receptor and G_S α subunit. Science 250, 121-123.

Kurjan J. (1993) The pheromone response pathway in *Saccharomyces cerevisiae*. Ann. Rev. Genet. 27, 147-179.

Leberer E., Dignard D., Harcus D., Thomas D.Y. and Whiteway M. (1992) The protein kinase homologue Ste20p is required to link the yeast pheromone response G protein $\beta\gamma$ subunits to downstream signalling components. EMBO J. 11, 4815-4824.

Lester H.A. (1988) Heterologous expression of excitability proteins: route to more specific drugs? Science 241, 1057-1063.

Luyten W.H.M.L. and Leysen J.E. (1993) Receptor cloning and heterologous expression - towards a new tool for drug discovery. Trends Biotechnol. 11, 247-254.

Manfredi J.P., Klein C., Herrero J.J., Byrd D.R., Trueheart J., Wiesler W.T., Fowlkes D.M. and Broach J.R. (1996) Yeast α mating factor structure-activity relationship derived from genetically selected peptide agonists and antagonists of Ste2p. Mol. Cell. Biol. 16, 4700-4709.

Marcus S., Polverino A., Barr M. and Wigler M. (1994) Complexes between STE5 and the components of the pheromone-responsive mitogen-activated protein kinase module. Proc. Natl. Acad. Sci. USA 91, 7792-7766.

Marjamäki A., Pohjanoksa K., Ala-Uotila S., Sizmann D., Oker-Blom C., Kurose H. and Scheinin M. (1994) Similar ligand binding in recombinant human α_2C2-adrenoceptors produced in mammalian, insect and yeast cells. Eur. J. Pharmacol. 267, 117-121.

Marsh L., Neiman A.M. and Herskowitz I. (1991) Signal transduction during pheromone response in yeast. Ann. Rev. Cell. Biol. 7, 699-728.

Neer E.J. (1995) Heterotrimeric G proteins: organizers of transmembrane signals. Cell 80, 249-257.

Payette P., Gossard F., Whiteway M. and Dennis M. (1990) Expression and pharmacological characterization of the human M1 muscarinic receptor in *Saccharomyces cerevisiae*. FEBS 266 (1, 2), 21-25.

Presland J.P. and Strange P.G. (1993) Expression of the rat D_2 dopamine receptor in the yeast *Saccharomyces cerevisiae*. Biochem. Soc. Transactions (21), 116S.

Price L.A., Kajkowski E.M., Hadcock J.R., Ozenberger B.A. and Pausch M.H. (1995) Functional coupling of a mammalian somatostatin receptor to the yeast pheromone response pathway. Mol. Cell. Biol. 15, 6188-6195.

Price L.A., Strnad J., Pausch M.H. and Hadcock J.R. (1996) Pharmacological characterization of the rat A_{2A} adenosine receptor functionally coupled to the yeast pheromone response pathway. Mol. Pharmacol. 50,829-837.

Sander P., Grünewald S., Bach M., Haase W., Reiländer H. and Michel H. (1994a) Heterologous expression of the human D_{2S} dopamine receptor in protease-deficient *Saccharomyces cerevisiae* strains. Eur. J. Biochem. 226, 697-705.

Sander P., Grünewald S., Maul G., Reiländer H. and Michel H. (1994b) Constitutive expression of the human D_{2S}-dopamine receptor in the unicellular yeast *Saccharomyces cerevisiae*. Biochimica et Biophysica Acta 1193, 255-262.

Sander P., Grünewald S., Reiländer H. and Michel H. (1994c) Expression of the human D_{2S} dopamine receptor in the yeasts *Saccharomyces cerevisiae* and *Schizosaccharomyces pombe*: a comparative study. FEBS Lett. 344, 41-46.

Strange P.G. (1991) Receptors for neurotransmitters and related substances. Curr. Opin. Biotechnol. 2, 269-277.

Strosberg A.D. and Marullo S. (1992) Functional expression of receptors in microorganisms. Trends Pharmacol. Sci. 13, 95-98.

Talmont F., Sidobre S., Demange P., Milon A. and Emmorine L.J. (1996) Expression and pharmacological characterization of the human μ-opioid receptor in the methylotrophic yeast *Pichia pastoris*. FEBS Lett. 394, 268-272.

Tate C.G. and Grisshammer R. (1996) Heterologous expression of G-protein-coupled receptors. Trends Biotechnol. 14, 426-430.

Weiβ H.M., Haase W., Michel H. and Reiländer H. (1995) Expression of functional mouse $5HT_{5A}$ serotonin receptor in the methylotrophic yeast *Pichia pastoris*: pharmacological characterization and localization. FEBS Lett. 377, 451-456.

Williams M. (1991) Receptor binding in the drug discovery process. Med. Res. Rev. 11, 147-184.

The Role of Small GTPases in Signal Transduction

J.L. Bos, P.D. Baas, B.M.Th. Burgering, B. Franke, M.P. Peppelenbosch, L. M'Rabet, M. Spaargaren, A.D.M. van Mansfeld. D.H.J. van Weeren, R.M.F. Wolthuis and F. Zwartkruis.
Dept. of Physiological Chemistry
Utrecht University
Universiteitweg 100
3584 CG Utrecht
The Netherlands

Introduction

Small GTPases are proteins of about 21kD that cycle between an active GTP-bound state and an inactive GDP-bound state. They can be considered as molecular switches in a variety of regulatory processes, including signal transduction (the Ras family), cytoskeletal reorganisation (Rac and Rho family) and vesicle transport (Rab family) (Figure 1).

Ras superfamily of small GTPases

Ras (Ras, R-ras, Rap, Ral, TC21) - *signal transduction*

Rho (Cdc42, Rac, Rho) - *cytoskeletal rearrangements*

Rad (Rad, Gem) - *?*

Rab-Arf - *vesicle transport*

Ran - *nuclear protein transport*

Figure 1: GTPases of the Ras superfamily are involved in the regulation of a variety of cell processes.

NATO ASI Series, Vol. H 101
Molecular Mechanisms of Signalling and Membrane Transport
Edited by Karel W. A. Wirtz

The best studied family is the Ras family. Ras is a product of the RAS proto-oncogene, which is mutated in a large variety of human tumors (Bos, 1989). The protein is isoprenylated at its C-terminus and localized at the innerside of the plasma membrane. The activity of the protein is regulated by guanine nucleotide exchange factors, which exchange GDP for GTP, and GTPase activating proteins (GAPs, e.g. RasGAP), which induce the hydrolysis of GTP. Recently, most of the components of a signalling pathway mediated by Ras have been identified. Receptor tyrosine kinases activate a guanine nucleotide exchange factor, which results in the activation of Ras. Active Ras binds to and activates the serine-threonine kinases Raf1, B-raf and A-raf, which in their turn phosphorylate and activate MEK1 and 2. These kinases, subsequently, phosphorylate and activate ERK1 and 2. These latter kinases are involved in the phosphorylation and activation of a variety of cellular targets. This pathway is under positive and negative control of other signalling events. For instance, protein kinase C activates Raf1 independent of an increase in RasGTP, and protein kinase A inactivates Raf1 and B-raf in certain cell types (Burgering and Bos, 1995) (Figure 2).

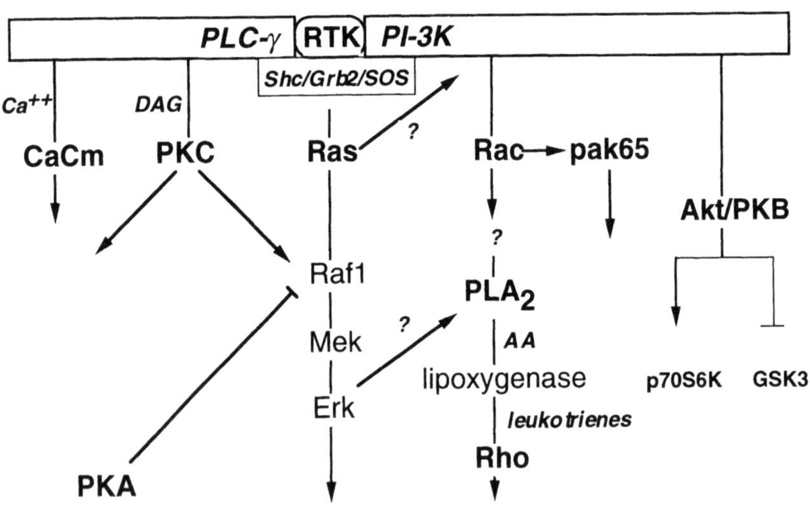

Figure 2: Major signalling events induced by receptor tyrosine kinases.
Most receptor tyrosine kinases activate phospholipase C-γ (PLC-γ). This leads to the release of diacylglycerol (DAG), which activates protein kinase C (PKC), and inositol-3-phosphate, which mobilizes calcium (Ca^{++}) from internal stores. Two other signalling molecules activated are phosphatidylinositol-3-kinase (PI-3K) and the guanine nucleotide exchange factor of Ras, SOS (see text)

In addition to the Rafs, several other putative downstream targets of Ras have been identified, including phosphatidylinositol-3-kinase (PI-3K) (Rodriguez-Viciana et al., 1994) and the RalGDS family members (Kikuchi et al., 1994; Spaargaren and Bischoff, 1994; Wolthuis et al., 1996). PI-3K plays a role in the regulation of the small GTPase Rac and in the activation of the serine/threonine kinase Akt/PKB. RalGDS is a guanine nucleotide exchange factor of the small GTPase Ral. The variety of putative effectors of Ras indicates that Ras mediates different signalling pathways. Whereas in the past most attention was given to the Raf-MEK-ERK pathway, currently the is on the additional signalling routes induced by Ras.

Rac and Rho mediate growth factor-induced cytoskeletal rearrangements

The action of growth factors coupled to receptor protein tyrosine kinases often includes the induction of morphological changes and cellular migration. These processes are brought about by a reorganization of the actin cytoskeleton, and are accompanied by phosphorylation and reorganization of other cytoskeletal components. The signal transduction pathways leading from receptor activation to actin remodelling are still unclear, but the small GTPases Rac and Rho have been implicated this process (Ridley and Hall, 1992; Ridley et al., 1992). For instance, in Swiss-3T3 cells epidermal growth factor (EGF) and platelet-derived growth factor (PDGF) first evoke rapid actin polymerization at the plasma membrane to produce lamellipodia and membrane ruffling, which is followed by the appearance of stress fibres. Introduction of dominant negative Rac proteins (Rac^{N17}) prior to growth factor addition abolishes these effects, while constitutively active Rac (Rac^{V12}) is sufficient to induce plasma membrane-localized actin polymerization and subsequent stress fibre formation in the absence of exogeneous factors.

Several reports indicate that activation of Rac by receptor tyrosine kinases is mediated by PI-3K (Hawkins et al., 1995). The "classical" PI-3K consists of a 85kD adaptor protein, which can bind to the activated receptor through SH2 domains, and a catalytic p110 protein. Thereby, most receptor tyrosine kinases directly activate PI-3K. In addition, the catalytic p110 subunit binds to active Ras (Rodriguez-Viciana et al., 1994). In both cases the association leads to activation of the p110 subunit. The contribution of each of these pathways in the activation of PI-3K is unclear, but the Ras-PI-3K interaction may explain the observation that introduction of active Ras^{V12} can induce the formation of lamellipodia. Recently, several additional PI-3Ks have been identified, some of which also interact with Ras (J. Downward, personal communication).

We have studied the connection between Ras and Rac in the neuroblastoma cell line SK-N-MC (van Weering et al., 1995). This cell line, which does not express EGF receptors, was stably transfected with a fusion receptor of the ligand binding domain of the EGF receptor and the tyrosine kinase domain of Ret. Ret is a receptor tyrosine kinase implicated in a variety of human syndromes, including multiple endocriene neoplasias 2A and 2B and Hirschsprung's desease. Activation of the Ret receptor tyrosine kinase with EGF leads, for instance, to MAP kinase activation and induces the formation of lamellipodia. In addition to EGF (thus Ret), this transfected cell line is also responsive to PDGF and fibroblast growth factor (FGF). Both EGF and PDGF induce the formation of lamellipodia. This formation was sensitive to dominant negative Rac^{N17} and to PI-3K inhibitors wortmannin and LY293002, but not to Ras^{N17}. In addition, both Rac^{V12} and Ras^{V12} can induce the formation of lamellipodia. Rac^{V12}-induced lamellipodia are insensitive to wortmannin, whereas Ras^{V12}-induced lamellipodia are inhibited by wortmannin (van Weering and Bos, in preparation).

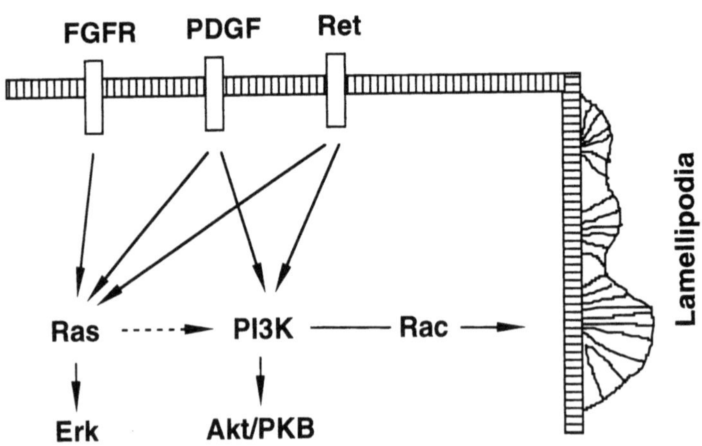

Figure 3: Differential effect of receptor tyrosine kinases on the induction of lamellipodia.

These results are compatible with the model that two signalling pathways induced by receptor tyrosine kinases lead to the activation of PI-3K, Rac and the formation of lamellipodia, a Ras-dependent and a Ras-independent pathway. However, stimulation of FGF did not lead to the formation of lamellipodia in SK-N-MC cells. This may imply that FGF does not activate PI-3K in these cells. Indeed, activation of PI-3K is hardly or not detectable in these cells when assayed by an immunecomplex assay. The result was rather surprising since in these cells FGF gave a very strong activation of Ras. This may imply that activation of endogenous Ras does not induce sufficient PI-3K activity to induce the formation of lamellipodia, in contrast to transfected active Ras^{V12} (Figure 3).
Rac induced stress fibre formation can be inhibited by C-3 transferase of Clostridium botulinum. This enzyme specifically inhibits the small GTPase Rho by ADP-ribosylation. Fetal calf serum (FCS), lysophosphatidic acid (LPA), and bombesin also lead to the formation of stress fibres in a Rho-dependent manner. However, in contrast to receptor tyrosine kinase-induced stress fibre assembly, these stimuli act independently of Rac (Ridley and Hall, 1992; Ridley et al., 1992). Therefore, at least two different signal transduction pathways exist, leading to the stimulation of Rho and subsequently to stress fibre formation, one Rac dependent and one Rac independent.
Besides Rac and Rho, metabolites of arachidonic acid (AA) are also implicated in growth factor-induced actin remodelling (Peppelenbosch et al., 1993). For instance, in A431 cells, EGF induced actin polymerisation is inhibited by inhibitors of phospholipase A_2 (PLA_2) and 5-lipoxygenase. In addition, cortical actin polymerization can be induced by leukotrienes (LTs) in the absence of added growth factors.
We have examined the relationship between Rac and Rho, and AA metabolism (Peppelenbosch et al., 1995). We first found that in Swiss 3T3 cells inhibitors of leukotriene synthesis, i.e. NDGA and MK886, abolished EGF-induced stress fibre formation, but not the formation of lamellipodia and membrane ruffles Furthermore, addition of LTs to serum-starved Swiss-3T3 cells is sufficient for the induction of Rho-dependent stress fibres. LPA-induced stress fibre formation is insensitive to inhibitors of the LT pathway, showing that LPA use a different, LT-independent route to activate Rho and to induce stress fibre formation. Since growth factor-induced, but not LPA-induced stress fibre formation is mediated by Rac, these results suggested that Rac may mediate the production of AA and LTs. Indeed, we found that dominant negative Rac^{N17} completely inhibited EGF-induced AA release and LT synthesis in Swiss-3T3, A14 and rat1 cell lines. Moreover, rat1 cells expressing active Rac^{V12} had elevated levels of AA and LTs which were largely insensitive to EGF-treatment. Together, these results showed

that EGF-induced AA release and subsequent LT synthesis, is mediated by Rac and that this signalling pathway leads to Rho-dependent stress fibre formation. The 85kD cytosolic PLA_2 is a likely candidate target that could mediate Rac dependent AA release. This protein is calcium dependent, needs translocation to the plasma membrane fraction and phosphorylation on Ser505 and Ser727 for full activation (de Carvalho et al., 1996). It has been shown that Ras and ERK are involved in the phosphorylation of Ser505. Perhaps Rac is involved in the phosphorylation of Ser727 or in the induction of calcium transients. We therefore studied the possible role of Rac in growth factor-induced Ca^{2+} influx. We found that expressing dominant negative Rac^{N17} abolished Ca^{2+} signalling in response to EGF but did not affect the Ca^{2+} transients induced by histamine, endothelin and ATP. PDGF-induced Ca^{2+} influx, however, was only partially inhibited by Rac^{N17} expression, showing that the PDGF receptor can activate Ca^{2+} channels via Rac-dependent and -independent signalling pathways (Peppelenbosch et al., 1996). From these results we concluded that Ca^{2+} influx is a Rac specific effector mechanism, and that this Rac-dependent Ca^{2+} influx is the major signalling pathway mediating the EGF-induced increase in $[Ca^{2+}]_i$. It is however unclear whether the inhibition of EGF-induced Ca^{2+} influx explains the inhibition of AA metabolism, since PDGF-induced AA release is also inhibited by Rac^{N17}, whereas Rac^{N17} has only a partial effect on Ca^{2+} influx.

A target for phosphatidylinositol 3-kinase: Akt/PKB

PI-3K phosphorylates the inositol ring of phosphoinositides at the D-3 position (PI-3-P, which includes PI-3-P, PI-3,4-P_2 and PI-3,4,5-P_3), but until recently it was unclear how these lipids direct their signal. It is likely that the entire phospholipid serves as a second messenger, since no phospholipase C is known that can cleave the inositol group with a D-3 phosphate. We (Burgering and Coffer, 1995) and others (Franke et al., 1995) have found that the 60kD serine/threonine kinase Akt/PKB is most likely a direct target of the D-3-phosphorylated lipids. The C-terminal catalytic domain of Akt/PKB is a kinase related to both protein kinase A and protein kinase C, whereas the N-terminal domain has a pleckstrin homology (PH) domain. PH domains are about 100 residues long and most likely involved in either protein-protein or protein-lipid interactions.

Akt/PKB is rapidly activated by a variety of different growth factors, such as PDGF, EGF and insulin, and three pieces of evidence show that this activation is mediated by PI-3K: (i) Activation of Akt/PKB by growth factors is inhibited by wortmannin, a rather specific inhibitor of PI-3K at low concentrations. (ii) PDGF receptor mutants that fail to activate PI-3K also failed to activate Akt/PKB. (iii) A dominant negative mutant of PI-3K inhibits PDGF-induced activation of Akt/PKB. (iv) Introduction of a constitutively active mutant of the p100 subunit of PI-3K activates Akt/PKB (Klippel et al., 1996).

A crucial question is whether Akt/PKB is a direct target for PI-3-P, as protein kinase A a target is for cAMP and protein kinase C for diacylglycerol. Indeed, Franke et al. (Franke et al., 1995) showed that PI-3-P, but not PI or PI-4,5-P_2 can activate Akt/PKB *in vitro*. Furthermore, they showed that the PH domain is essential for Akt/PKB activation and, thus, the model could be that PI-3-P binds to the PH domain, for instance, to the hydrophobic pocket formed by the seven antiparallel β-sheets characteristic for PH domains. As a consequence a conformational change in the protein is induced, which results in the activation of Akt/PKB. Although this model is very provocative, PI-3 P is not the lipid that is formed by PI-3K, but rather PI-3,4 P_2.

Several putative downstream targets of Akt/PKB have been identified. (i) GSK3 has been found to be phosphorylated and inactivated by Akt/PKB (Cross et al., 1996). GSK3 is involved in a variety of different cellular processes including the regulation of glycogen synthase and the regulation of β-catenin stability. (ii) $p70^{S6}$ kinase, responsible for the phosphorylation of the ribosomal S6 protein is activated by constitutive active Akt/PKB *in vivo* . This, together with the finding that PI-3K mediates growth factor-induced activation of $p70^{S6}$ kinase (Ming et al., 1994), indicates that Akt/PKB directs the activation of $p70^{S6}$ kinase. However, it is unlikely that $p70^{S6}$ is a direct target for Akt/PKB (Burgering and Coffer, 1995). (iii) Another downstream target of Akt/PKB could be the Rac signalling pathway, since Rac was shown to be a downstream target of PI-3K as well (Hawkins et al., 1995). However, this is unlikely, since constitutively active Akt/PKB did not induce lamellipodia formation.

Rlf, a guanine nucleotide exchange factor as putative effector of Ras and Rap

A very close relative of Ras is Rap1. Rap1 was first identified as a protein, Krev-1, that could revert the effect of active Ras in cell transformation. Ras and Rap1 are very similar, in particular in the guanine nucleotide binding domain and in the domain that binds putative effectors. The function of this small GTPase is, however, completely unclear (Noda, 1994).

To identify proteins that may interact with Rap1 in vivo, we have performed a yeast-two hybrid screen using an embryonal mouse cDNA library. We identified three different proteins, ralGDS, RGL (ralGDS-like) and a novel protein which we called Rlf (RalGDS-like factor). Rlf is a 778 residue protein that is ubiquitously

expressed and a distant relative of RalGDS (29% identity). The similarity is particularly striking in the Rap-binding domain (RBD) and the region of the putative guanine nucleotide exchange activity (Wolthuis et al., 1996).
In vitro, Rlf has guanine nucleotide exchange active for small GTPase Ral. In this respect Rlf is also a functional homologue of RalGDS. Ral has been implicated in the activation of phospholipase D (Jiang et al., 1996) and in the regulation of the small GTPase CDC42 (Cantor et al., 1995). Via its RBD, Rlf binds specifically to the GTP-bound form of Rap1 and thus Rlf is a putative effector of Rap1. In addition to Rap1, Rlf also binds to the GTP-bound form of Ras. This association can be observed *in vivo* in Cos cells transfected with both Rlf and active Ras. Therefore Rlf could be an effector of Ras as well. Importantly, White et al, (White et al., 1996) recently showed that the putative pathway induced by RalGDS, and thus probably also by Rlf, is essential for the induction of cell transformation by active Ras. This conclusion was reached from the observation that effector domain mutants of Ras that fail to bind RalGDS also fail to transform. In addition, RalGDS in combination with active Raf1, the additional pathway induced by Ras, can induce cell transformation. Although suggestive, more experiments are needed to establish the function of RalGDS, Rlf and Ral in Ras mediated signalling. Also the binding of Rlf and RalGDS to Rap needs further investigation.
Most studies indicate Rap1 (or K-rev1) as an antagonist of active Ras. Indeed introduction of Rap1 reverts Ras-induced cell transformation and Ras-induced germical vesicle breakdown in Xenopus oocytes. However, the molecular mechanism behind this antagonistic effect is unknown. One of the possible mechanisms could be a competitive inhibition of direct effectors of Ras, such as RalGDS and Raf.

Rap1 activation is an early events in platelet activation

To study the function of Rap1 in more detail we have investigated the role of Rap1 in human platelets. We have choosen for this system, since Rap1 is abundantly expressed in platelets. We developed an assay system to measure activation of Rap1. This assay is based on the high affinity of $Rap1^{GTP}$ for the Rap-binding domain of RalGDS. This 97 aa domain binds with a K_D of 10 nM to $Rap1^{GTP}$, whereas no affinity could be detected for $Rap1^{GDP}$ (Herrmann et al., 1996). We found that Rap1 is very rapidly activated by platelet agonists like thrombin, the thromboxane A_2 (TxA_2) analogue U46619, ADP and collagen. This activation is mediated by an increase in intracellular calcium, which is both necessary and sufficient. PGI_2, a strong platelet antagonist, inhibits both thrombin- and calcium-induced activation of Rap1. From these results we conclude that in platelets Rap1 functions in calcium-mediated signalling. How calcium activates Rap1 is unknown. Also the response that is mediated by Rap1 is still elusive. However, one of the major downstream events in which calcium-induced signalling is involved, is the activation of integrin $\alpha_{IIb}\beta_3$ resulting in the exposure of binding sites for fibrinogen and, subsequently, platelet aggregation. Perhaps Rap1 functions in inside-out signalling to activate integrin $\alpha IIb\beta 3$ (Franke et al., submitted for publication).

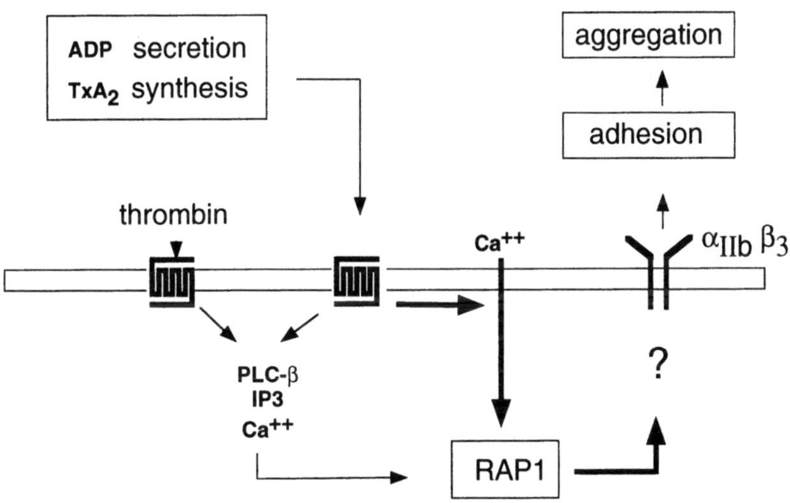

Figure 4: Schematic representation of the signalling pathway used to activate Rap1 in human platelets. (For explanation, see text).

Acknowledgements. Our work is supported by grants from the Dutch Cancer Society, the Netherlands Heart Foundation, the Diabetes Foundation Netherlands and the Netherlands Organisation for Scientific Research.

Bos, J. L. (1989). ras oncogenes in human cancer: a review. Cancer Res. 49: 4682-4689.

Burgering, B. M. T., and Bos, J. L. (1995). Regulation of ras-mediated signalling: more than one way to skin a cat. Trends Bioch. Sci. 20: 18-22

Burgering, B. M. T., and Coffer, P. J. (1995). Protein kinase B (c-Akt) in phosphatidylinositol-3-OH kinase signal transduction. Nature 376: 599-602

Cantor, S. B., Urano, T., and Feig, L. A. (1995). Identification and characterization of Ral-binding protein 1, a potential downstream target of Ral GTPases. Mol. Cell. Biol. 15: 4578-4584

Cross, D. A., Alessi, D. R., Cohen, P., Andjelkovich, M., and Hemmings, B. A. (1996). Inhibition of glycogen synthase kinase-3 by insulin mediated protein kinase B. Nature 378: 785-789

de Carvalho, M. G. S., McCormack, A. L., Olson, E., Ghomashchi, F., Gelb, M. H., Yates III, J. R., and Leslie, C. C. (1996). Identification of phosphorylation sites of human 85-kDa cytosolic phospholipase A2 expressed in insect cells and present in human monocytes. J. Biol. Chem. 271: 6987-6997

Franke, T. F., Yang, S. I., Chan, T. O., Datta, K., Kazlauskas, A., Morrison, D. K., Kaplan, D. R., and Tsichlis, P. N. (1995). The protein kinase encoded by the proto-oncogene Akt is a target of the PDGF-activated phosphatidylinositol 3-kinase. Cell 81: 727-736

Hawkins, P. T., Eguinoa, A., Qui, R. G., Stokoe, D., Cooke, F. T., Walters, R., Wennström, S., Claesson-Welsh, L., Evans, T., Symons, M., and Stephens, L. (1995). PDGF stimulates an increase in GTP-Rac via activation of phosphoinositide 3-kinase. Curr. Biol. 5: 393-403

Herrmann, C., Horn, G., Spaargaren, M., and Wittinghofer, F. (1996). Differential interaction of the Ras family GTP-binding proteins H-ras, Rap1A, and R-Ras with the putative effector molecules Raf kinase and Ral-guanine nucleotide exchange factor. J. Biol. Chem. 271: 6794-6800

Jiang, H., Luo, J. Q., Urano, T., Frankel, P., Lu, Z., Foster, D. A., and Feig, L. A. (1996). Involvement of Ral GTPase in v-src-induced phospholipase D activation. Nature 378: 409-412

Kikuchi, A., Demo, S. D., Ye, Z.-H., Chen, Y.-W., and Williams, L. T. (1994). RalGDS family members interact with the effector loop of ras p21. Mol. Cell. Biol. 14: 7483-7491

Klippel, A., Reinhard, C., Kavanaugh, W. M., Apell, G., Escobedo, M. A., and Williams, L. T. (1996). Membrane localization of phosphatidylinositol 3-kinase is sufficient to activate multiple signal-transducing kinase pathways. Mol. Cell. Biol. 16: 41117-4127

Ming, X.-F., Burgering, B. M. T., Wenström, S., Claesson-Welsh, L., Heldin, C. H., Bos, J. L., Kozma, S. C., and Thomas, G. (1994). Activation of p70/p85 S6 kinase is a pathway independent of p21ras. Nature 371: 426-429

Noda, M. (1994). Structures and functions of the Krev-1 transformation suppressor gene and its relatives. Biochim. Biophys. Acta 1155: 97-109

Peppelenbosch, M. P., Qiu, R.-G., De Vries-Smits, A. M. M., Tertoolen, L. G. J., De Laat, S. W., McCormick, F., Hall, A., Symons, M. H., and Bos, J. L. (1995). Rac mediates growth factor-induced arachidonic acid release. Cell 81: 849-856

Peppelenbosch, M. P., Tertoolen, L. G. J., de Vries-Smits, A. M. M., Qui, R. G., Symons, M. H., de Laat, S. W., and Bos, J. L. (1996). Rac-dependent and -independent pathways mediate growth factor-induced Ca^{2+} influx. J. Biol. Chem. 271: 7883-7886

Peppelenbosch, M. P., Tertoolen, L. G. J., Hage, W. J., and de Laat, S. W. (1993). Epidermal growth factor-induced actin remodelling is regulated by 5-lipoxygenase and cyclooxygenase products. Cell 70: 389-399

Ridley, A. J., and Hall, A. (1992). The small GTP-binding protein rho regulates the assembly of focal adhesions and actin stress fibers in response to growth factors. Cell 70: 389-399

Ridley, A. J., Paterson, H. F., Johnston, C. L., Diekmann, D., and Hall, A. (1992). The small GTP-binding protein rac regulates growth factor-induced membrane ruffling. Cell 70: 401-410

Rodriguez-Viciana, P., Warne, P. H., Dhand, R., Vanhaesebroeck, B., Gout, I., Fry, M. J., Waterfield, M. D., and Downward, J. (1994). Phosphatidylinositol 3-OH kinase as a direct target of Ras. Nature 370: 527-532

Spaargaren, M., and Bischoff, J. R. (1994). Identification of the guanine nucleotide dissociation stimulator for Ral as a putative effector molecule of R-ras, H-ras, K-ras and Rap. Proc. Natl. Acad. Sci. USA 91: 12609-12613

van Weering, D. H. J., Medema, J. P., van Puijenbroek, A., Burgering, B. M. T., Baas, P. D., and Bos, J. L. (1995). Ret receptor tyrosine kinase activates extracellular signal-regulated kinase 2 in SK-N-MC cells. Oncogene 11: 2207-2214

White, M. A., Vale, T., Camonis, J. H., Schaefer, E., and Wigler, M. H. (1996). A role for the Ral guanine nucleotide dissociation stimulator in mediating Ras-induced transformation. J. Biol. Chem. 271: 16439-16442

Wolthuis, R. M. F., Bauer, B., van 't Veer, L. J., de Vries-Smits, A. M. M., Cool, R. H., Spaargaren, M., Wittinghofer, A., Burgering, B. M. T., and Bos, J. L. (1996). RalGDS-like factor (Rlf) is a novel Ras and Rap1A-associating protein. Oncogene 13: 353-362

Small GTPases in the morphogenesis of yeast and plant cells

Viktor Žárský[1] and Fatima Cvrčková
Department of Plant Physiology
Faculty of Sciences
Charles University
Viničná 5
CZ 128 44 Praha 2
Czech Republic

Abstract

Small GTPases play a central part in cell morphogenesis of both yeast and animal cells. Intracellular vesicle transport, essential for cell growth, requires the function of Rab-like GTPases, while the Rac/Cdc42-like GTPases control cytoskeletal organisation. Since relatively little is known about analogous regulation of cell morphogenesis in higher plants, we started a systematic search for plant homologues of several known yeast and animal morphogenetic genes. So far, we identified a number of small GTPases (including Rab and Rac homologues) expressed in the tobacco pollen, i.e. in a cell type capable of extremely rapid polar cell growth, and cloned an *Arabidopsis thaliana* homologue of the Rab-associated GDP dissociation inhibitor (Rab-GDI) by complementation of the *Saccharomyces cerevisiae sec19-1* mutation.

Introduction

What determines the shape of multicellular organisms? In plants, whose cells as a rule do not migrate, morphogenesis results from two processes - oriented cell division and differential cell growth. Unlike its animal counterpart, a typical plant (or fungal) cell has a rigid shape determined by its cell wall. Cell shape can change only slowly, since any reshaping requires remodeling of the exocellular matrix. Multiple factors affect the cell shape. Diffusible molecules (phytohormones), as well as mechanical forces and signals originating from the surrounding tissue, influence the cell morphogenesis "from outside". Plant cells are not only

---

[1] and Institute of Experimental Botany, Academy of Sciences of the Czech Republic, Prague, Czech Republic

NATO ASI Series, Vol. H 101
Molecular Mechanisms of Signalling
and Membrane Transport
Edited by Karel W. A. Wirtz

passively formed by extracellular forces, but also actively responding to environmental signals. Indeed, a considerable part of cellular morphogenesis seems to be driven "from inside", perhaps by intracellular cytoskeletal movements, as documented by the existence of structures such as root hairs, trichomes and pollen tubes. Here we shall focus on some of the intracellular molecular mechanisms contributing to plant cell morphogenesis, and show that participating proteins are, despite the major differences in cell structure and behaviour, very similar to those found in the fungal and animal kingdoms.

Yeast buds and pollen tubes: simple models of plant cell morphogenesis

Intracellular processes determining the cell shape in higher eukaryotes are nowadays only beginning to be understood. However, considerable progress has been made in the study of cell morphogenesis in a "simple" model eukaryote - the budding yeast *Saccharomyces cerevisiae.*

Like plant cells, yeasts posses a rigid cell wall. Being unicellular for most of their lives, and having a distinctive cell shape, the yeasts provide a good model for the study of morphogenesis on the single cell level. However, yeasts can also produce multicellular structures - colonies and pseudomycelia, where we can study mechanisms responsible for non-random orientation of cell division, a process of extreme importance in the development of plant tissues.

Vegetative cells of *S. cerevisiae* divide by budding, which is based on oriented cell growth. A number of genes involved in various steps of budding and cytokinesis were identified; we shall focus on some of them in the next section. Both genetical and cytological data lead to the following model of the sequence of events related to budding and cytokinesis (see Lew and Reed, 1995, Mischke and Chant, 1995): The first step towards budding is the polarization of cell growth and establishment of the future bud site. Once the bud site is chosen, subsurface actin aggregates in its vicinity. In parallel, a ring of 10 nm microfilaments (the bud neck filament or septin ring) assembles under the cytoplasmic membrane around the site of bud emergence (see Chant, 1994). This ring consists of structurally related proteins collectively termed the septins. Members of the septin gene family were found also in *Drosophila, Caenorhabditis* and vertebrates (Neufeld and Rubin, 1994, Sanders and Field, 1994, Fares et al., 1996; see also Anonymous, 1996a,b).

The next step in normal bud development is the establishment of a "border" that limits cell wall growth only to the bud and seems to act as a barrier for surface diffusion. This border is somewhat reminiscent e.g. of the apical/basolateral border of epithelial cells.Position of the border is marked by the septin ring, which does not necessarily mean that the ring and the border are identical. Indeed, yeast mutants forming normal-looking septin rings but unable to restrict growth to the bud have been found (Cvrčková, 1994; Cvrčková et al., 1995). Bud formation and growth, as well as subsequent cytokinesis, are subjected to a temporal control ensuring that these processes are kept in phase with other cell cycle events such as DNA replication and mitosis. This control is apparently accomplished by the general cell cycle regulators - cyclin-dependent kinases (CDKs) and cyclins (see Lew and Reed, 1995).

Plant cells exhibit a variety of morphogenetic processes that may have their counterparts in yeasts. Formation of cell protuberances such as unicellular trichomes and root hairs, whose development formally resembles budding in yeast. These structures arise as a result of localized growth of a distinct part of the cell surface, and usually develop their shape by restricting the growing area to the tip of the elongating protuberance (Hülskamp et al., 1994, Peterson and Farquhar, 1996, Obermeyer and Bentrup, 1996).

In most seed plants, formation of a highly polar cell protuberance - the pollen tube, formed by the vegetative cell of a mature pollen grain - even represents an essential step in the sexual life cycle (see Bedinger et al., 1994). Pollen tubes are an interesting model for the study of plant cell morphogenesis. Large populations of pollen grains can be easily brought to germination in vitro, providing ample material for biochemical, physiological and - in species with sufficiently large pollen tubes - even electrophysiological analysis, which is not technically feasible in small yeast cells (Obermeyer and Bentrup 1996, Feijó et al., 1995). The behaviour of actin and tubulin cytoskeleton (Joshi and Palevitz, 1996, Staehelin and Hepler 1996) as well as the overall pattern of cell surface growth in pollen tubes exhibits apparent similarities to that in the yeast buds. A germinating pollen grain chooses a single location of its surface to start the polarized growth, and forms a sharp border between the growing and non-growing surface, reminiscent of the yeast bud neck (although no septins were found in plant cells so far). This border is determined by the preexisting structure of the cell wall, which at least partly reflects cytoskeletal organization during pollen maturation (Schmid et al., 1996). Moreover, the observation that germinating vegetative pollen cells are in the G1 phase of their cell cycle (Žárský et al., 1992) may be more than a coincidence - also yeast cells grow in a polar manner only in this phase of the cell cycle (Lew and Reed, 1993). The question whether similarities could also be detected on the level of underlying regulatory mechanisms arises as an obvious consequence of comparing the fungal and plant cells.

Cell growth, intracellular transport and the Rab GTPases

Which molecular mechanisms are contributing to the shape of a plant or fungal cell? In cells with a rigid cell wall, any change of the cell shape requires remodeling of a preexisting wall structure. This is accomplished by enzymes localized in the periplasmic space, which locally loosen the wall structure and insert new wall material. Both the enzymes and the wall material to be inserted are secreted from within the cell in a carefully regulated manner, since in the usual hypotonic environment the wall is required for physical stability of cells whose vacuole maintains turgor pressure (Harold et al., 1995, Heath, 1995).

Our recent understanding of secretion is based mostly on studies in yeast and mammalian cells. Novick et al. (1980) identified a group of 23 yeast genes (the *SEC* genes) required for protein secretion; nowadays, the number of known secretion-related genes from various organisms exceeds 200 (Rothblatt et al., 1994). The first step of secretion is the (usually cotranslational) transport of secretory proteins into the endoplasmic reticulum (ER). From the ER, the proteins continue through the cisternae of the Golgi complex to the cell surface, being usually at the same time processed by proteolytic cleavage and (often complex) glycosylation. Each step of the secretory transport is accomplished by fission of vesicles from a donor compartment (ER or Golgi), their transport to the target site and fusion with the target compartment (Golgi or plasmalemma). The formation, transport and delivery of secretory vesicles is a part of a complex network of vesicle traffic which is responsible for the maintenance of proper protein composition of all compartments of a eukaryotic cell (see Rothman and Wieland, 1996, Rothblatt et al., 1994).

How do the secretory (and other) vesicles find their targets? Mutual recognition of specific proteins on the vesicles (v-SNAREs) and their targets (t-SNAREs) is required for the final step of vesicle delivery - its binding to and fusion with the target membrane (see Rothman and Wieland, 1996). The irreversibility of vesicle docking is ensured by a conformational change in a small GTP-binding protein from the Ras superfamily - the Rab GTPase - that is associated with the vesicle and responds to a proper vSNARE-tSNARE contact by GTP hydrolysis. This conformational change causes detachment of the GTPase from the membrane. The Rab protein remains in the cytosol in a GDP-bound state, stabilized by a GDP dissociation inhibitor (GDI), and reattaches to the membrane upon acquisition of a new GTP molecule, facilitated by a guanosine nucleotide exchange factor (GEF). The hydrolysis of GTP also requires a protein cofactor, the GTPase activating protein (GAP) that is probably provided by the acceptor compartment (Takai et al., 1993). Rab proteins are not the only small GTPases associated with vesicle transport. Budding of vesicles from the donor membrane requires other members of the Ras superfamily - the Arf GTPases (Rothmann and Wieland, 1993).

Multiple genes for Rab proteins were found in both animal and yeast cells. They include one of the first isolated yeast *SEC* genes - *SEC4*, and the family of *YPT* genes (Novick et al. 1993). Different Rab proteins are associated with distinctive stages of the vesicle trafficking processes. Rab1 and Rab2 are associated with vesicle transport between ER and Golgi, Rab6 has been found in the middle and trans Golgi cisternae, acting presumably in intra-Golgi transport. The yeast Sec4 and mammalian Rab3 and Rab8 proteins accompany the vesicles traveling from Golgi to the plasmalemma. Other GTPases of the Rab family are involved in endocytosis, a process that could be viewed as the opposite of secretion (see Rothblatt et al., 1994).

Small GTPases from the Rab family were recently found also in plants (Ma, 1994; Verma et al., 1994, Jonak et al., 1995, Moore et al., 1996). Only very few of them have been characterized functionally. In some cases, plant genes were shown to complement yeast mutations (e.g. Fabry et al., 1995). Cheon et al. (1993) showed that Rab homologues are participating in the formation of peribacteroid membranes in symbiotic root nodules. Apparently, components of the well-conserved secretion apparatus were recruited for a process that appeared relatively late in the evolution of higher plants.

Secretion could be viewed as a "housekeeping" process essential for the survival of any cell. However, it has been found that components of the secretion machinery have an important role in plant morphogenesis. Two *A. thaliana* genes, identified on the basis of mutations disturbing embryogenesis, code for proteins involved in secretion: *EMB30/GNOM* is related to the yeast *SEC7* gene, while *KNOLLE* codes for a t-SNARE (Shevell et al., 1994, Lukowitz et al., 1996).

Regulation of cytoskeletal behaviour and the Rac/Cdc42 GTPases

Yeast studies were also instrumental for our recent understanding of additional regulatory aspects of cell morphogenesis. Mutants defective in the temporal or spatial control of budding and cytokinesis were found in screens for cell division cycle (*cdc*) mutants, as well as in other mutant hunts (see Hartwell et al., 1974, Adams et al., 1990, Lew and Reed, 1995, Mischke and Chant, 1995). Small GTPases from the Rho subfamily were found to contribute to the control of cell shape not only in yeast but also in animal cells (see Nobes and Hall, 1994, 1995, Chant and Stowers, 1995).

Establishment of the bud site in yeast requires the Rho-like GTPase Cdc42p (Johnson and Pringle, 1990). This GTPase communicates with protein cofactors - a GEF (Cdc24p - Hart et al., 1991) and a GAP (Bem3p - Zheng et al., 1994). Subsequent steps in budding - rearrangement of the actin cytoskeleton, formation of the septin ring and establishment of a border between the growing and non-growing surface - require, directly or indirectly, proper function of Cdc42.

Mechanisms by which Cdc42 controls these events are apparently different. Mutants with defects in Cdc42p cannot perform any of the three events. However, mutants lacking two functionally redundant protein kinases Cla4p and Ste20p localize actin normally, assemble normal-looking septin rings but fail to restrict the growth to the bud side of the ring. The Cla4p kinase is likely to be directly regulated by the active, GTP-bound form of Cdc42p (Cvrčková, 1994; Cvrčková et al., 1995). Proteins related to Cdc42p as well as interacting kinases corresponding to Cla4p/Ste20p (the PAK-type kinases) have also been found in mammalian cells (Manser et al., 1994, Martin et al., 1995).

Cdc42p and its relatives (the Rho and Rac GTPases) have multiple roles. They play a central part in the organization of the subcortical, especially actin, cytoskeleton, and therefore in the control of cell shape in both yeast and higher eukaryotes. In animal cells, Rho-related GTPases are required for processes such as membrane ruffling, the formation of lamellipodia and filopodia, and formation of actin stress fibers (Nobes and Hall, 1995). In addition, mammalian Cdc42p and Rac are involved in a signaling pathway controlling transcription of certain genes (Minden et al., 1995). The yeast Cdc42p GTPase has also a role in transduction of the pheromone signal that induces differentiation of vegetative cells into gametes, a process involving also transcriptional control (Simon et al., 1995). An additional, recently discovered, function of yeast Rho proteins may be especially relevant to plant cell morphogenesis: the Rho1p GTPase was found to regulate the activity of the glucan synthase, an enzyme involved in cell wall construction (Drgoňová et al., 1996).

Recently, first small GTPases of the Rho family (denoted Rop - *R*ho *of p*lant, although they are structurally closer to Rac/Cdc42 than to Rho) were identified in pea and *Arabidopsis* (Yang and Watson, 1993; Anonymous, 1996b). The intracellular localization of Rop proteins in the cortical cytoplasm of growing pea pollen tube tips suggests that they may have a role in tip growth (Lin et al., 1996).

A search for small GTPases involved in pollen development

An essential step in unraveling the mechanism of plant cell morphogenesis would be the identification of components of the responsible molecular apparatus. For this reason, we decided to start a systematic search for small GTPases expressed in the tobacco pollen, using a modification of RACE-PCR (Martin-Parras and Zerial, 1995). Template cDNA was prepared by reverse transcription of total RNA from a population of *Nicotiana tabacum* cv. Samsun pollen grains at different stages of development. PCR amplification was performed with a non-specific 5' end primer and a degenerate intragenic primer that should detect a broad spectrum of small GTPases (described in Martin-Parras and Zerial, 1995). 5'- terminal portions of at least 12 different, mostly novel GTPase-encoding cDNAs were found among 30 randomly chosen sequenced clones (Fig. 1). Among them were sequences encoding several members of the Rab family (Rab1, Rab6, Rab8 and Rab11), a Rho-like GTPase (see below), two Arf homologues, and a protein related to the Ran GTPase involved in the transport of mRNA from the nucleus (Kadowaki et al., 1993). We do not know yet how many of these genes are expressed in a pollen-specific manner; at least one of them (the Ran homologue) is identical with the previously cloned tobacco RanA1 gene expressed in somatic tissues (GenBank Acc. No. L16767). Characterization of a larger number of PCR products would also be required to saturate the screen; for example, a homologue of Rab5 is transcribed during tobacco pollen development (V.Ž. and F. Nagy, unpublished data), although no such clone was obtained among the cDNAs sequenced so far.

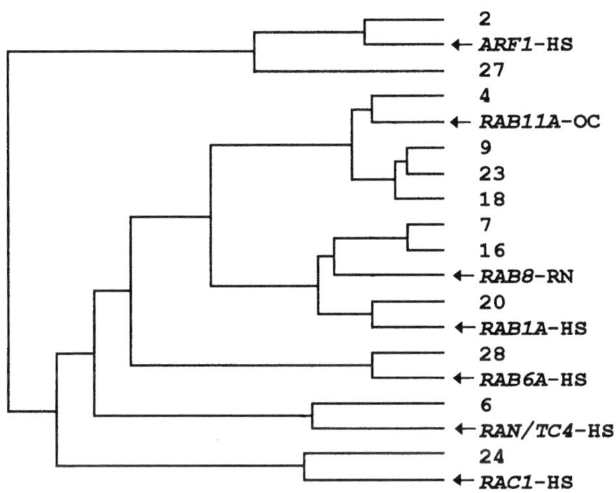

Fig. 1. A dendrogram representing the diversity of tobacco pollen GTPases identified in a PCR screen. Branches denoted by numbers represent clones from the screen, those denoted by gene names correspond to known mammalian GTPases. The dendrogram was produced from an alignment of predicted protein sequences by the CLUSTAL program (Higgins and Sharp, 1988; settings: K-tuple 1, gap penalty 5, window size 10, open gap cost 10, unit gap cost 10, filtering load 2.5); longer sequences were cut at the position of the degenerate primer. HS - *Homo sapiens*, RN - *Rattus norvegicus*, OC = *Oryctolagus cuniculus*.

Isolation of the *Arabidopsis thaliana* Rab-associated GDI

Identification of a GDP dissociation inhibitor (GDI) associated with the Rab proteins acting in pollen development would be extremely interesting not only *per se* (since there are no plant GDIs known so far), but also for a technical reason. Functional analysis of the Rab-like

GTPases in pollen is likely to be hampered by the functional overlap between the multiple isoforms of Rab proteins in every cell. Inactivation of a single gene e.g. by a mutation or by antisense RNA could be compensated by the activity of a closely related gene product. A possibility to interfere with the function of a whole subgroup of GTPases would be desirable. Since one or a few GDIs are likely to be shared by all proteins of the Rab subfamily (Rothblatt et al., 1994), an isolated plant GDI gene could serve as an interesting tool in the future studies of the biological function of Rab proteins.

We decided to search for functional homologues of the well-characterized yeast Rab-GDI, the *SEC19/GDI1* gene product (Garrett et al., 1994), by complementation of the thermosensitive growth defect of a *S. cerevisiae* strain carrying the *sec19-1* mutation. The mutant yeast strain was transformed by a cDNA library derived from whole *A. thaliana* plants of different developmental stages, constructed in the yeast expression vector λYES (Elledge et al., 1991), using the method of Dohmen et al. (1991). Transformants were plated on standard media (see Nasmyth and Dirick, 1991) lacking uracil (to select for the plasmid) and containing galactose (to induce expression of the cDNAs), allowed to recover for 24 h at permissive temperature and transferred to the restrictive temperature. Plasmids from colonies formed under such conditions were examined for their ability to complement *sec19-1* after repeated transformation. At least two plasmids complementing *sec19-1* were found after screening approximately 40 000 cDNAs. A weakly complementing clone, apparently a bypass suppressor of the GDI mutation, was unrelated to *SEC19* or other known genes in available databases[2]. A strongly complementing cDNA coded for an obvious Sec19 homologue (53 % amino acid identity throughout the protein) and therefore most likely represents a genuine *A. thaliana* Rab GDI (V. Ž., F. C., F. Bischoff and K. Palme, manuscript in preparation; EMBL database Acc. No. Y07961).

How conserved are the regulatory aspects of plant cell morphogenesis?

There are reasons to believe that molecular mechanisms responsible for cell morphogenesis are evolutionally old. Small GTPases involved in secretion seem to be relatively well conserved between fungi, animals and higher plants. At least a part of the diversity of Rab proteins apparently evolved already in the ancient progenitor common to the recent plant and animal kingdoms (see Fig. 1).

---

[2] There was significant similarity between this clone and several plant ESTs of unknown function (Anonymous, 1996b).

However, the small GTPases that control the cytoskeletal organization might present a different case. In both yeasts and animal cells, these Rho-related GTPases form three distinct subfamilies - Rho, Rac and Cdc42. Do plants have genuine homologues of all these GTPases? Our attempt to identify plant *CDC42* homologues by PCR amplification with primers chosen from the regions where this GTPase differs from Rho, Rac and Rop using *A. thaliana* genomic DNA and *A. thaliana* and alfalfa (*Medicago sativa*) cDNA libraries as templates yielded only sequences unrelated to the *CDC42* gene, and screening approximately 200 000 *M. sativa* cDNAs for suppressors of the yeast *cdc42-1* mutation produced no complementing clones (F.C. and M. Zachlederová, unpublished observations). The only Rho-like GTPase found in our PCR-based screen for GTPases expressed in the tobacco pollen is an evident member of the Rop protein family (Fig. 2; Lin et al., 1996).

```
24            MSAPRFIKCVTVGDGAVGKTCLLISYTXNTFP
ROP1          MSASRFIKCVTVGDGAVGKTCLLISYTSNTFP
RAC1             MQAIKCVVVGDGAVGKTCLLISYTTNAFP
CDC42            MQTLKCVVVGDGAVGKTCLLISYTTNQFP
RHO1      MSQQVGNSIRRKLVIVGDGACGKTCLLIVFSKGQFP
                     * * ***** *******      **

24            MDYVPTVFDNFSANVVVNGSTVNLGLWDTXGQ
ROP1          TDYVPTVFDNFSANVVVNGSTVNLGLWDTAGQ
RAC1          GEYIPTVFDNYSANVMVDGRPINLGLWDTAGQ
CDC42         ADYVPTVFDNYAVTVMIGDEPYTLGLFDTAGQ
RHO1          EVYVPTVFENYVADVEVDGRRVELALWDTAGQ
                * **** *    *        * * *****
```

Fig. 2. Partial sequence of the predicted protein product of the tobacco pollen Rop-related cDNA (clone 24 from Fig. 1) aligned to the sequences of pea Rop1, *Caenorhabditis elegans* Rac1 and yeast Cdc42p and Rho1p proteins (sequences from Anonymous, 1996a,b; the alignment was made using the program MACAW - Schuler et al., 1991). Asterisks denote completely conserved residues.

The plant Rop proteins may well represent a special subfamily of GTPases that does not directly correspond to any of the animal or yeast proteins but performs functions that are shared by two or more proteins in fungi or animal cells. While yeast and animal Rho (or Cdc42) proteins are genuine homologues of each other, plant Rop proteins may represent a paralogous family that evolved to accomplish similar functions by different means. This would suggest that the GTPase system responsible for cytoskeletal organization has evolved after the separation of the plant and animal/fungal kingdoms, and is therefore much younger than the GTPase apparatus associated with secretion.

Conclusion

Last decades of molecular biology produced an image of biological systems that are based on common molecular principles despite their (phenotypic) variability. Genes were viewed as "written" perhaps in different languages but in the same alphabet - the genetic code. The important question was which genes does an organism need to produce a phenotype. Nowadays it becomes obvious that even the sentences of the "genetic language", such as signal transduction pathways or protein secretion mechanisms, are surprisingly well conserved not only between ourselves and cows, but also between cows and grass. It is perhaps time to ask which phenotypes can a gene produce in different organismal backgrounds. If we were to study this question, conserved protein families, such as the small GTPases and their cofactors, could serve as a good model system.

Acknowledgements

We thank L. Nečasová for expert technical assistance, A. Ragnini and R. Schekman for the yeast strains, R. Davis for the cDNA library, F. Bischoff, S. Fabry and T. Schmülling for helpful discussion, and V. Vondrejs for critical reading of the manuscript. This work has been supported by the Grant Agency of the Czech Republic Grants 204/95/1069 and 204/95/1296; part of the work has been done in the laboratory of K. Palme (MPI Köln), supported by the European Commission S & T Cooperation with Central and East Europe Grant 11051 to V.Ž.

References

Adams AM, Johnson DI, Longnecker RM, Sloat BF, Pringle JR (1990) *CDC42* and *CDC43*, two additional genes involved in budding and in the establishment of cell polarity in the yeast *Saccharomyces cerevisiae*. J Cell Biol 111:131-142

Anonymous (1996a) Collection of non-redundant sequence databases accessible through the NCBI Entrez server. URL http://www.ncbi.nlm.nih.gov

Anonymous (1996b) Plant EST databases. URL http://www.tigr.org and http://genome-www.stanford.edu

Bedinger PA, Hardeman KJ, Loukides CA (1994) Travelling in style: the cell biology of pollen. TICB 4:132-138

Chant J (1994) Cell polarity in yeast. TIG 10:328-333

Chant J, Stowers L (1995) GTPase cascades choreographing cellular behavior: movement, morphogenesis and more. Cell 81:1-4

Cheon C, Lee N-G, Siddique A-BM, Bal AK, Verma DP (1993) Roles of plant homologues of Rab1p and Rab7p in the biogenesis of the peribacteroid membrane, a subcellular compartment formed *de novo* during root nodule symbiosis. EMBO J 12:4125-4135

Cvrčková F (1994) From "start" to finish: G1 cyclins have a role in yeast cytokinesis. Thesis, Universitaet Wien

Cvrčková F, De Virgilio C, Manser E, Pringle JR, Nasmyth K (1995) Ste20-like protein kinases are required for normal localization of cell growth and for cytokinesis in budding yeast. Genes Dev 9:1817-1830

Dohmen JR, Strasser AW, Höner CB, Hollenberg CP (1991) An efficient transformation procedure enabling long-term storage of competent cells of various yeast genera. Yeast 7:691-692

Drgoňová J, Drgoň T, Tanaka K, Kollár R, Chen G-C, Ford RA, Chan CSM, Takai Y, Cabib E (1996) Rho1p, a yeast protein at the interface between cell polarization and morphogenesis. Science 272:277-281

Elledge SJ, Mulligan T, Ramer SW, Spottswood M, Davis RW (1991) λYES: a multifunctional cDNA expression vector for the isolation of genes by complementation of yeast and *Escherichia coli* mutations. PNAS USA 88:1731-1735

Fabry S, Steigerwald R, Bernklau C, Dietmaier W, Schmitt R (1995) Structure-function analysis of small G proteins from *Volvox* and *Chlamydomonas* by complementation of *Saccharomyces cerevisiae YPT/SEC* mutations. Mol Gen Genet 247:265-274

Fares H, Goetsch L, Pringle JR (1996) Identification of a developmentally regulated septin and involvement of the septins in spore formation in *Saccharomyces cerevisiae.* J Cell Biol 132:399-411

Feijó JA, Malhó R, Obermeyer G (1995) Ion dynamics and its possible role during in vitro pollen germination and tube growth. Protoplasma 187:155-167

Garrett MD, Zahner JE, Cheney CM, Novick PJ (1994) *GDI1* encodes a GDP dissociation inhibitor that plays an essential role in the yeast secretory pathway. EMBO J 13:1718-1728

Harold FM, Harold RL, Money NP (1995) What forces drive cell wall expansion? Can J Bot 73(Suppl. 1):S379-S383

Hart MJ, Eva A, Evans T, Aaronson SA, Cerione RA (1991) Catalysis of guanine nucleotide exchange on the CDC42Hs protein by the dbl oncogene product. Nature 354:311-314

Hartwell LH, Culotti J, Pringle JR, Reid BJ (1974) Genetic control of the cell division cycle in yeast. Science 183:46-51

Heath IB (1995) Integration and regulation of hyphal tip growth. Can J Bot 73(Suppl. 1):S131-S139

Higgins DG, Sharp PM (1988) CLUSTAL: a package for performing multiple sequence alignment on a microcomputer. Gene 73:237-244

Hülskamp M, Miséra S, Jürgens G (1994) Genetic dissection of trichome cell development in *Arabidopsis*. Cell 76: 555-566

Johnson DI, Pringle JR (1990) Molecular characterization of *CDC42*, a *Saccharomyces cerevisiae* gene involved in the development of cell polarity. J Cell Biol 111:143-152

Joshi HC, Palevitz BA (1996) γ-tubulin and microtubule organization in plants. TICB 6:41-44

Jonak C, Heberle-Bors E, Hirt H (1995) A cDNA from *Medicago sativa* encodes a protein homologous to small GTP-binding proteins. Plant Physiol 107:263-264

Kadowaki T, Goldfarb D, Spitz LM, Tartakoff AM, Ohno M (1993) Regulation of RNA processing and transport by a nuclear guanine nucleotide release protein and members of the Ras superfamily. EMBO J 92:2927-2937
Lew DJ, Reed SI (1993) Morphogenesis in the yeast cell cycle: regulation by Cdc28 and cyclins. J Cell Biol 120:1305-1320
Lew DJ, Reed SI (1995) Cell cycle control of morphogenesis in budding yeast. Curr Opin Genet Dev 5:17-23
Lin Y, Wang Y, Zhu J, Yang Z (1996) Localization of a Rho GTPase implies a role in tip growth and movement of the generative cell in pollen tubes. Plant Cell 8:293-303
Lukowitz W, Mayer U, Jürgens G (1996) Cytokinesis in the *Arabidopsis* embryo involves the syntaxin-related *KNOLLE* gene product. Cell 84:61-71
Ma H (1994) GTP-binding proteins in plants: new members of an old family.
Manser E, Leung T, Salihuddin H, Zhao Z, Lim L (1994) A brain serine-threonine protein kinase activated by cdc42 and rac. Nature 367:40-46.
Martin GA, Bollag G, McCormick F, Abo A (1995) A novel serine kinase activated by Rac/Cdc42Hs-dependent autophosphorylation is related to PAK65 and STE20. EMBO J 14:77-81 (corrigenda: EMBO J 14:4385)
Martin-Parras L, Zerial M (1995) Using degenerate oligonucleotides for cloning of rab proteins by polymerase chain reaction. Meth Enzymol 257-189-199
Minden A, Lin A, Claret FX, Abo A, Karin M (1995) Selective activation of the JNK signaling pathway and c-Jun transcriptional activity by the small GTPases Rac and Cdc42Hs. Cell 81:1147-1157
Mischke B, Chant J (1995) The shape of things to come: morphogenesis in yeast and related patterns in other systems. Can J Bot 73 (Suppl. 1):S234-S242
Moore I, Diefenthal T, Žárský V, Schell J, Palme K (1996) Homologues of mammalian rab2 from *Arabidopsis*. PNAS USA, in press
Nasmyth K, Dirick L (1991) The role of *SWI4* and *SWI6* in the activity of G1 cyclins in yeast. Cell 62:631-647
Neufeld TP, Rubin GM (1994) The *Drosophila peanut* gene is required for cytokinesis and encodes a protein similar to yeast putative bud neck filament proteins. Cell 77:371-379
Nobes C, Hall A (1994) Regulation and function of the Rho-subfamily of small GTPases. Curr Opin Genet Dev 4:77-81
Nobes C, Hall A (1995) Rho, Rac and Cdc42 GTPases regulate the assembly of multimolecular focal complexes associated with actin stress fibers, lamellipodia and filopodia. Cell 81:53-62
Novick P, Field Ch, Schekman R (1980) Identification of 23 complementation groups required for post-translational events in the yeast secretory pathway. Cell 21:205-215
Novick P, Brennwald P, Walworth NC, Kabcenell AK, Garrett M, Moya M, Roberts D, Müller H, Govindan B, Bowser R (1993) The cycle of *SEC4* function in vesicular transport. Ciba Foundation Symposium 176:218-232
Obermeyer G, Bentrup FW (1996) Developmental physiology: regulation of polar cell growth and morphogenesis. Progr Bot 57:54-67
Peterson RL, Farquhar ML (1996) Root hairs: specialized tubular cells extending root surfaces. Bot Rev 62:2-40
Rothblatt J, Novick P, Stevens TH (eds) (1994) Guidebook to the secretory pathway. A Sambrook and Tooze Publication. Oxford University Press, Oxford
Rothman JE, Wieland FT (1996) Protein sorting by transport vesicles. Science 272:227-234
Sanders SL, Field CM (1994) Septins in common? Curr Biol 4:907-910

Simon MN, De Virgilio C, Souza B, Pringle JR, Abo A, Reed SI (1995) Role for the Rho-family GTPase Cdc42 in yeast mating-pheromone signal pathway. Nature 376:702-705

Schmid A-MM, Eberwein RK, Hesse M (1996) Pattern morphogenesis in cell walls of diatoms and pollen grains: a comparison. Protoplasma 193:144-173

Schuler GD, Altschul SF, Lipman DJ (1991) A workbench for multiple alignment construction analysis. Prot Struct Funct Genet 9:180-190

Shevell D, Leu W-M, Gilmour CS, Xia G, Feldmann KA, Chua N-H (1994) *EMB30* is essential for normal cell division, cell expansion, and cell adhesion in *Arabidopsis* and encodes a protein that has similarity to Sec7. Cell 77:1051-1062

Staehelin LA, Hepler PK (1996) Cytokinesis in higher plants. Cell 84:821-824

Takai Y, Kaibuchi K, Kikuchi A, Sasaki T, Shirataki H (1993) Regulators of small GTPases. Ciba Foundation Symposium 176:128-146

Verma DPS, Cheon C, Hong Z (1994) Small GTP-binding proteins and membrane biogenesis in plants. Plant Physiol 106:1-6

Yang Z, Watson JC (1993) Molecular cloning and characterization of rho, a ras-related small GTP-binding protein from the garden pea. PNAS USA 90:8732-8736

Zheng Y, Cerione R, Bender A (1994) Control of the yeast bud-site assembly GTPase Cdc42. Catalysis of guanine nucleotide exchange by Cdc24 and stimulation of GTPase activity by Bem3. J Biol Chem 269:2369-2372

Žárský V, Garrido D, Říhová L, Tupý J, Vicente O, Heberle-Bors E (1992). Derepression of the cell cycle by starvation is involved in the induction of tobacco pollen embryogenesis. Sex Plant Reprod 4:204-207

A role for G-protein βγ-subunits in the secretory mechanism of rat peritoneal mast cells

Jef A. Pinxteren, Antony J. O'Sullivan and Bastien D. Gomperts
Secretory Mechanisms Group
Physiology Departement - Rockefeller Building
University College London
London WC1E 6JJ
UK

Introduction

Our aim is to advance knowledge and understanding in the field of regulated exocytotic secretion. This is the process by which substances packaged in membrane-bound vesicles are released to the exterior. In its final stages, the exocytotic mechanism involves fusion of the vesicle membrane and the plasma membrane and it occurs without any leakage or even a transient insult to the integrity of either. Regulated exocytosis is crucial to the functioning of complex organisms in which communication takes place between cells and between tissues, and it is our major interest. Depending on the system, exocytosis may occur in response to elevations in intracellular Ca^{2+} or alternative second messengers such as cyclicAMP. We focus our attention on the granulocytic cells of the immune system (mast cells, eosinophils, neutrophils, T-cells etc) in which the regulation of exocytosis involves the activation of a GTP-binding protein that we have called G_E (Gomperts, 1990).

While Ca^{2+} clearly plays an important role in the stimulus-secretion pathway of these cells, the late events determining exocytotic fusion can be triggered in permeabilised cells in the effective absence of Ca^{2+} (at concentrations $<10^{-9}$M) by application of GTP-γ-S alone. Neither Ca^{2+} nor ATP are needed, though when provided both have important modulatory effects. We use the technique of cell permeabilisation with streptolysin-O as a biochemical approach to detect

NATO ASI Series, Vol. H 101
Molecular Mechanisms of Signalling
and Membrane Transport
Edited by Karel W. A. Wirtz

proteins which may play a role in exocytosis. Following on the loss of cytosol proteins, secretory cells lose their capacity to secrete upon stimulation though this can be delayed and retarded by provision of exogenous proteins. We have called this phenomenon "run–down" and we use the modulation of "run–down" as the basis of a cellular bioassay to identify protein regulators of exocytosis in mast cells and eosinophils (O'Sullivan *et al.*, 1996).

For mast cells, it has been clear for some time that, dependent on the identity of the (cell surface) stimulus, different GTP-binding proteins mediate exocytosis. Agents stimulating mast cells fall into two classes. There are those which act to crosslink the cell surface receptors for IgE, initiating a series of phosphorylations on tyrosine residues and leading to activation of phospholipase-C_γ and then there are the receptormimetic agents (mastoparan, compound 48/80, substance P etc) which appear to activate GTP-binding proteins directly bypassing conventional receptors and activating PLC_β and other targets. While the late stages of the secretory pathway initiated by both forms of stimulation, and indeed that due to Ca^{2+}-ionophores as well, require the presence of GTP within the cells (Wilson *et al.*, 1989; Marquardt *et al.*, 1987), only the pathway initiated by the receptormimetic agents is inhibitable by pertussis toxin (Saito *et al.*, 1987; Aridor *et al.*, 1990). This indicates that different G_E proteins mediate exocytosis dependent on the nature of the cell surface stimulus and that the G_E active in the pathway initiated by the receptormimetics is a pertussis toxin substrate.

Work from this laboratory has shown that the complex of Rac (a monomeric GTP-binding protein) with RhoGDI certainly has characteristics of a G_E protein mediating exocytosis (O'Sullivan *et al.*, 1996). RhoGDI itself accelerates the rundown and is also inhibitory to mast cells in the whole cell patch-clamp configuration (Mariot *et al.*, 1996) so we can be sure that one of the Rho-related proteins is a contender for the G_E function. However, RhoGDI is only capable of inhibiting secretion up to 80%, implying that at least one other GTP-binding protein, not of the Rho class, is capable of inducing secretion. It is unlikely, that this GTP-binding protein is a Rab as RabGDI has no effect on secretion from permeabilized or patch-clamped cells.

The heterotrimeric G-protein G_{i3} has been recognised as another regulator having functional characteristics of a G_E protein. The site of action of G_{i3} in the regulation of exocytosis from mast cells is understood to occur subsequent to the elevation of intracellular Ca^{2+} in the pathway activated by receptormimetic (i.e. non-immunological) agents such as mastoparan and compound 48/80 (Aridor *et al.*, 1993; Aridor *et al.*, 1990). Although we cannot rule out the possibility of further downstream GTP-binding proteins, the two pathways could be served by the two GTP-binding proteins so far discovered, namely G_{i3} for non-immunological stimuli, and Rac for allergens and other IgE crosslinkers. Whether they operate in parallel or in series has yet to be determined. Importantly, because the identification of G_{i3} as a G_E was based on the use of reagents specific to the C-terminus of the α-subunit, this must have prevented activation of the G-protein, dissociation and any further downfield events related to the independent actions of the α and the βγ-subunits.

G-protein βγ-subunits enhance secretion

We have thus been testing the possibility that the βγ-subunits of heterotrimeric G-proteins (prepared essentially according to the method of Sternweis and Pang (1990)) might be able to modulate exocytosis from permeabilised mast cells stimulated by Ca^{2+}-plus-GTP-γ-S. At the start, we considered all possibilities i.e. that βγ-subunits might modulate exocytosis positively, negatively, or not at all. We have used the "run–down" assay to determine the effects of βγ-subunits which we isolate from cholate extracts of bovine brain membranes. The cells, permeabilised with streptolysin-O are incubated in the presence of ATP and a Ca^{2+} buffer to maintain Ca^{2+} at 10^{-8}M, are allowed to run down for various times in the presence of proteins or appropriate controls. The cells are then stimulated by addition of GTP-γ-S (100μM) and concomitent elevation of free Ca^{2+} to 10^{-5}M and after a further 10min incubation they are sedimented and the supernatants sampled for measurement of secreted hexosaminidase.

We have found that these βγ-subunits (mainly G_i and G_o-derived) retard run–down and this suggests that they support exocytosis (Fig. 1.).

Since the βγ-subunits are unstable in cholate, and since this is also detrimental to secretion, we

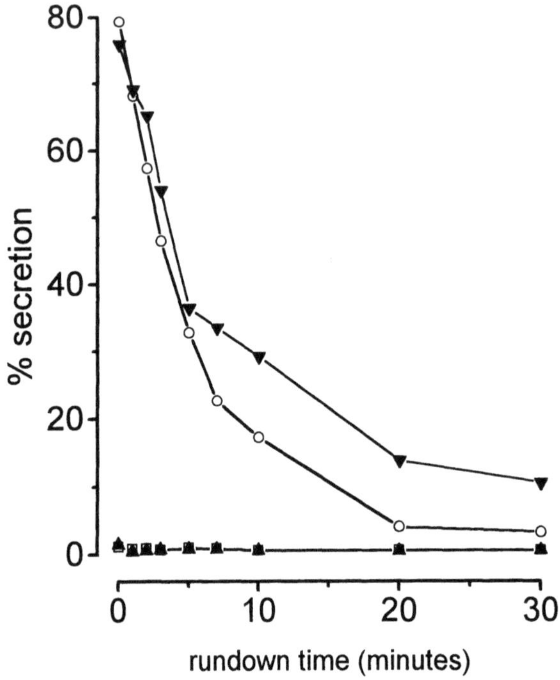

Fig.1. Example of a run down experiment, illustrating the effect of βγ subunits (10^{-8}M). If stimulated at the time of permeabilisation, mast cells can release 100% of their hexoaminidase by exocytosis. If the stimulus is applied at times after permeabilisation, then there is progressive decline in their responsiveness: this is called "run down". By timing the application of the stimulus with care, it is possible to introduce exogenous proteins into the cells during the run down period, and so test their effect on the regulation of secretion. Boiled protein was used as control (○,□). Buffer with low calcium concentration (pCa7; □,▲) was used as control for stimulation (pCa5 with GTP-γ-S, 100μM;▼,○).

have transferred them to Lubrol-PX in which they can be stored. However, while this preserves the integrity of the protein in the long term, it is also unsuitable for the secretion measurements. Perhaps this should come as no surprise but it is only recently discovered that low concentrations of detergent are inhibitory to all forms of membrane fusion (Chernomordik

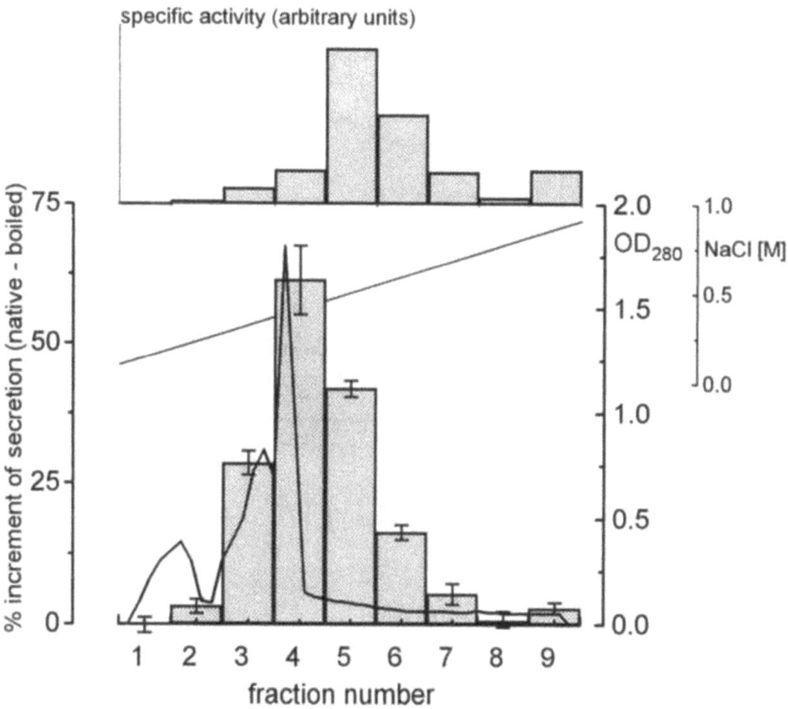

Fig.2. Subfractionation of βγ subunits. The exchange of detergent into CHAPS using a MonoQ column (Lower panel), causes some fractionation of the βγ subunits. After allowing unbound material to wash through the column, a NaCl gradient (0-1M) was imposed and the fractions collected. Analysis of the fractions in the run down experiment indicates that the profile of activity is shifted with respect to the main peak of eluted protein (Top panel).

et al., 1993). CHAPS turned out to be the most useful detergent so far, in a way that it seems not to disturb secretion and keeps the proteins in solution, at least during the time of the experiment. This stage was however only reached using a Mono Q column to exchange detergents and not by using a DEAE Sephacel column as described earlier (Logothetis *et al.*, 1988). In Figure 2 we illustrate the elution profile of βγ-subunits from the Mono Q column used to exchange detergents (from cholate to CHAPS). After allowing unbound material to wash through, we imposed a

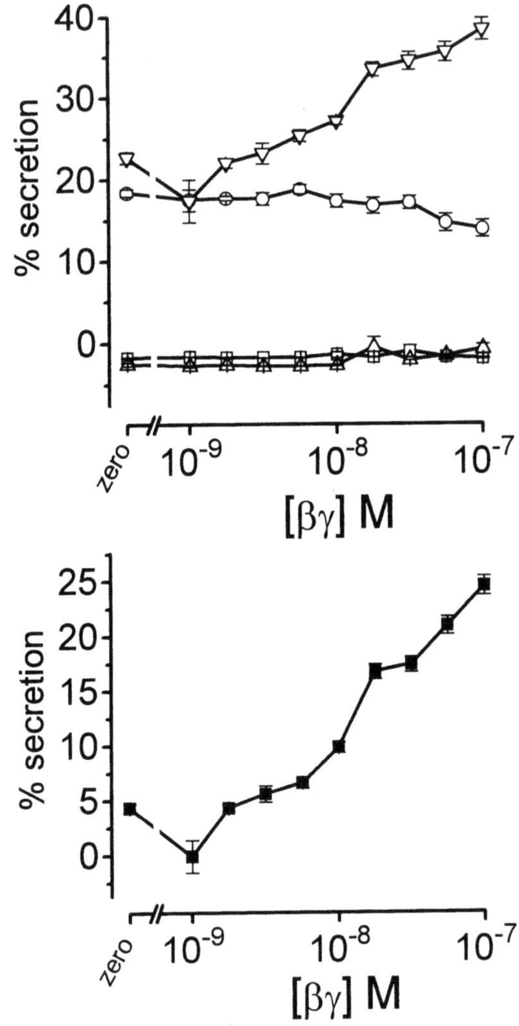

Fig.3. Concentration-effect relationship for mixed brain βγ subunits. Cells were stimulated (pCa5 with GTP-γ-S, 100μM) after allowing 10min for the protein to penetrate the cells. The effect of the protein in enhancing the extent of secretion is clearly manifest at concentrations below 10^{-8}M and is not fully saturated at 10^{-7}M. Boiled protein was used as control (O,□). Buffer with low calcium concentration (pCa7; □,▲) was used as control for stimulation (pCa5 with GTP-γ-S, 100μM;▼,O). Error bars represent SEM (n=4).

NaCl gradient (0-1M) and collected the fractions. SDS-PAGE and immunoblotting indicates that the high resolution column is also causing some slight separation of different βγ-subtypes, with enrichment of β_1 over β_2 in the fractions leading up to the peak and relative enrichment of β_2 relative to β_1 in the later fractions. Analysis of the fractions in the rundown experiment indicates that the profile of activity is shifted with respect to the main peak of eluted protein. From this experiment we conclude that βγ-subunits can induce secretion, and furthermore, that certain pairs of βγ-subunits seem more active than others (nb boiled protein was used as a control to compensate for possible artefacts arising from the presence of the detergent). In Figure 3, the concentration-effect relationship for mixed brain βγ-subunits in the rundown experiment is illustrated. Once again, the cells were stimulated by Ca^{2+} and GTP-γ-S after allowing 10min for the protein to penetrate the cells.

There is no secretion in the absence of the stimulus (pCa7), and in this experiment the secretion that occurred in the absence of protein was 20% (normally close to 100% in cells stimulated immediately at the time of permeabilisation). The effect of the protein in enhancing the extent of secretion is clearly manifest at concentrations below 10^{-8}M and is not fully saturated at 10^{-7}M. We get the impression that the slope of the activation curve is rather shallow, and this could be interpreted to mean that the βγ-subunits are activating the system through interaction with different targets (possibly indicating the activities of different βγ-pairs).

Possible mechanisms of action of G-protein βγ-subunits

Recent work has indicated that βγ-subunits can bind to pleckstrin homology (PH) domains present on other proteins (Pitcher *et al.*, 1995). The fact that these domains are found on several key enzymes in cell signalling (e.g. PLC (all β and at least the types γ1, γ4, δ1 and δ4), MAP kinase, Raf1, PK, cdc24, βARK, PKC_{μ}, Tsk and Btk tyrosine kinases), suggests that the PH-domains are of potential importance in the operation of signalling pathways (Saraste and Hyvonen, 1995). Competition studies have revealed that the binding of βγ-subunits to PH-

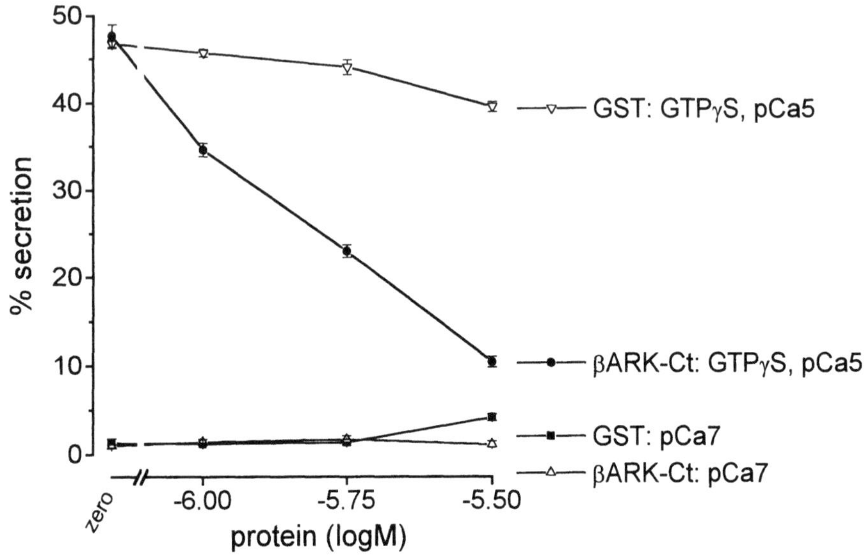

Fig.4. The effect of βARK C-terminal peptide on secretion of permeabilized mast cells. A very clear concentration-dependent relationship between the βARK peptide and inhibition of secretion can be found, indicating that a PH domain is involved in the secretory mechanism. Buffer with low calcium concentration (pCa7; ■,Δ) was used as control for stimulation (pCa5 with GTP-γ-S, 100μM;∇,●). Error bars represent SEM (n=4).

domains has about the same affinity as the binding to α-subunits, underlining the importance of this interaction. One method to investigate if the βγ-subunits act through a PH-domain would be by using the β-adrenergic receptor kinase (βARK) C-terminal peptide. This peptide constitutes a PH domain that has been shown to bind βγ-subunits with high affinity. However, PH-domain interactions are not limited to proteins. It has been shown that PH-domains bind to PIP_2 (K_d=30μM) as does IP_3 (Musacchio *et al.*, 1993). These inositol derivatives compete for the same binding site, suggesting that these PH-domain interactions are regulated by the ratio of PIP_2/IP_3. The interaction of the PH-domain is through the positive and negative charged ends of the domain. It has been suggested that PH-domain bound PIP_2 can serve as an activating ligand for heterotrimeric G-proteins. Again, studies with the βARK peptide reveal that the dissociation of the $\beta\gamma/\beta$ARK complex is possibly regulated by the level of PIP_2. The interaction between $\beta\gamma$-subunits and some PLCs is well recognised but it is remarkable that PLC_γ , which shows the strongest binding to $\beta\gamma$-subunits, is actually not activated or regulated by them, while PLC_β is.

In Figure 4 we illustrate the effect of the βARK C-terminal peptide on secretion of permeabilized mast cells after 10 min run-down. There is a very clear concentration-dependent relationship between the βARK peptide and inhibition of secretion, which shows that at some point, a PH domain is involved in secretion. At this point it is of course not possible to distinguish between a possible involvement of PIP_2 or a purely $\beta\gamma$-effect. Experiments with point mutations of the βARK peptide will show how much the effect is dependent on the $\beta\gamma$-subunits. Preliminary data show that the βARK C-terminal peptide also inhibits the stimulation of secretion that is induced by the $\beta\gamma$-subunits. This gives a clue to the mechanism but this still does not rule out a PIP_2 effect.

Conclusions

Collectively, these experiments give direct credence to the idea of a heterotrimeric G_E protein mediating exocytosis, and that the activating component is the $\beta\gamma$ subunit. More than this, it appears that when working at low concentrations, a considerable degree of specificity between different $\beta\gamma$ pairs is apparent.

References

Aridor, M., Traub, L.M. and Sagi-Eisenberg, R. (1990), Exocytosis in mast cells by basic secretagogues: Evidence for direct activation of GTP-binding proteins. J. Cell Biol. 111, 909-917.

Aridor, M., Rajmilevich, G., Beaven, M.A. and Sagi-Eisenberg, R. (1993), Activation of exocytosis by the heterotrimeric G protein Gi3. Science 262, 1569-1572.

Chernomordik, L.V., Vogel, S.S., Sokoloff, A., Onaran, H.O., Leikina, E.A. and Zimmerberg, J. (1993), Lysolipids reversibly inhibit Ca^{2+}-, GTP- and pH-dependent fusion of biological membranes. FEBS Lett. 318, 71-76.

Gomperts, B.D. (1990), G_E: A GTP-binding protein mediating exocytosis. Annu. Rev. Physiol. 52, 591-606.

Logothetis, D.E., Kim, D.H., Northup, J.K., Neer, E.J. and Clapham, D.E. (1988), Specificity of action of guanine nucleotide-binding regulatory protein subunits on the cardiac muscarinic K+ channel. Proc. Natl. Acad. Sci. U. S. A. 85, 5814-5818.

Mariot, P., O'Sullivan, A.J., Brown, A.M. and Tatham, P.E.R. (1996), Rho-guanine nucleotide dissociation inhibitor protein (Rho-GDI) inhibits exocytosis in mast cells. EMBO J. (in press).

Marquardt, D.L., Gruber, H.E. and Walker, L.L. (1987), Ribavirin inhibits mast cell mediator release. J. Pharmacol. Exp. Therap. 240, 145-149.

Musacchio, A., Gibson, T., Rice, P., Thompson, J. and Saraste, M. (1993), The PH domain: a common piece in structural patchwork of signalling proteins. Trends in Biochem. Sci. 18, 343-348.

O'Sullivan, A.J., Brown, A.M., Freeman, H.N.M. and Gomperts, B.D. (1996), Purification and identification of FOAD-II, a cytosolic protein that regulates secretion in streptolysin-O permeabilised mast cells, as a rac/rhoGDI complex. Mol. Biol. Cell 7, 397-408.

Pitcher, J.A., Touhara, K., Payne, E.S. and Lefkowitz, R.J. (1995), Pleckstrin homology domain-mediated membrane association and activation of the beta-adrenergic receptor kinase requires coordinate interaction with G beta gamma subunits and lipid. J. Biol. Chem. 270, 11707-11710.

Saito, H., Okajima, F., Molski, T.F.P., Sha'afi, R.I., Ui, M. and Ishizaka, T. (1987), Effects of ADP-ribosylation of GTP-binding protein by pertussis toxin on immunoglobulin E-dependent and -independent histamine release from mast cells and basophils. J. Immunol. 138, 3927-3934.

Saraste, M. and Hyvonen, M. (1995), Pleckstrin homology domains: A fact file. Curr. Biol. 5, 403-408.

Sternweis, P.C. and Pang, I. (1990), Preparation of G-proteins and their subunits. In:. "Receptor-Effector. Coupling:. A. practical. approach". Hulme. E. C. ed. IRL. Press;. Oxford. UK. pp. 1-30, 30.

Wilson, B.S., Deanin, G.G., Standefer, J.C., Vanderjagt, D. and Oliver, J.M. (1989), Depletion of guanine nucleotides with mycophenolic acid suppresses IgE receptor-mediated degranulation in rat basophilic leukemia cells. J. Immunol. 143, 259-265.

Subcellular localisation of ARF1-regulated phospholipase D in HL60 cells.

Clive Paul Morgan
Jacqueline Whatmore and
Shamshad Cockcroft
Department of Physiology
University College London
London WC1E 6JJ
United Kingdom

INTRODUCTION

Phospholipase D (PLD) is a ubiquitous enzyme found in cells and tissues from a wide variety of species. PLD is activated in response to the occupation of many cell-surface receptors including those of the heterotrimeric G-protein, and tyrosine kinase families (Cockcroft, 1992; Billah, 1993; Exton, 1994). It is one of a family of phospholipases which include isozymes of the phospholipase A_2 (PLA_2), and phospholipase C (PLC) families. PLD catalyses the hydrolysis of phosphatidylcholine (PC) to form phosphatidic acid (PA) and free choline. Alternatively, in the presence of a short chain alcohol, such as ethanol, a phophatidylalcohol is produced in preference to PA. PA has been implicated as an activator of PKCξ (PKCξ) (Limatola et al., 1994), type II phosphatidylinositol 4-phosphate 5-kinase (PI4P 5-K) (Jenkins et al., 1994), along with a novel form of serine-threonine kinase (Khan et al., 1994). PA can also readily be converted to diacylglycerol (DAG) by the enzyme phosphatidate phosphohydrolase (PAP) and may thus also function to activate other PKC isoforms although there is some evidence that the fatty acid composition of PA derived from PC is not suitable as an activator of PKC (Leach et al., 1991).

It has been established by a number of groups that GTPγS is a potent activator of phospholipase D activity (Olson et al., 1991; Geny and Cockcroft, 1992; Dubyak et

NATO ASI Series, Vol. H 101
Molecular Mechanisms of Signalling
and Membrane Transport
Edited by Karel W. A. Wirtz

al., 1993). Independently using either permeabilised HL60 cells, or HL60 membranes a cytosolic factor was identified which conferred this GTPγS sensitivity of PLD, a known protein termed ADP ribosylation factor (ARF) (Brown et al., 1993; Cockcroft et al., 1994). ARF had previously been identified as a cofactor necessary for the ADP ribosylation of Gs by cholera toxin. There are six mammalian ARF isotypes (ARF1-6, Tsuchiya et al., 1991) of which five have been shown to activate PLD (Cockcroft et al., 1994; Massenburg et al., 1994).

Since this work another small GTP-binding protein, RhoA, has been implicated as an activator of PLD. Work with Rho-GDI, which keeps Rho in an inactive GDP bound form, has shown that membranes stripped of endogenous Rho have substantially depleted PLD activity (Bowman et al., 1993). In addition to Rho, the regulatory subunit of PKCα has been shown to activate PLD in an ATP independent manner (Conricode et al., 1992; Singer et al., 1996). However, it seems that in cells the major regulator of phospholipase D is ARF, and it is not yet clear what the role of Rho and PKC are in a physiological context.

Several attempts have been made to purify PLD from a wide variety of sources. However, those attempts have so far not been successful. Hammond *et al* (1995) have recently cloned a mammalian isoform of PLD from HeLa cells based on sequence similarity of the recently cloned yeast PLD and plant PLD sequence. The clone (termed hPLD1) has been shown to be regulated by ARF, Rho, and PKC and from its behaviour would seem to be highly related to the granulocyte PLD previously studied.

In addition to the activation of PLD, ARF has been shown to be a requirement for coatomer (COP) binding to the Golgi stacks during vesicle budding (Serafini et al., 1991; Rothman and Orci, 1992; Donaldson and Klausner, 1994; Zhang et al., 1994). Isolated Golgi stacks *in vitro* do not make coated vesicles unless ARF is included in the reaction mixture. It has also been shown that ARF-GAP is located at the Golgi- and this strongly suggests the major site of ARF activity is at the Golgi (Cukierman et al., 1995). Recently Ktistakis *et al* (1996) have shown that PA, the product of PLD activity, promotes coatomer binding on the Golgi stack and may be the site of coatomer binding. This suggests that PLD may function at the Golgi and be involved in protein trafficking in addition to the suggested role in signal transduction.

It has also been shown that ARF is a requirement for regulated secretion supporting the protein trafficking role of ARF/PLD in cells (Fensome et al., 1996). The activation of PLD is also regulated by phophatidylinositol 4,5-bisphosphate (PIP_2), which has been shown to be necessary for ARF stimulated PLD activity *in vitro*, and in intact cells (Whatmore et al., 1994; Siddiqi et al., 1995). The recently cloned hPLD1 is also PIP_2 dependent. It is therefore of interest to locate the site(s) of PIP_2 synthesis as previous reports have suggested it is confined to the plasma membrane (Rawler et al., 1982; Cockcroft et al., 1985; Helms et al., 1991), however the evidence for PLD being at the Golgi is quite strong and this would suggest a requirement for PIP_2 for optimal PLD activity.

In this study we have studied the localisation of ARF1 regulated PLD in HL60 cells. A continuous sucrose gradient was used which allowed complete separation of plasma membrane from the denser endomembranes (ER/ Golgi/ lysozymes). The site of PLD activity was additionally monitored in FMLP stimulated intact HL60 cells (in the presence of ethanol) to determine the physiologically relevant PLD activity. In addition we monitored the localisation of the inositol lipids (PI/ PIP/ PIP_2), and the inositol lipid kinases (PI4K, PI4P 5-K) to determine PIP_2 localisation in these cells. We found that ARF1 regulated phospholipase D in HL60 cells was located at intracellular membranes (endomembranes), as well as at the plasma membrane.

RESULTS

Characterisation of the sucrose gradient

To identify the subcellular localisation of the ARF1-regulated PLD activity in HL60 cells, a cell fractionation procedure based on linear sucrose gradients was used (Lewis et al., 1992). The sucrose gradient was analysed by assaying for marker enzyme activities and specific proteins by Western blotting. Fractions 2-3 are enriched in markers for the Golgi complex (galactosyl transferase), endoplasmic reticulum (E.R.) (NADPH-cytochrome c reductase) as well as lysosomal granules (arylsulphatase and β-hexosaminidase). Plasma membrane was found to be highly enriched in fractions 6-9 as assessed by the distribution of HLA class I antigen across the gradient. This localisation for plasma membrane is supported by the characteristic phospholipid composition of these fractions which were enriched in sphingomyelin, and phosphatidylserine. The contamination of the plasma membrane fractions with endomembranes is minimal judging by the low levels of marker enzymes for Golgi, E.R. and lysosomal granules.

Western blotting was performed with antibodies specific to ARF, PI-TP, p102 (β`-COP), $G_{i\alpha 2}$, $G_{i\alpha 3}$, β subunit (common), RhoA, and PLCβ2. Protein immunoreacivity confirmed the enzyme localisation data for the gradient. ARF and Rho are believed to be the major regulators of phospholipase D. ARF was detected in two regions of the gradient, fractions 1-4 (endomembranes) showed a relatively weak signal while fractions 15-17 (cytosol) contained the bulk of the ARF immunoreactivity. RhoA was localised mainly in the cytosol with immunoreactivity detected at both the plasma membrane and endomembranes.

Subcellular Localisation of ARF1-regulated PLD activity using endogenous lipid substrate

In order to monitor PLD acitivity in the fractions the cellular lipids were prelabelled to equilibrium by growing HL60 cells in the presence of [^{14}C]acetate. PLD activity was measured in the individual fractions by the production of radiolabelled phosphatidylethanol (PEt) (Fig. 1A). GTPγS alone stimulated PLD activity in fraction 2 which most likely utilises the endogenous ARF proteins present in this fraction. Addition of both GTPγS and rARF1 revealed a major peak of PLD activity in fractions 2-3 and a second smaller peak in fractions 7-8.

Analysis of the gradient fractions using exogenous PC as substrate for PLD

The subcellular fractions were also assayed for rARF1-stimulated PLD activity using vesicles containing PIP_2:PE:PC ([^{3}H]dipalmitoyl-PC) (Fig. 1B). GTPγS alone gave a small stimulation of PLD activity in fractions 2, and in fractions 14-15. This stimulation by GTPγS most likely utilises the endogenous ARF/Rho proteins since these fractions

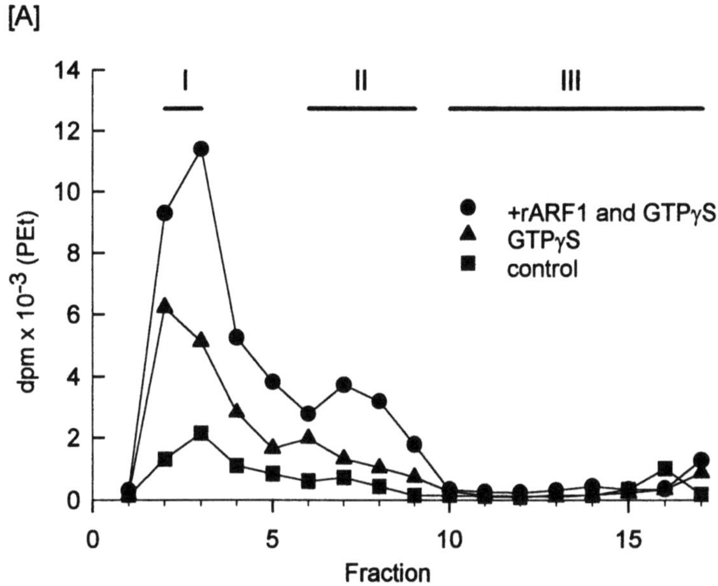

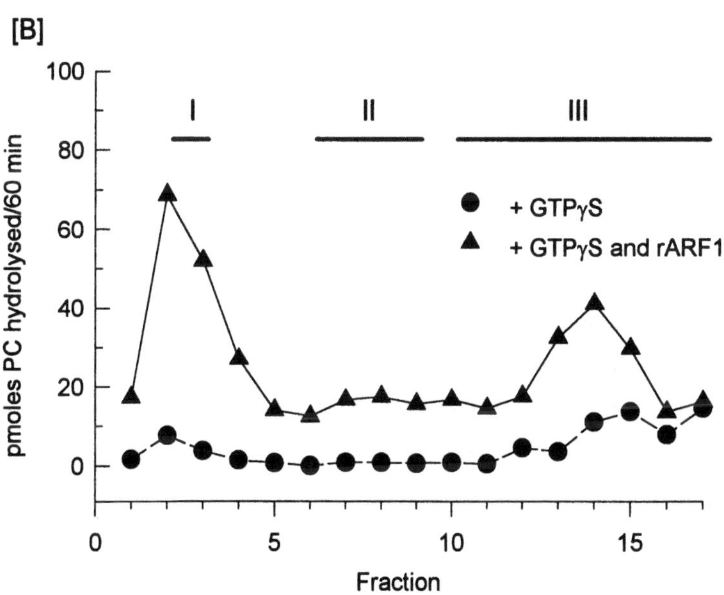

Figure 1. Subcellular localisation of rARF1-stimulated PLD activity using [A] endogenous, or [B] exogenous lipids as substrate.

(I) endomembranes, (II) plasma membrane, and (III) cytosolic regions are indicated by horizontal lines.

Data reproduced from Whatmore et al. (1996)

are found to contain them. When the fractions were analysed with rARF1 alone no activity was seen. Addition of both GTPγS and rARF1 increased PLD activity in all the fractions, with two major peaks of activity. One peak was in fractions 2-3 and the second in fractions 13-15 (Fig. 1B). A third region of activity was also observed in fractions 7-10. In three out of six experiments this activity was observed as a distinct peak.

Subcellular localisation of FMLP-stimulated PLD activity

It is clear from the results described above that ARF-PLD activity is present in both endomembranes and the plasma membrane. To examine which of the activities are responsible for PEt production in intact cells, we analysed the distribution of PEt in the subcellular fractions from control and FMLP-stimulated cells. Differentiated HL60 cells prelabelled with [^{14}C]acetate were stimulated with FMLP in the presence of ethanol, fractionated after disruption, and PEt levels determined in each fraction. The absolute DPM in PEt was greatest in fractions 2-3 (Fig. 2A) with a distinct shoulder seen in fractions 6-8. The shoulder seen in fractions 6-8 is observed as a distinct peak when the data are expressed as a function of the amount of PC present in the individual fractions (Fig. 2B).

MgATP requirement for the rARF1-stimulated PLD activity

It has been previously demonstrated that rARF1-stimulated PLD activity is enhanced by MgATP (Geny and Cockcroft, 1992; Kusner et al., 1993; Whatmore et al., 1994). This MgATP requirement may represent the availability of PIP_2 produced by sequential phosphorylation of PI by the PI 4-kinase and PI4P 5-kinase. We have analysed the MgATP dependence of the ARF1-regulated PLD activity from [^{14}C]acetate-labelled HL60 cells. The endomembrane-enriched fractions 2-3 were pooled and similarly the plasma membrane-enriched fractions 6-7. The pooled fractions were incubated in the presence or absence of both GTPγS and rARF1 (Table 1). In endomembranes, a small level of rARF1-stimulated PLD activity was observed which was enhanced 8-fold by the presence of 1mM MgATP. In contrast, a robust rARF1-dependent activity was found at

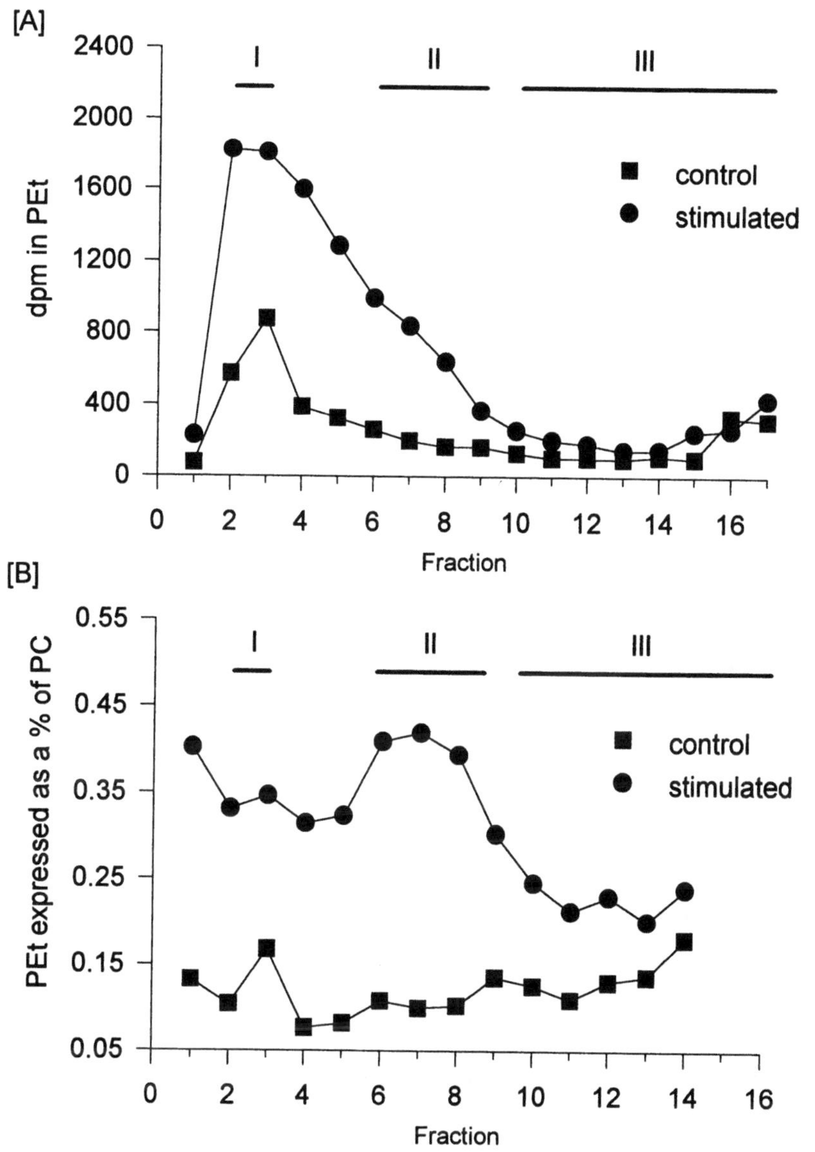

Figure 2. Subcellular distribution of FMLP-stimulated PLD activity in intact HL60 cells.

[A] Expressed as dpm in PEt, [B] expressed as a percentage of PC in each fraction. (I) endomembranes, (II) plasma membrane, and (III) cytosolic regions.
Data reproduced from Whatmore et al. (1996)

the plasma membrane which was only increased 2-fold by MgATP (Table 1). Neomycin sequesters PIP_2 and has been recently found to be a potent inhibitor of rARF1-reconstituted activity in human neutrophils (Whatmore et al., 1994). In the presence of MgATP, the rARF1-stimulated PLD activity associated with the endomembranes was inhibited by neomycin more strongly compared to the activity at the plasma membrane (Table 1).

	[^{14}C]PEt (d.p.m.)	
Endomembranes	**Control**	**rARF1 + GTPγS**
No additions	651 (100%)	1586 (244%)
MgATP (1mM)	1962 (301%)	12220 (1877%)
MgATP + neomycin (1mM)	774 (119%)	1045 (160%)
Plasma Membranes	**Control**	**rARF1 + GTPγS**
No additions	177 (100%)	738 (417%)
MgATP (1mM)	276 (156%)	1571 (888%)
MgATP + neomycin (1mM)	177 (100%)	561 (317%)

Table 1. Comparison of MgATP dependence of ARF-regulated phospholipase D activity at plasma membrane versus endomembranes.

Distribution of the polyphosphoinositides in the subcellular fractions

Since the assay for ARF1-regulated PLD activity using exogenous or endogenous substrate has identified a role for PIP_2 as a cofactor, we have examined the subcellular localisation of the inositol lipids (Fig. 3). The distribution of [^{3}H]inositol-labelled lipids in HL60 cells was examined after fractionation on the sucrose gradient. The fractions obtained from [^{3}H]inositol-labelled cells were incubated with MgATP to ensure that the endogenous inositol lipids were phosphorylated to their maximal capacity by the endogenous lipid kinases. The inositol lipids PI, and PIP were found mainly in fractions

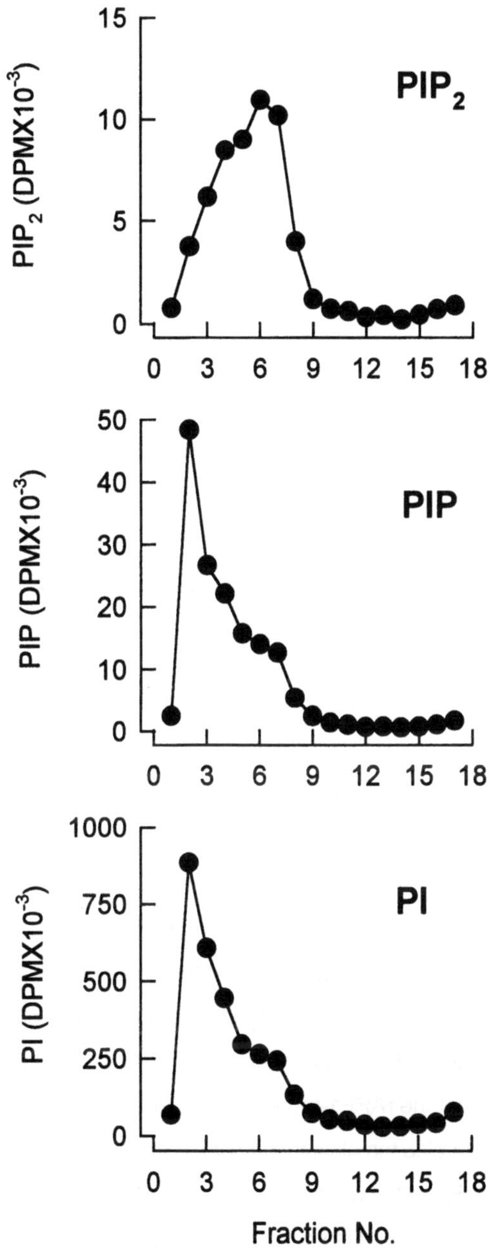

Figure 3. Distribution of inositol-containing phospholipids in fractionated [^{3}H]inositol-labelled cells.

Data reproduced from Whatmore et al. (1996)

2 and 3 which reflected the distribution of the total lipids in the fractions. In contrast, the distribution of PIP_2 does not reflect the distribution of the total lipids but is highly enriched in fractions 6 and 7, the region identified as plasma membrane. The endomembrane region contained some PIP_2 corresponding with the presence of some PIP 5 kinase activity (see below).

The fractions were additionally assayed for PI 4-kinase, and PI4P 5-kinase activities to locate the site of PIP_2 synthesis (Fig. 4). The fractions were analysed using both the endogenous lipids as substrate and also using exogenous lipid in the presence of 1% Triton. PI 4-kinase was localised to fractions 2-3 and additionally in fractions 6-8, in agreement with previous reports that this enzyme is localised at the Golgi, lysosomes and plasma membranes (Cockcroft et al., 1985) (Fig. 4A). When assayed using the endogenous substrate (Fig 4A, see inset), PIP was formed mainly in fraction 2 with some in the fractions 5-7. PI4P 5-kinase activity was found to be primarily localised in fractions 6-9 when assayed using both exogenous and endogenous substrate (Fig. 4B), although a small amount of activity was consistently detected in fractions 2-3.

DISCUSSION

HL60 cells and neutrophils are two of the most extensively studied cell types with respect to the regulation of PLD activity. Previous studies have established that the PLD activity in these cells is located in the membrane fraction and that its stimulation by guanine nucleotides is mediated by small GTP binding proteins belonging to the ARF family (Brown et al., 1993; Cockcroft et al., 1994; Whatmore et al., 1994). The mammalian ARF family consists of 6 proteins, of which five are able to stimulate PLD activity (Cockcroft et al., 1994; Massenburg et al., 1994). Using subcellular fractionation, we demonstrate that the rARF1-regulated PLD activity is localised not only in intracellular compartments but also the plasma membrane and the cytosol.

Subcellular fractionation of HL60 cells on continuous sucrose gradients allowed for a clean separation of the plasma membranes from other intracellular compartments. The markers for the Golgi complex, endoplasmic reticulum and the lysosomal granules co-sedimented at the bottom of the gradient. In addition, approximately 10-15% of the plasma membrane markers were detected in these dense fractions. Attempts were

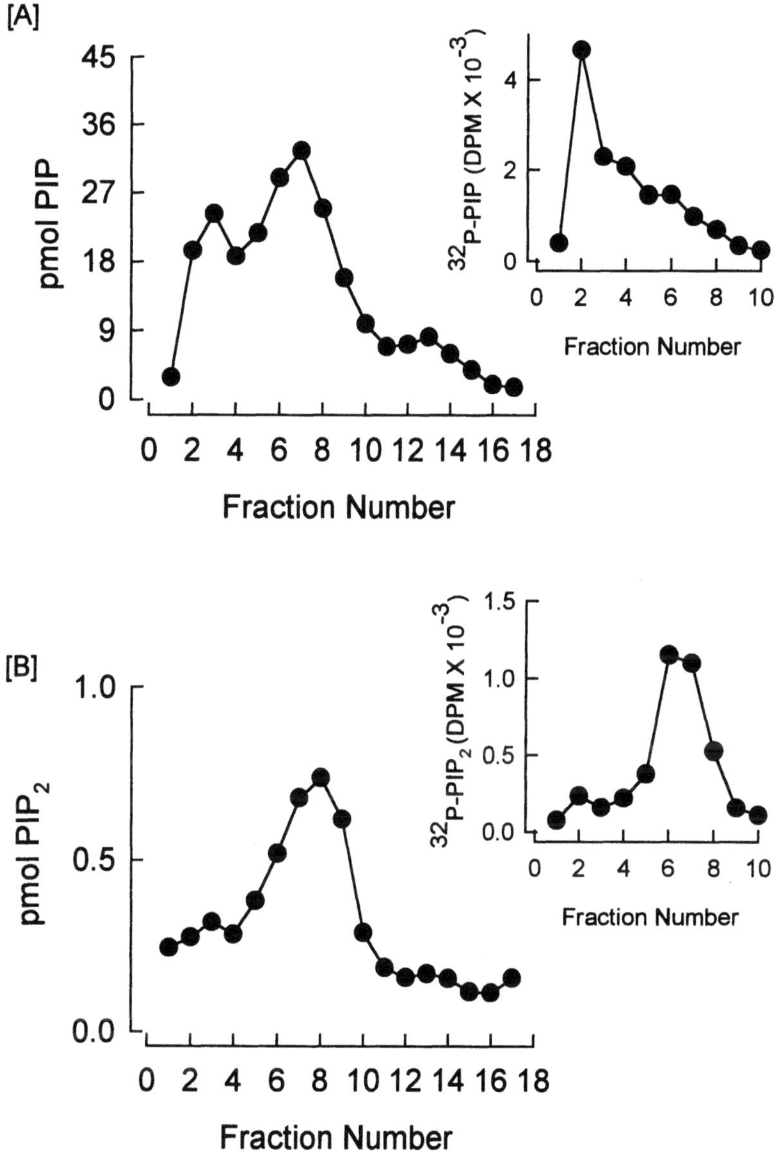

Figure 4. Distribution of [A] PI 4-kinase, and [B] PI4P 5-kinase in the sucrose gradient.

The main figures show the localisation of enzyme when exogenous substrate was used.
Inset shows the kinase activity with endogenous substrate.
Data reproduced from Whatmore et al. (1996)

made to further separate these membrane compartments using Nycodenz gradients but this was unsuccessful in HL60 cells. Ktistakis *et al* (1995) have recently shown that a brefeldin A-sensitive ARF-activated PLD activity was localised at the Golgi complex but not in membranes enriched in E.R.

The presence of rARF1-regulated PLD activity at the plasma membrane indicates that ARF proteins are functional not only at intracellular membranes but also at the plasma membrane. Recently, ARF6 when overexpressed was found to be localised at the plasma membrane and has been suggested to be involved in endocytosis (D'Souza-Schorey et al., 1995). It is possible that ARF1 is able to replace ARF6 *in vitro*.

PIP_2 is necessary for the demonstration of ARF-stimulated PLD activity. The majority of PIP_2 is reported to be localised at the plasma membranes (Cockcroft et al., 1985; Helms et al., 1991; Rawler et al., 1982). The subcellular distribution of the inositol lipids and the associated kinases responsible for inositol lipid phosphorylation was examined in fractionated HL60 cells. PIP as well as PI 4-kinase activity was present in intracellular compartments and the plasma membrane. As expected, PIP_2 as well as the PI4P 5-kinase activity was predominantly found at the plasma membrane. However, a small, but consistent amount of activity was also observed in endomembranes. The PI4P 5-kinase data was supported by an examination of the distribution of PIP_2 in the gradient. These results are consistent with previous reports that PIP_2 synthesis occurs in a microsomal fraction of CHO cells distinct from the plasma membrane (Helms et al., 1991). The presence of PIP_2 in endomembranes is also supported by the identification of a PIP_2 5-phosphatase at the Golgi (Olivos-Glander et al., 1995; Zhang et al., 1995).

CONCLUSION

In summary, we demonstrate that ARF-regulated PLD activity is present not only at endomembranes but also at the plasma membrane. Both these activities are stimulated upon receptor occupation with FMLP as judged by the formation of PEt in these membrane compartments. A previous study had reported that PEt was formed exclusively at the plasma membrane upon stimulation with FMLP (Gelas et al., 1989). However, this study relied on the use of lyso-PAF to prelabel the cells and thus it is not clear whether the label had fully distributed to the intracellular compartments. The labelling procedure used here was designed to label all the phospholipids to near equilibrium. Our study clearly indicates that FMLP stimulates a substantial degree of PEt production in intracellular compartments. The degree of PEt formed in the different membrane compartments upon receptor stimulation reflects the distribution of the PLD activity. The intracellular localisation of the ARF-regulated PLD activity is very indicative of it playing a prominent role in vesicular transport. The identification of ARF-regulated PLD activity at the plasma membranes poses interesting questions concerning PA metabolism. The production of both PA and diacylglycerol (DAG) at the plasma membrane, and their ready interconversion suggests there may be the need for subtle differences in their mode of action.

REFERENCES

Billah, M.M. (1993) Curr. Opin. Immunol. **5**, 114-123

Bowman, E.P., Uhlinger, D.J. & Lambeth, J.D. (1993) J. Biol. Chem. **268**, 21509-21512

Brown, H.A., Gutowski, S., Moomaw, C.R., Slaughter, C. & Sternweis, P.C. (1993) Cell **75**, 1137-1144

Cockcroft, S. (1992) Biochem. Biophys. Acta **1113**, 135-160

Cockcroft, S., Taylor, J.A. & Judah, J.D. (1985) Biochim. Biophys. Acta **845**, 163-170

Cockcroft, S., Thomas, G.M.H., Fensome, A., Geny, B., Cunningham, E., Gout, I., Hiles, I., Totty, N.F., Troung, O. & Hsuan, J.J. (1994) Science **263**, 523-526

Conricode, K.M., Brewer, K.A., & Exton, J.H. (1992) J. Biol. Chem. **267**, 7199-7202

Cukierman, E., Huber, I., Rotman, M. & Cassel, D. (1995) Science **270**, 1999-2002

Donaldson, J.G. & Klausner, R.D. (1994) Curr. Opin. Cell Biol. **6**, 527-532

Dubyak, G.R., Schomisch, S.J., Kusner, D.J. & Xie, M. (1993) Biochem. J. **292**, 121-128

Exton, J.H. (1994) Biochem. Biophys. Acta **1212**, 26-42

Fensome, A., Cunningham, E., Prosser, S., Tan, S.K., Swigart, P., Thomas, G., Hsuan, J., Cockcroft, S. (1996) Curr. Biol. **6**, 730-738

Gelas, P., Ribbes, G., Record, M., Terce, F. & Chap, H. (1989) FEBS Lett. **251**, 213-218

Geny, B. & Cockcroft, S. (1992) Biochem. J. **284**, 531-538

Hammond, S.M., Altshuller, Y.M., Sung, T., Rudge, S.A., Rose, K., Engebrecht, J., Morris, A.J. & Frohman, M.A. (1995) J. Biol. Chem. **270**, 29640-29643

Helms, J.B., De Vries, K.J. & Wirtz, K.W.A. (1991) J. Biol. Chem. **266**, 21368-21374

Jenkins, G.H., Fisette, P.L., & Anderson, R.A. (1994) J. Biol. Chem. **269**, 11547-11554

Khan, W.A., Blobe, G.C., Richards, A.L., & Hannun, Y.A. (1994) J. Biol. Chem. **269**, 9729-9735
Ktistakis, N.T., Brown, A., Sternweis, P.C. & Roth, M.G. (1995) Proc. Natl. Acad. Sci. U. S. A. **92**, 4952-4956
Ktistakis, N.T., Brown, H.A., Waters, M.G., Sternweis, P.C., & Roth, M.G. (1996) J. Cell Biol. **134**, 295-306
Kusner, D.J., Schomisch, S.J. & Dubyak, G.R. (1993) J. Biol. Chem. **268**, 19973-19982
Leach, K.L., Ruff, V.A., Wright, T.M., Pessin, M.S., & Raben, D.M. (1991) J. Biol. Chem. **266**, 3215-3221
Lewis, V.A., Hynes, G.M., Zheng, D., Saibil, H. & Willison, K. (1992) Nature **358**, 249-252
Limatola, C., Schaap, D., Moolenaar, W.H., & van Blitterswijk, W.J. (1994) Biochem. J. **304**, 1001-1008
Massenburg, D., Han, J., Liyanage, M., Patton, W.A., Rhee, S.G., Moss, J. & Vaughan, M. (1994) Proc. Natl. Acad. Sci. U. S. A. **91**, 11718-11722
Olivos-Glander, I.M., Janne, P.A. & Nussbaum, R.L. (1995) Am. J. Hum. Genet. **57**, 817-823
Olson, S.C., Bowman, E.P. & Lambeth, J.D. (1991) J. Biol. Chem. **266**, 17236-17242
Pai, J., Siegel, M.I., Egan, R.W. & Billah, M.M. (1988) J. Biol. Chem. **263**, 12472-12477
Rawler, A.J., Roelofsen, B., Wirtz, K.W.A. & Op den Kamp, J.A.F. (1982) FEBS Lett. **148**, 140-144
Rothman, J.E. & Orci, L. (1992) Nature **355**, 409-415
Serafini, T., Orci, L., Amherdt, M., Brunner, M., Kahn, R.A. & Rothman, J.E. (1991) Cell **67**, 239-253
Siddiqi, A.R., Smith, J.L., Ross, A.H., Qiu, R., Symons, M. & Exton, J.H. (1995) J. Biol. Chem. **270**, 8466-8473
Singer, W.D., Brown, H.A., Jiang, X., & Sternweis, P.C. (1996) J. Biol. Chem. **271**, 4504-4510
Tsuchiya, M., Price, S.R., Tsai, S., Moss, J., & Vaughan, M. (1991) J. Biol. Chem. **266**, 2772-2777
Whatmore, J., Cronin, P. & Cockcroft, S. (1994) FEBS Lett. **352**, 113-117
Whatmore, J., Morgan, C.P., Cunningham, E., Collison, K.S., Willison, K.R., & Cockcroft, S. (1996) Biochem. J. **320**, 785-794
Zhang, C., Rosenwald, A.G., Willingham, M.C., Skuntz, S., Clark, J. & Kahn, R.A. (1994) J. Cell Biol. **124**, 289-300
Zhang, X., Jefferson, A.B., Auethavekiat, V. & Majerus, P.W. (1995) Proc. Natl. Acad. Sci. USA **92**, 4853-4856

Electrophysiological Aspects of Growth Factor Signaling in NRK Fibroblasts: The Role of Intercellular Communication, Membrane Potential and Calcium

A.D.G. de Roos, E.J.J. van Zoelen, and A.P.R. Theuvenet
University of Nijmegen
Department of Cell Biology
Toernooiveld 1
NL-6525 ED Nijmegen
The Netherlands

Introduction

Normal, non-transformed cells become limited in their growth when cultured at high cell densities. This density-dependent growth inhibition, which is also known as contact-inhibition, is one of the most prominent characteristics of non-transformed cells and it is lost upon tumorigenic transformation (Van Zoelen, 1991b). In our laboratory, normal rat kidney (NRK) fibroblasts are used as a model system to study density-dependent growth regulation. These cells are usually grown to confluence in the presence of serum after which they are serum-deprived and become quiescent. When these quiescent cells are then cultured in the presence of epidermal growth factor (EGF) as the only growth factor present, these cells undergo density-dependent growth inhibition. Several polypeptide growth factors are capable to reversibly induce the loss of density-dependent growth inhibition in the presence of EGF and thus phenotypically transform these cells (Van Zoelen et al., 1988). This loss of growth inhibition is strongly inhibited by the nonapeptide bradykinin (Afink et al., 1994, Van Zoelen et al., 1994).

The mechanisms underlying density-dependent inhibition of growth are not well understood and several hypotheses to explain this phenomenon have been put forward. These include the release of growth-inhibitory factors, reduction of growth factor receptor numbers, the expression of growth inhibitory plasma membrane glycoproteins (contact-inhibins), the

NATO ASI Series, Vol. H 101
Molecular Mechanisms of Signalling
and Membrane Transport
Edited by Karel W. A. Wirtz

activation of tyrosine phosphatases and regulation of growth by intercellular communication and membrane potential (Rijksen et al., 1993; reviewed in Van Zoelen, 1991a).

In NRK fibroblasts, downregulation of EGF receptors appears to be an important growth regulatory mechanism (Rizzino et al., 1990, Van Zoelen, 1991b). However, it has been postulated that also electrophysiological mechanisms are involved. An increase in intercellular communication via gap junctions at high cell densities might inhibit proliferation of cells (Loewenstein, 1981; Yamasaki, 1990). Moreover, It has been suggested that membrane hyperpolarization at high cell densities is a general mechanism for density-dependent growth inhibition (Binggeli and Weinstein, 1986). These results initiated electrophysiological studies on NRK cells in our laboratory and in this paper we focus on these studies.

Intercellular coupling has been shown to be involved in the growth control of normal cells and in cellular transformation (Trosko et al., 1990; Ruch, 1994). It is thought that intercellular communication inhibits the growth of normal cells either by diluting out effects of cytosolic growth stimulatory factors or by the intercellular diffusion of growth inhibitory factors (Ruch, 1994). A loss of intercellular communication may, therefore, stimulate cells to grow or permit transformed cells to proliferate. Indeed, inhibition of gap junctional communication often results in the oncogenic transformation of cells. Many transformed cells show a decreased intercellular communication and many growth factors also decrease intercellular communication (Ruch, 1994, Trosko et al., 1990).

Mitogenic activation of cells by growth factors is generally accompanied by changes in membrane potential. One of the earliest events after serum stimulation of cells is a prolonged depolarization (Moolenaar and Jalink, 1992). It was recently discovered that activation of Cl^- channels by the serum component lysophosphatidic acid (LPA) contributes to this depolarization (Postma et al., 1996). However, growth factors can also hyperpolarize the membrane mainly due to a Ca^{2+}-activated K^+ conductance, as has been shown with EGF (Peppelenbosch et al., 1991; Pandiella and Meldolesi, 1992). These observations show that ionic events may play an important role in the action of growth factors.

In electrically coupled cells, changes in membrane potential can be transduced to neighboring cells, which gives the opportunity for intercellular signaling. For instance, in excitable cells such as nerve and muscle cells, action potentials caused by the regenerative opening of voltage-dependent ion channels provide a fast (0.1-100 m/s) mechanism for signaling over long distances. Action potentials can be easily transduced to other cells via gap junctions for which the propagation of the heart action potential is a classical example (De Mello, 1994). Also in other cells, membrane potential has been shown to be involved in the coordination of cellular activities. For instance, synchronous Ca^{2+} oscillations in pancreatic islets are mediated by changes in membrane potential (Santos et al., 1991). In excitable cells, the depolarizations associated with the action potentials open voltage-dependent Ca^{2+} channels in the plasma membrane, causing a fast influx of Ca^{2+} (Putney, 1993).

In non-excitable cells, membrane potential does not seem to play a prominent role in the coordination of intercellular signaling. However, gap junctions do play an important role in intercellular Ca^{2+} signaling in these cells. In electrically inexcitable cells, Ca^{2+} signalling is believed to be regulated via inositoltrisphosphate formation, subsequent Ca^{2+} release from intracellular stores, followed by Ca^{2+} entry from the extracellular medium via store-regulated Ca^{2+}-influx channels (Putney, 1993). Many growth factors, including PDGF, LPA, and bradykinin, cause the production of inositoltrisphosphate with a concomitant release of Ca^{2+} from intracellular stores. It has been shown that the regenerative generation of inositoltrisphosphate can induce Ca^{2+} waves that slowly propagate (50-100 μm/s) from cell to cell via gap junctions (Sanderson et al., 1994). Thus, also in non-excitable cell, gap junctions play an important role in the coordination of cooperative cellular responses.

Here, we show that NRK fibroblasts are electrically well-coupled, which provides a way for the transduction of membrane potential changes. Our results also show that monolayers of NRK fibroblasts, which are considered to be inexcitable cells, can behave like an excitable syncytium. Increases in $[Ca^{2+}]_i$ by agonists such as bradykinin can depolarize the cells by activation of Ca^{2+}-dependent Cl^- channels, translating the initial Ca^{2+}signal into an electrical signal (depolarization). The opening of voltage-dependent Ca^{2+} channels by a depolarization can trigger a Ca^{2+} action potential that propagates through the monolayer. The consequence of this action potential is an almost synchronous increase in $[Ca^{2+}]_i$ in large numbers of cells. In this way, a locally evoked electrical signal (action potential) is converted into synchronized long-range Ca^{2+} signals.

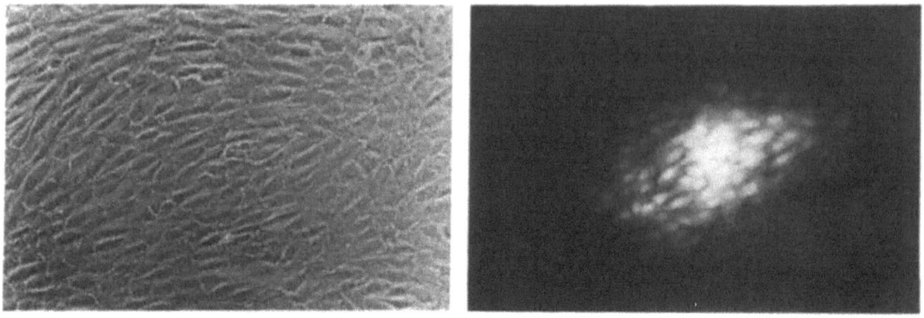

Fig. 1 Dye coupling in NRK fibroblasts. The left picture shows a phase-contrast image of quiescent NRK fibroblasts. One cell was microinjected with the fluorescent dye Lucifer Yellow. The right picture shows a fluorescent image taken 5 minutes after the injection. Coupling is demonstrated by the diffusion of the dye from the injected cell (in the middle of the fluorescent cells) to almost 100 cells.

Intercellular communication in NRK cells

Several methods have been described to measure gap junctional communication. These include techniques to assay biochemical communication, in which the diffusion of tracer molecules was followed (Stewart, 1978; Wade et al., 1986), and electrical communication, using voltage clamp techniques (Neyton and Trautman, 1985; Spray et al., 1981; Maldonado et al., 1988). Figure 1 shows the extensive biochemical intercellular communication in quiescent NRK fibroblasts assayed by the microinjection of the fluorescent dye Lucifer Yellow. After injection into a single cell, the dye diffuses within minutes to neighboring cells demonstrating that the cells are dye-coupled. Earlier studies did not detect significant dye coupling in NRK fibroblasts (van Zoelen and Tertoolen, 1991), which may depend on the dye used.

Biochemical communication may not be a good measure for electrical coupling. In several instances electrical coupling of cells has been demonstrated in the absence of dye coupling (Steinberg et al., 1994; Traub et al., 1994). Also, a reduction in gap junctional coupling may abolish transfer of, for instance, fluorescent dyes, but the remaining electrical coupling may be sufficient to transduce electrical signals.

Measurements of electrical communication between cells are usually performed using double voltage clamp techniques on cell pairs, but these measurements are hampered by the

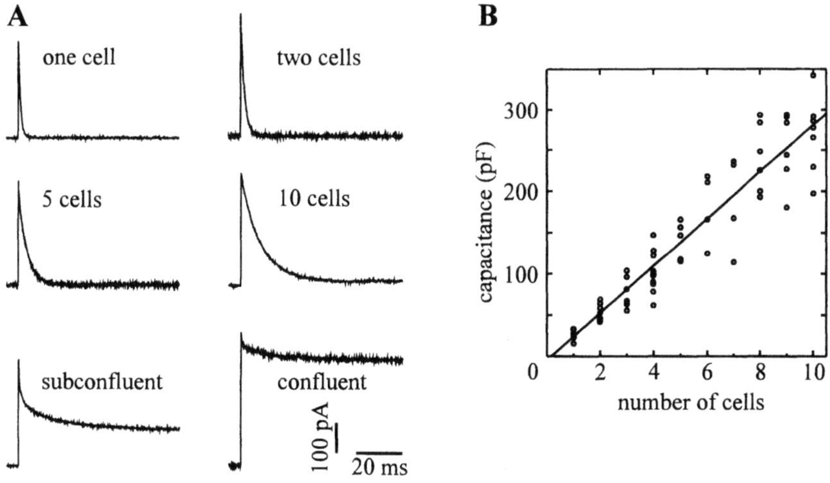

Fig. 2 Capacitance measurements in NRK cells. **A**. Current responses to a voltage pulse in NRK cells. One cell of a cluster of cells was patched in the whole-cell configuration and a voltage pulse of 10 mV was given. Pipette capacitance was canceled and transients represent capacitive transients of the cell membrane. **B**. Capacitance as a function of cell number. By integrating the current transient, the accumulated charge could be determined and the total membrane capacitance was obtained by dividing the charge by the applied voltage pulse. Thus, with increasing cluster size, the capacitance of the cluster of cells increased linearly, implying that the cells are electrically coupled.

fact that they are technically difficult and require two patch clamp setups. Recently, we developed a simple single patch clamp technique, that is not limited to cell pairs, to study electrical intercellular communication (de Roos et al., 1996). This technique is based on the principle that the plasma membrane acts as a capacitor (Kado, 1993). With increasing cell number, the total membrane surface, and thus the capacitance, will increase when the cells are electrically coupled.

In clusters of NRK cells, the capacitive transients increase with increasing cluster size (fig. 2A), indicating electrical coupling. Capacitive transients in (sub)confluent cells are determined by the large steady-state currents due to a large total membrane conductance, characteristic for coupled cells. Uncoupling of the cells, for instance by halothane or octanol, reduces these capacitive currents to single cells levels. From the area under the curve in figure 2A the capacitance can be calculated and the fact that capacitance increases linearly with cell number (fig. 2B) shows that NRK cells are electrically well-coupled. Since it was found that even small clusters of cells are well-coupled, upregulation of gap junctional intercellular communication at high cell densities does not seem to be implicated in density-dependent growth inhibition of NRK cells.

The fact that NRK fibroblasts are electrically well-coupled implicates that the membrane potential is shared between cells and that differences in membrane potential between cells are averaged. We found that membrane potential is more stable in coupled cells than in single cells (de Roos et al., unpublished). Also, changes in membrane potential, for instance depolarizations, will not be limited to single cells, but will be passively transduced to other cells, resulting in an averaged, coordinate response.

Bradykinin depolarizes NRK fibroblasts via a Ca^{2+}-activated Cl^- conductance

Many extracellular stimuli affect the membrane potential of cells. One of the earliest events after addition of serum is a sustained depolarization (Moolenaar and Jalink, 1992). In quiescent NRK cells bradykinin also induces a depolarization, from a resting membrane potential of about -70 mV to -17 mV, that generally lasts up to 10 minutes[1] (fig. 3A). This depolarization is due to an increased Cl^- conductance, as demonstrated by the fact that depolarization is enhanced in media containing a low concentration of Cl^- (fig. 3B). The receptors of bradykinin belong to the class of G-protein-coupled receptors possessing seven membrane-spanning domains which, upon binding of their ligand, activate among others phospholipase C with consequent Ca^{2+} mobilization from internal stores by released inositoltrisphosphate (Afink et al., 1994). When $[Ca^{2+}]_i$ is buffered by loading the cells with BAPTA, the depolarization by bradykinin is abolished[1], showing that the depolarization is secondary to the increase in $[Ca^{2+}]_i$. Additional experiments with blockers of Ca^{2+}-activated Cl^- channels and Ca^{2+}-releasing agents confirmed the presence of a Ca^{2+}-activated Cl^- conductance in NRK cells[1].

By inducing a depolarization through an increase in $[Ca^{2+}]_i$, a chemical signal is translated into an electrical signal. Transmission of the depolarization to neighboring cells via gap junc-

---

[1] de Roos et al., J Cell Physiol, *in press*

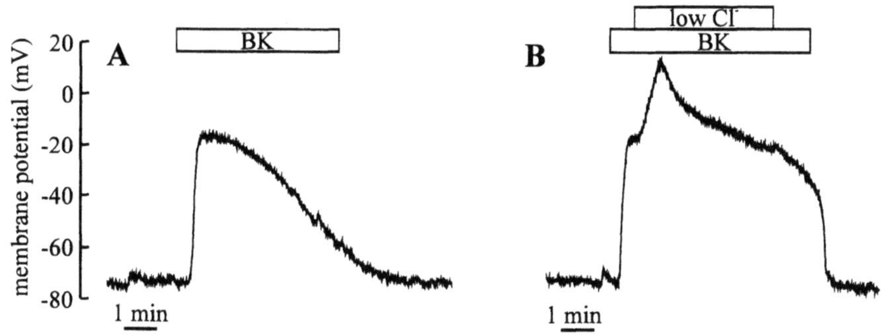

Fig. 3 Bradykinin induces a Ca^{2+}-activated Cl^- conductance. **A**. Typical depolarization in response to 100 nM bradykinin measured in the current clamp mode of the whole cell patch clamp configuration. Cells were continuously perfused and bradykinin was included in the perfusion medium during the indicated time. **B**. Effect of changing to a solution that contained 8 mM Cl^- after the cells were depolarized with 100 nM bradykinin. The enhancement of the depolarization by bradykinin in low-Cl^- media demonstrates that the depolarization is mediated by a Cl^- conductance.

tions can depolarize cells that are not themselves exposed to bradykinin and therefore may represent a mechanism for fast intercellular transduction of locally evoked responses.

The membrane potential of quiescent NRK cells is around -70 mV, close to the equilibrium potential of K^+. However, a hyperpolarization of the membrane at density-arrest, as has been shown in epithelial and other cells (Binggeli and Weinstein, 1986), was not observed. This excludes a role of membrane potential in density-dependent growth inhibition in NRK fibroblasts.

Transforming growth factor β, retinoic acid or LPA cause, in the presence of EGF, a loss of density-dependent growth arrest in NRK cells, resulting in a phenotypic transformation of the cells (Van Zoelen et al., 1988). It has been shown that bradykinin is able to block this growth stimulus-induced phenotypic transformation (Afink et al., 1994; Van Zoelen et al., 1994). This effect of bradykinin was not mimicked by prostaglandin $F_{2\alpha}$, which depolarizes quiescent as well as density-arrested cells as effectively as bradykinin. Therefore, it is unlikely that the block of phenotypic transformation of NRK cells by bradykinin is related to depolarization of the cells.

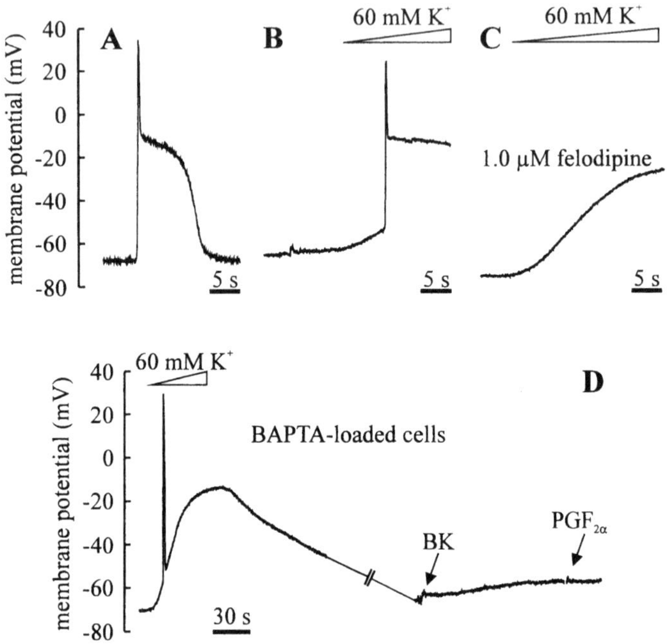

Fig. 4 Generation of action potentials in monolayers of NRK cells. **A**. Measurement of an action potential that was induced by application of a small volume (5 μl) of 10 nM bradykinin at 10 cm distance from the site of measurement. By this mode of application, it was ensured that only a small part of the monolayer was exposed to stimulatory concentrations of the agonist and that the action potential was recorded in cells that were not exposed to bradykinin. **B**. Generation of an action potential by depolarization with perfusion of medium with a high K^+ concentration. **C**. Abolishment of the action potential induced by depolarization in the presence of 1.0 μM felodipine, an L-type Ca^{2+} channel blocker. **D**. Lack of effect of buffering $[Ca^{2+}]_i$ with BAPTA on the depolarization and action potential generation by high K^+, while depolarization and action potential generation by bradykinin and $PGF_{2\alpha}$ are abolished. Media contained 3 mM Sr^{2+}, with no Ca^{2+} added.

Propagating Ca^{2+} action potentials in monolayers of NRK fibroblasts

Depolarization by bradykinin was sometimes preceded by a fast action potential-like spike depolarization. Subsequently, we discovered that depolarization by bradykinin can evoke action potentials in monolayers of NRK cells. Depolarization of only a small part of the monolayer can induce a propagating action potential throughout the entire monolayer (fig. 4A).

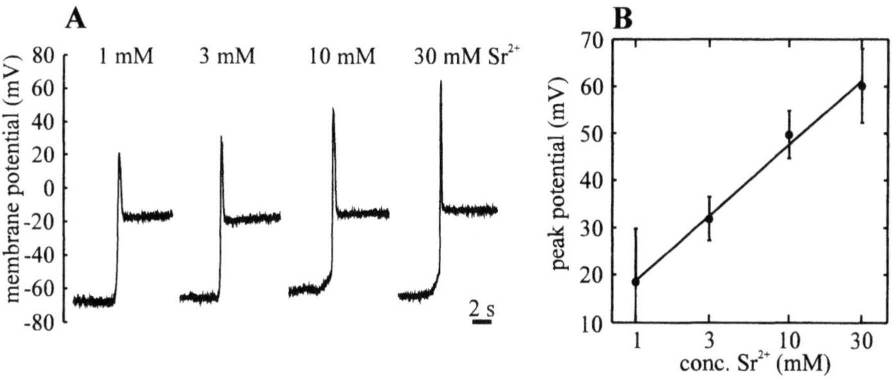

Fig. 5 Action potentials are mediated by Ca^{2+} influx. **A**. Action potentials in media containing 1, 3, 10 or 30 mM Sr^{2+}. **B**. Relation between peak value of the action potential and the extracellular Sr^{2+} concentration (mean ± SD, n=6). Since Sr^{2+} can readily permeate Ca^{2+} channels, Sr^{2+} was used instead of Ca^{2+} because of the better solubility of Sr^{2+} in bicarbonate-based buffers. The peak potential increased linearly with the logarithm of Sr^{2+} with a slope of 29 mV per 10-fold increase (r^2=0.996) in the extracellular Sr^{2+} concentration, as expected for a Sr^{2+}-electrode (Hille, 1992). This indicates that the spike depolarization is solely caused by an increased Ca^{2+} (Sr^{2+}) permeability and demonstrates the presence of Ca^{2+} action potentials

This action potential propagates over large distances in an all-or-none fashion at a velocity of 6.0 mm/s, resulting in a depolarization of cells that have not been exposed to bradykinin (de Roos et al., submitted). The role of an initial depolarization in the generation of the action potential is demonstrated by the fact that depolarization by perfusion of a high extracellular K^+ concentration also induces an action potential (fig. 4B). This action potential can be blocked by L-type Ca^{2+} channel blockers (fig. 4C), indicating that these channels mediate the action potential. Buffering $[Ca^{2+}]_i$ with BAPTA does not abolish the action potential evoked by high extracellular K^+, but does block the action potential evoked by bradykinin or prostaglandin $F_{2\alpha}$. These results show that depolarization is the trigger for the action potential. The response to bradykinin is dependent on a rise in $[Ca^{2+}]_i$, since depolarization by bradykinin is mediated by Ca^{2+}-dependent Cl^- channels.

Action potentials are caused in excitable cells by the regenerative opening of voltage-dependent ion channels. The action potential in axon and neuronal cells is caused by the regenerative opening of Na^+ channels (Hille, 1992), but also Ca^{2+} action potentials can be generated by opening of voltage-dependent Ca^{2+} channels (Hagiwara and Byerly, 1981). The magnitude of the fast action potential in NRK cells is increased by an increase in the extracellular Sr^{2+} concentration (fig. 5A and B). Sr^{2+} can permeate Ca^{2+} channels (Hille, 1992), and therefore, these results show that Ca^{2+} channels mediate the action potential. Since in addition the action potentials continue in Na^+-free media, and can be blocked by (L-type) Ca^{2+} channel antagonists (de Roos et al., submitted), it is concluded that they are Ca^{2+} action potentials.

Although it has been reported that fibroblasts, which are considered to be inexcitable cells, possess voltage-dependent Ca^{2+} channels (Lovisolo et al., 1988; Baumgarten et al., 1992; Chen et al., 1988; Peres et al., 1988; Harootunian et al., 1991) that would in principle enable these cells to generate action potentials, a function for such channels in these cells has so far been unclear. Our results suggest that they may function in the generation and propagation of action potentials.

Action potentials can not only be evoked by depolarization with bradykinin or high K^+ in monolayers of quiescent NRK cells, but are also observed to occur spontaneously in monolayers of density-arrested NRK cells (fig. 6A). These action potentials, just as the ones evoked in quiescent cells, are paralleled by synchronous increases in $[Ca^{2+}]_i$ (fig. 6B) and result from the opening of voltage-dependent L-type Ca^{2+} channels during the action potential. Although bradykinin increases $[Ca^{2+}]_i$ mainly by a release of Ca^{2+} from intracellular stores, the presence of voltage-dependent Ca^{2+} channels in these cells suggests that the depolarization resulting from the Ca^{2+} release opens these Ca^{2+} channels causing an additional Ca^{2+} influx. Although both quiescent and density-arrested NRK cells are able to generate action potentials, the occurrence of spontaneous action potentials in quiescent cells could point at a different, growth status dependent, regulation of the Ca^{2+} channels involved.

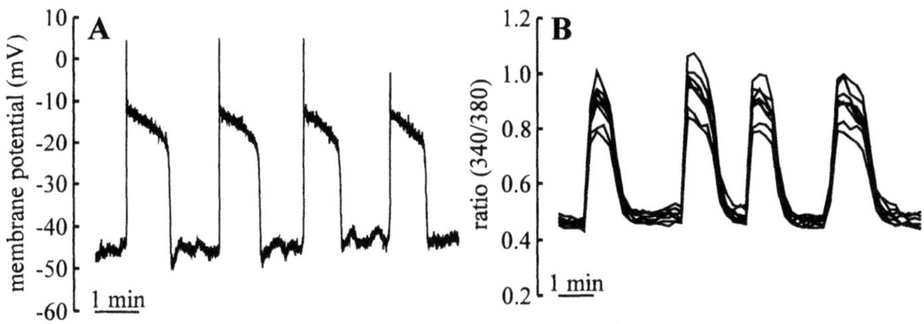

Fig. 6 Spontaneous repetitive action potentials in density-arrested monolayers of NRK cells. A. Spontaneous action potentials recorded in density-arrested NRK cells. **B**. Synchronous repetitive increases in $[Ca^{2+}]_i$ in Fura loaded cells. The average response in $[Ca^{2+}]_i$ of seven individual cells from a monolayer of density-arrested NRK cells is shown. These results show that each action potential is accompanied by an increase in $[Ca^{2+}]_i$ that is virtually synchronized in a large number of cells. Simultaneous measurements of membrane potential with bisoxonol and Ca^{2+} measurements confirmed that increases in $[Ca^{2+}]_i$ occurred during each action potential (data not shown). Calcium measurements were performed as described in Willems et al. (1993).

Summary and conclusions

It is shown here that in monolayers of NRK fibroblasts, local depolarization with BK can induce an intercellularly propagating Ca^{2+} action potential which results in a fast transient increase in $[Ca^{2+}]_i$ in large numbers of cells. Thus, bradykinin acts by first increasing $[Ca^{2+}]_i$, thereby inducing a depolarization via Ca^{2+}-dependent Cl^- channels. This depolarization opens voltage-dependent Ca^{2+} channels which induces an intercellularly propagating action potential and an increase in $[Ca^{2+}]_i$ by an influx of Ca^{2+}. The Ca^{2+} action potentials reported here provide a mechanism for intercellular communication which is several magnitudes faster than the reported intercellular IP_3-mediated Ca^{2+} waves in non-excitable cells (Jaffe, 1991; Sanderson et al., 1994).

Fibroblasts can be electrically coupled to each other (Hashizume et al., 1992; Jester et al., 1995; de Roos et al., 1996) and in several tissues, such as kidney, intestine and dermis, these cells form three-dimensional communicating networks (Komuro, 1990). Fibroblasts can also be electrically coupled to other cells (Hunter and Pitts, 1981), and in this way may play a role in the coordinated behavior of tissues. It is hypothesized that fibroblasts play an active role in the initiation and transduction of electrical signals involved in the coordination of multicellular activities.

In conclusion, membrane potential and Ca^{2+} play an important role in intercellular signaling in NRK fibroblasts. However, we did not find evidence for the involvement of gap junctional intercellular communication or membrane potential in growth control of NRK fibroblasts.

Acknowledgments: This work was funded by The Netherlands Foundation for Life Sciences. We thank dr. Peter Willems, dept. of Biochemistry, University of Nijmegen for helpful discussions. Dye-coupling experiments were performed at the dept. of Toxicology, Agricultural University of Wageningen, The Netherlands

References

Afink GB, Van Alewijk DCGJ, De Roos ADG, Van Zoelen EJJ (1994) Lysophosphatidic acid and bradykinin have opposite effects on phenotypic transformation of normal rat kidney cells. J Cell Biochem 56: 480-489

Baumgarten LB, Toscas K, Villereal ML (1992) Dihydropyridine-sensitive L-type Ca^{2+} channels in human foreskin fibroblast cells. J. Biol. Chem. 267: 10542-10530

Binggeli R, Weinstein RC (1986) Membrane potentials and sodium channels: hypotheses for growth regulation and cancer formation based on changes in sodium channels and gap junctions. J Theor Biol 123: 377-401

Chen C, Corbley MJ, Roberts TM, Hess P (1988) Voltage-sensitive calcium channels in normal and transformed 3T3 fibroblasts. Science 239: 1024-1026

De Roos ADG, van Zoelen EJJ, Theuvenet APR (1996) Determination of gap junctional intercellular communication by capacitance measurements. Pflügers Arch 431: 556-563

De Mello WC (1994) Gap junctional communication in excitable tissues: the heart as paradigma. Prog Biophys Molec Biol 61: 1-35

Hagiwara S, Byerly L (1981) Membrane biophysics of calcium currents. Fed Proc 40: 2220-2225

Harootunian AT, Kao JPY, Paranjape S, Tsien RY (1991) Generation of calcium oscillations in fibroblasts by positive feedback between calcium and IP_3. Science 251: 75-77

Hashizume T, Imayama S, Hori Y (1992) Scanning electron microscopic study on dendritic cells and fibroblasts in connective tissue. J of Electron Microsc 41: 434-437

Hille B (1992) Ionic channels in excitable membranes. Sinauer, Sunderland, MA, USA

Hunter GK, Pitts JD (1981) Non-selective junctional communication between some different mammalian cell types in primary culture. J Cell Sci 49: 163-175

Jaffe LF (1991) The path of calcium in cytosolic calcium oscillations: a unifying hypothesis. Proc Natl Acad Sci USA 88: 9883-9887

Jester JV, Petroll WM, Barry PA, Cavanagh HD (1995) Temporal, 3-dimensional, cellular anatomy of corneal wound tissue. J Anat 186: 301-311

Kado RT (1993) Membrane area and electrical capacitance. Methods in Enzymol 221: 273-299

Komuro T (1990). Re-evaluation of fibroblasts and fibroblast-like cells. Anat Embryol 182: 103-112

Loewenstein WR (1981) Junctional intercellular communication: the cell-to-cell membrane channel. Physiol Rev 61: 829-913

Lovisolo D, Alloatti G, Bonelli G, Tessitori L, Baccino FM (1988) Potassium and calcium currents and action potentials in mouse Balb/c 3T3 fibroblasts. Pflügers Arch 412: 530-534

Maldonado PE, Rose B, Loewenstein WR (1988) Growth factors modulate junctional cell-to-cell communication. J Membr Biol 106: 203-210

Moolenaar WH, Jalink K (1992) Membrane potential changes in the action of growth factors. Cell Physiol Biochem 2: 189-195

Neyton J, Trautmann A (1985) Single-channel currents of an intercellular junction. Nature 317: 331-335

Pandiella A, Meldolesi J (1992) Phosphoinositide hydrolysis and ensuing calcium and potassium fluxes: Role in the action of EGF and other growth factors. Cell Physiol Biochem 2: 196-2121

Peppelenbosch MP, Tertoolen LGJ, de Laat SW (1991) Epidermal growth factor-activated calcium and potassium channels. J Biol Chem 266: 19938-19944

Peres A, Sturani E, Zippel R (1988) Properties of the voltage-dependent calcium channel of mouse Swiss 3T3 fibroblasts. J Physiol 401: 639-655

Postma FR, Jalink K, Hengeveld T, Bot AGM, Alblas J, de Jonge HR, Moolenaar WH (1996) Serum-induced membrane depolarization in quiescent fibroblasts: activation of a chloride conductance through the G protein-coupled LPA receptor. EMBO J 15: 63-72

Putney JW (1993) Excitement about calcium signaling in inexcitable cells. Science 262: 676-678

Rijksen G, Völler MCW, van Zoelen EJJ (1993) Orthovanadate both mimics and antagonizes the transforming growth factor ß action on normal rat kidney cells. J Cell Physiol 154: 393-401

Rizzino A, Kazakoff P, Nebelsick J (1990) Density-induced down regulation of epidermal growth factor receptors. In Vitro Dev Biol 26: 537-542

Ruch RJ (1994) The role of gap junctional intercellular communication in neoplasia. Ann Clin Lab Sci 24: 216-231

Sanderson MJ, Charles AC, Boitano S, Dirksen ER (1994) Mechanisms and function of intercellular calcium signaling. Mol Cell Endocrinol 98: 173-187

Santos RM, Rosario LM, Nadal A, Garcia-Sancho J, Soria B, Valdeolmillos M (1991) Widespread synchronous $[Ca^{2+}]_i$ oscillations due to bursting electrical activity in single pancreatic islets. Pflügers Arch 418: 417-422

Steinberg TH, Civitelli R, Geist ST, Robertson AJ, Hick E, Veenstra RD, Wang H-Z, Warlow PM, Westphale EM, Laing JG, Beyer EC (1994) Connexin 43 and connexin 45 form gap junctions with different molecular permeabilities in osteoblastic cells. EMBO J 13: 744-750

Spray DC, Harris AL, Bennett MVL (1981) Equilibrium properties of a voltage-dependent junctional conductance. J Gen Physiol 77: 77-93

Stewart WW (1978) Gap junctional connections between cells as revealed by dye-coupling with a highly fluorescent naphthalimide tracer. Cell 14: 741-759

Traub O, Eckert R, Lichtenberg-Fraté H, Elfgang C, Bastide B, Scheidtmann KH, Hülser DF, Willecke K (1994). Immunochemical and electrophysiological characterization of murine connexin40 and -43 in mouse tissues and transfected human cells. Eur J Cell Biol 64: 101-112

Trosko JE, Chang CC, Madhukar BV, Klaunig JE (1990) Chemical, oncogene and growth factor inhibition of gap junctional intercellular communication: an integrative hypothesis of carcinogenesis. Pathobiol 58: 265-278

Van Zoelen EJJ, Van Oostwaard TMJ, De Laat SW (1988) The role of polypeptide growth factors in phenotypic transformation of normal rat kidney cells. J Biol Chem 263: 64-68

Van Zoelen EJJ (1991a) Density-dependent control of cell proliferation: Molecular mechanisms involved in contact-inhibition. In: Growth Regulation and Carcinogenesis 1: 91-93. WR Paukovitz, ed. CRC Press, Inc, Boca Raton, USA

Van Zoelen EJJ (1991b) Phenotypic transformation of normal rat kidney cells: a model for studying cellular alterations in oncogenesis. Crit Rev Oncogenesis 2: 311-333

Van Zoelen EJJ, Tertoolen LGJ (1991) Transforming growth factor-β enhances the extent of intercellular communication between normal rat kidney cells. J Biol Chem 266: 12075-12081

Van Zoelen EJJ, Peters PHJ, Afink GB, van Genesen S, de Roos ADG, van Rotterdam W, Theuvenet APR (1994) Bradykinin-induced growth inhibition of normal rat kidney cells supports a direct role for epidermal growth factor receptor expression in density-dependent growth control. Biochem J 298: 335- 340

Wade MH, Trosko JE, Schindler MA (1986) A fluorescent photobleaching assay of gap junction mediated communication between human cells. Science 232: 525-528

Willems PHGM, Van Emst-De Vries SE, Van Os CH, De Pont JJHHM (1993) Dose-dependent recruitment of pancreatic acinar cells during receptor-mediated calcium mobilization. Cell Calcium 14: 145-159

Yamasaki H (1990) Gap junctional intercellular communication and carcinogenesis. Carcinogenesis 11: 1051-1058

Cross-talk between calmodulin and protein kinase C

Arndt Schmitz, Enrico Schleiff[1], and Guy Vergères
Department of Biophysical Chemistry
Biozentrum, University of Basel
Klingelbergstr. 70
CH-4056 Basel, Switzerland

Introduction

Today a new picture of signal transduction events is emerging due to our better understanding of cellular behavior. The idea of linear signaling cascades is substituted by the new paradigm of networks in which different strains of signaling are balanced and tuned to a uniform reaction of the cell against the multitude of signals from the outside and the inside of the cell (Nishizuka, 1992a). For this purpose, the individual states of each of these pathways can potentially be cross-linked and compared to each other. In this context, "cross-talk" was defined as "a modulatory interaction between two distinct channels of information transfer" (Liscovitch, 1992).

It is intriguing to look for the molecular realization of such systems in the example of signaling cascades involving calmodulin (CaM) and protein kinase C (PKC) for different reasons. First, the individual pathways involving these proteins are relatively well understood, such that there is a wealth of information available. Second, both pathways share a common regulator, calcium. Third, at least 40 proteins interact with both CaM and PKC, suggesting that cross-talk might be a general mechanism.

In this review, we briefly describe the signal transduction cascades resulting in activation of CaM and PKC. The proteins, which are targets of both proteins, and which might thus mediate cross-talk are listed (see Table I). Three of these proteins, namely neuromodulin, neurogranin and the myristoylated alanine-rich C kinase substrate (MARCKS), for which sufficient data is available to support a role in cross-talk, are discussed. Finally, we present new data from our laboratory on MARCKS-related protein (MRP) to further demonstrate the potential of the members of the MARCKS family in linking the CaM and PKC signal transduction pathways. The article is concluded by an outlook.

---

[1] Present address: Department of Biochemistry, McIntyre Medical Sciences Building, McGill University, 3655 Drummond St., Montreal, Quebec H3G 1Y6, Canada.

NATO ASI Series, Vol. H 101
Molecular Mechanisms of Signalling
and Membrane Transport
Edited by Karel W. A. Wirtz

The calmodulin and protein kinase C signal transduction pathways

Calmodulin

CaM is a ubiquitous protein recognized as the prototype mediator of calcium signaling. The cytoplasmic concentration of calcium is highly regulated and maintained at a level of *ca.* 0.1 μM in the resting cell (Taylor and Marshall, 1992). Influxes from intracellular stores or from the extracellular space, which contain calcium in millimolar concentrations, result in an increase in the cytosolic calcium concentration of several hundred nanomolars. At these concentrations calcium ions bind to CaM and induce a conformational change which allows binding of CaM to most of its targets. Although the mechanisms of recognition of CaM substrates are well understood from a structural point of view (O'Neil and DeGrado, 1990; Török and Whitaker, 1994; Crivici and Ikura, 1995; James et al., 1995), the molecular events leading from CaM to the regulation of gene expression, cell growth and cell cycle progression (Rasmussen and Means, 1989; Means, 1994) are much less defined. In this respect the targets of CaM need to be thoroughly characterized.

One category of targets contains proteins which bind directly to CaM. Within this category, a group of targets contains regulatory enzymes whose activation depends on CaM and which can be regarded as parts of signal transduction cascades. These enzymes can be divided into four subgroups (Klee, 1991). The first subgroup contains enzymes regulating the metabolism of cyclic nucleotides, such as adenylate cyclase, NO synthase, guanylate cyclase and phosphodiesterase (Beltman et al., 1993). The second subgroup, which contains protein kinases, may be further divided into two classes, according to substrate specificity. Some kinases phosphorylate mainly one substrate and, consequently, link CaM to a well defined cellular response. These enzymes are myosin light chain kinase (involved in muscle contraction), phosphorylase kinase (glycogen catabolism) and elongation factor-2 kinase (protein synthesis). Other kinases, such as CaM kinases Ia, Ib, II and IV, are much less specific and thus affect a broad range of substrates. The functions of these kinases, even of CaM kinase II which has been intensively studied, are poorly understood (Schulman, 1993). The third subgroup of CaM-dependent enzymes contains phosphatases and is represented by a single member, calcineurin (Klee, 1991). The fourth subgroup contains plasma membrane Ca^{2+}-ATPase and $InsP_3$ 3-kinase, two CaM-dependent enzymes involved in the regulation of intracellular calcium. These enzymes allow a CaM-mediated feedback regulation of calcium concentration (Williams, 1992). Furthermore, the list of CaM targets is not restricted to enzymes which require CaM for activity. A second group of CaM targets contains proteins such as MAP-2, neuromodulin, and MARCKS, which do not have an absolute requirement for CaM to be functional, but whose activities may be modulated by CaM.

Finally, a second category of targets, which will not be discussed in great detail, contains proteins acting further downstream of the CaM-mediated signaling cascade, *i.e.* which are modified by the CaM-dependent enzymes mentioned above.

Protein kinase C

PKC consists of a family of closely related phospholipid-dependent serine/threonine kinases which are implicated in numerous cellular processes, including growth, differentiation, secretion and metabolism. These kinases transduce their activating signals, calcium and lipid second messengers, into phosphorylation of their protein substrates.

The levels of the second messengers calcium and diacylglycerol (DAG) are highly regulated. Activation of phospholipase C by G-protein coupled receptors induces the hydrolysis of phosphatidylinositol 4,5-biphosphate to inositol 1,4,5-triphosphate ($InsP_3$) and DAG. $InsP_3$, in turn, increases the intracellular calcium concentration by opening calcium channels on the surface of intracellular storage organelles. Calcium and DAG released *via* these pathways induce a transient activation of PKC which is typical of early cellular responses such as secretion. A second wave of DAG is also produced from dephosphorylation of phosphatidic acid resulting from the hydrolysis of phosphatidylcholine by phospholipase D. This pool of DAG induces a sustained activation of PKC which, independently from the early transient increase in calcium, is responsible for late cellular responses such as proliferation and differentiation. Experimentally, PKC can be activated by the tumor-promoting phorbol esters. These widely used compounds mimic the action of DAG, but are metabolically stable and therefore lead to a more persistent activation of PKC than achieved by DAG whose turnover occurs rapidly. Note that other lipid metabolites may also activate PKC; for example, the reaction products of phospholipase A_2, lysophospholipid and free fatty acid (*e.g.* arachidonic acid), activate PKC and take part in the propagation of inflammatory response (for reviews see: Bell, 1986; Asaoka et al., 1992; Nishizuka, 1992b; 1995).

The members of the PKC family can be divided into three categories based on their requirements for cofactors: 1) conventional PKCs (α, β I, β II, and γ) require phosphatidyl-serine (PS), DAG and calcium as cofactors; 2) novel PKCs (δ, ε, η, θ, and μ) require PS and DAG but not calcium; 3) atypical PKCs (ζ and λ (ι)) only require PS for activation. These isozymes share a common structure. Binding sites for the cofactors are found in the amino-terminal regulatory domain. The carboxyl-terminus contains the catalytic domain that binds substrates and ATP. Calpain cleaves between these two domains and releases a constitutively active form of PKC, protein kinase M (PKM), which is independent of calcium and lipids and which is thus often used for *in vitro* phosphorylation (see section "MARCKS-related protein"). In the resting state, the substrate-binding site of PKC is occupied by a pseudosubstrate autoinhibitory site contained in the amino-terminal regulatory domain; this interaction is believed to inhibit basal kinase activity. Activation of PKC by its cofactors results in a complex reaction, including its translocation from the cytosol to the membrane and removal of the pseudosubstrate autoinhibitory site from the catalytic domain which may now be recognized by a substrate molecule (for reviews see: Huang, 1989; Azzi et al., 1992).

Although a precise substrate consensus sequence cannot be defined, the threonine and serine residues phosphorylated by PKC are always flanked by one or several basic residues at positions -1 to -3 and/or +1 to +3 (Kemp and Pearson, 1990; Kennelly and Krebs, 1991). The multiplicity of PKC isozymes suggests that each isozyme phosphorylates specific substrates and, consequently, is responsible for a specific cellular response. *In vitro* studies have shown, however, that each isozyme can efficiently phosphorylate a broad range of substrates and that one substrate can be phosphorylated by several isozymes. It has therefore become obvious that the tissue distribution as well as the cellular and subcellular localization of PKC isozymes must be important to determine the selective activity of these enzymes *in vivo*. In this respect, the discovery of proteins, such as receptors for activated C kinase, which might be involved in targeting PKC to the correct membrane localization, has highlighted the importance of subcellular targeting for the specificity of PKC isozymes (for reviews see: Nishizuka, 1988; Hug and Sarre, 1993; Dekker and Parker, 1994; Jaken, 1996; Faux and Scott, 1996).

Over the last two decades many PKC substrates, including membrane-bound proteins such as channels and receptors, contractile and cytoskeletal proteins as well as enzymes, have been identified (Nishizuka, 1986; Azzi et al., 1992). Although the mechanisms leading to the activation of PKC have been relatively well investigated, the consequences of phosphorylating these substrates are significantly less understood and the physiological significance of many of these phosphorylation reactions remains to be explored. One heterogeneous category of PKC substrates is, however, central to the theme of this article since it is composed of proteins which also participate in the CaM-dependent signal transduction pathway. These substrates can therefore potentially allow cross-talk between the CaM- and PKC-dependent signal transduction pathways.

Cross-talk between CaM and PKC

Numerous pharmacological studies have shown that cross-talk between the CaM and PKC signal transduction pathways occurs. Only rarely have the molecular mechanisms underlying these observations been revealed. In an effort to identify the proteins which can potentially mediate cross-talk downstream of CaM and PKC, we have listed the PKC substrates which either bind to CaM or are modified by CaM-dependent enzymes. A survey of the literature published during the last 10 years shows that *ca.* 40 proteins fulfill these criteria (Table I).

The proteins listed in Table I are classified in two categories according to the level at which cross-talk might take place. The first category contains PKC substrates which bind directly to CaM. Within this category, three different groups of proteins are distinguished. The first group (group 1. a) contains the CaM-dependent enzymes mentioned in the section about calmodulin. Note that, with the exception of the subgroup composed of the phosphatase calcineurin, the three other subgroups of enzymes described in that section, *i.e.* enzymes involved in the metabolism of cyclic nucleotides, in phosphorylation (kinases), and in the regulation of calcium, are represented. The second group of PKC substrates binding directly to CaM contains neuronal proteins without enzymatic activity which are involved in the regulation of the cytoskeleton (group 1. b). We have separated these proteins from a heterogeneous group (group 1. c), whose members, apparently, do not share common properties with the proteins of group 1. b.

A second category of PKC substrates contains proteins acting further downstream of CaM, *i.e.* which are targets for CaM-dependent enzymes. Interestingly, the majority of these PKC substrates (group 2. a) are transmembrane proteins which are also phosphorylated by CaM-dependent kinases and which constitute important elements of other signal transduction pathways. The signaling function of these membrane-bound proteins can be modulated by PKC

TABLE I

Potential mediators of cross-talk downstream of CaM and PKC.

Category 1: PKC substrates which bind to CaM.

a) Enzymes which require CaM for activation. b) Neuronal proteins, without enzymatic activity, involved in regulation of the cytoskeleton. c) Other proteins.

Category 2: PKC substrates which are modified by CaM-dependent enzymes.

a) Transmembrane proteins involved in signal transduction, phosphorylated by CaM-dependent kinases. b) Other proteins.

This table was prepared by searching the MedLine Databank for abstracts containing the terms "calmodulin" and "PKC". Note that the groups within the categories 1 and 2 are not mutually exclusive.

	PKC substrate	Reference
1. a)	CaMKII	Waxham MN et al., Biochemistry 32 (1993) 2923-30
	cAMP phosphodiesterase PDE1	Spence S et al., Biochem. J. 310 (1995) 975-82
	$InsP_3$ 3-kinase	Communi D et al., Cell. Signal. 7 (1995) 643-650
	MLCK	Jiang H et al., Mol. Cell. Biochem. 135 (1994) 1-9
	multifunctional CaMK	MacNicol M et al., J. Biol. Chem. 267 (1992) 12197-201
	NO synthase (brain)	Bredt DS et al., J. Biol. Chem. 267 (1992) 10976-81
	plasma membrane Ca^{2+}-ATPase	Smallwood JI et al., J. Biol. Chem. 263 (1988) 2195-202
1. b)	adduccin	Matsuoka Y et al., J. Biol. Chem. 271 (1996) 25157-66
	AKAP-79	Klauck TM et al., Science 271 (1996) 1589-92
	caldesmon	Stasek JE et al., J. Cell. Physiol. 153 (1992) 62-75
	calponin	Fraser ED et al., Biochemistry 32 (1993) 13327-33
	MARCKS	see text
	MRP	see text
	neurogranin	see text
	neuromodulin	see text
	tau	Steiner B et al., EMBO J. 9 (1990) 3539-44
	troponin I	Bazzi MD et al., Biochim. Biophys. Acta 931 (1987) 339-46
1. c)	kinesin	Matthies HJ et al., J. Biol. Chem. 268 (1993) 11176-87
	MyOD transcription factor	Baudier J et al., Biochemistry 34 (1995) 7834-46
	myosin light chain	Nakabayashi H et al., FEBS Lett. 294 (1991) 144-48
	NAP-22	Maekawa S et al., J. Biol. Chem. 269 (1994) 19462-65
	NMDA receptor / channel	Ehlers MD et al., Cell 84 (1996) 745-55
	phospholamban	Allen BG et al., Mol. Cell. Biochem. 155 (1996) 91-103
	pp65 (macrophages)	Shinomiya H et al., J. Immunol. 154 (1995) 3471-78
	p68 RNA helicase	Buelt MK et al., J. Biol. Chem. 269 (1994) 29367-70
	p75 (A10 cells)	Zhao D et al., Biochem. J. 277 (1991) 445-50
	SR Ca^{2+}-channel (cardiac)	Takasago T et al., J. Biochem. 109 (1991) 163-70
2. a)	AMPA-type Glu receptor	Tan SE et al., J. Neurosci. 14 (1994) 1123-29
	cystis fibrosis TR	Picciotto MR et al., J. Biol. Chem. 267 (1992) 12742-52
	$GABA_A$-receptor γ 2L	Machu TK et al., J. Neurochem. 61 (1993) 375-77
	$InsP_3$ receptor	Ferris CD et al., Proc. Natl. Acad. Sci. USA 88 (1991) 2232-35
	L-type Ca^{2+}-channel α	Hell JW et al., J. Biol. Chem. 268 (1993) 19451-57
	SR Ca^{2+}-ATPase	Qu Y et al., Can. J. Physiol. Pharmacol. 70 (1992) 1230-35
	SR Ca^{2+}-channel (skeletal)	Mayrleitner M et al., Cell Calcium 18 (1995) 197-206
2. b)	choline acetyltransferase	Bruce G et al., Neurochem. Res. 14 (1989) 613-620
	connexin32	Saez JC et al., Eur. J. Biochem. 192 (1990) 263-273
	cytochrome P450 2E1	Menez JF et al., Alcohol Alcohol. 28 (1993) 445-51
	20K-Mr (cerebral cortex)	Suzuki T et al., J. Neurochem. 53 (1989) 1751-1762
	5-HT 1C receptor	Boddeke HW et al., Naunyn Schmiedebergs Arch. Pharmacol. 348 (1993) 221-24

or by CaM-dependent kinases. This suggests that these proteins are important modules on which several signal transduction pathways converge. In this second category we have also listed a group of several other proteins without apparent similarities (group 2.b).

The group 1. b in Table I contains neuronal proteins that have been proposed to act at the end of signaling cascades as regulators of the cytoskeleton. These proteins may ultimately translate the incoming information into a cellular response; important clues regarding this essential aspect of signal transduction should therefore be obtained by investigating their molecular properties. Interestingly, the most convincing data on cross-talk between the CaM and PKC pathways have been obtained with some members of this group. These proteins, namely neuromodulin, neurogranin and the proteins of the MARCKS family, will be treated in more detail in the following sections.

Neuromodulin and neurogranin

Neuromodulin (also known as GAP-43, B-50 and F-1) and neurogranin (RC3 or BICKS) are two major PKC substrates in the central nervous system (Huang and Huang, 1993). In neurons these proteins have opposite subcellular localizations: Neuromodulin is located in presynapses, whereas neurogranin is found at the posterior side of synapses. Although these proteins have unrelated overall primary structures, they share a conserved domain which contains the PKC phosphorylation site and an adjacent CaM binding motif (Apel and Storm, 1992).

Most interestingly, neuromodulin and neurogranin bind to CaM in the absence of calcium (Baudier et al., 1991), in contrast to the vast majority of CaM targets (see section "Calmodulin"). The structural basis for this mechanism is not fully understood. While neuromodulin and neurogranin lack secondary structure elements, circular dichroic spectroscopy reveals that CaM increases the α-helical content of both proteins in the absence, but not in the presence, of calcium (Gerendasy et al., 1995). These results were confirmed by NMR spectroscopy with a peptide corresponding to the CaM-binding domain of neuromodulin (Urbauer et al., 1995).

Cross-talk - A mutual influence of the signal transducers CaM and PKC was observed on neuromodulin and neurogranin. Phosphorylation by PKC inhibits adsorption of neuromodulin to CaM-modified sepharose (Apel et al., 1990). Also a mutated neuromodulin, in which Ser41 was substituted by Asp to mimic a phospho-Ser residue, does not bind to CaM-sepharose (Chapman et al., 1991). *In vitro*, CaM significantly inhibits PKM-catalyzed phosphorylation of neuromodulin and neurogranin in the absence, but not in the presence of 100 μM $CaCl_2$ (Sheu et al., 1995). These studies were confirmed by an *in vivo* study which showed that stimulation of primary cultures of hippocampal neurons with a calcium ionophor or a phorbol ester increases the amount of free neuromodulin at the expense of its complex with CaM (Gamby et al., 1996).

Establishment of new signaling networks - The ability of neuromodulin and neurogranin to sequester CaM in resting cells and to increase the pool of free CaM upon activation suggests unusual modes of signal transduction. First, neuromodulin and neurogranin facilitate the activation of calcium-dependent CaM targets, such as MARCKS, MRP (see below) and CaM-dependent kinase II (see section "Calmodulin"), instead of competing with these targets for Ca^{2+}/CaM (Estep et al., 1990; Gerendasy et al., 1994). Second, neuromodulin and neurogranin not only regulate these CaM-dependent enzymes, but are also targets of some of them, allowing a highly controlled feedback regulation of the posttranslational modifications of both proteins as well as the establishment of new signaling networks; this phenomenon has been documented for several proteins, including phosphorylase kinase, calcineurin and, presumably, NO synthase.

In the presence of calcium, phosphorylase kinase is activated 2- to 7-fold by addition of exogenous calmodulin (Pickett-Gies and Walsh, 1986). This raises the possibility that, under cellular conditions stimulating PKC and releasing CaM from neuromodulin and neurogranin, *i.e.* at high calcium concentrations, phosphorylase kinase could be activated. Since neuromodulin and neurogranin are also phosphorylated by phosphorylase kinase as judged from *in vitro* experiments (Paudel et al., 1993) and since phosphorylase kinase and PKC phosphorylate the same residues in neuromodulin and neurogranin, both kinases could eventually compete for their common substrates.

Phosphorylation is typically a transient modification which is strictly controlled and often reversed by dephosphorylation. Phospho-neuromodulin is dephosphorylated by the CaM-dependent phosphatase calcineurin (Liu and Storm, 1989). Neurogranin is an even better substrate for calcineurin and both proteins are also dephosphorylated by protein phosphatases 1 and 2A, at least *in vitro* (Seki et al., 1995). These observations indicate that the levels of free CaM in neurons may be regulated by cycles of phosphorylation and dephosphorylation of neuromodulin and neurogranin (Estep et al., 1990; Klee, 1991).

Another CaM-dependent enzyme found in brain is NO synthase, which is subject to regulation by different pathways (Bredt et al., 1992). Its product, nitric oxide, is an important physiologic messenger molecule (Bredt and Snyder, 1994). Interestingly, neuromodulin (Slemmon and Martzen, 1994) and neurogranin (Martzen and Slemmon, 1995) depress NO synthase activity by competing for CaM at low calcium concentrations. At elevated concentrations (1-6 μM), NO synthase regains its activity, presumably due to the decreased affinity of neuromodulin and neurogranin for CaM. These inhibitory effects are eliminated by phosphorylation of neuromodulin and neurogranin by PKC and restored by subsequent treatment with phosphatases, further supporting the disruptive influence of phosphorylation on the complexation of neuromodulin and neurogranin by CaM.

The regulation of NO synthase becomes even more interesting in the light of two recent articles showing that NO-mediated oxidation of cysteine residues in neurogranin leads to the formation of intramolecular disulfide bridges and, consequently, to a conformational change in neurogranin which hinders its phosphorylation by PKM as well as its binding to CaM (Sheu et al., 1996; Mahoney et al., 1996). These reports suggest that NO synthase may interrupt phosphorylation/dephosphorylation cycles by modifying neurogranin with NO. Note that chemical NO-donors were used in these studies; whether similar results can be achieved with endogenous NO *via* stimulation of the NO synthase, remains to be demonstrated. NO also inhibits palmitoylation of neuromodulin *in vivo*, most probably by modifying the normally acylated cysteine residues (Hess et al., 1993). This inhibition could affect the subcellular localization of neuromodulin by decreasing its affinity for membranes (Taniguchi et al., 1994).

Physiological function - It has been proposed that neuromodulin and neurogranin constitute a part of the molecular basis for long term potentiation (Meberg et al., 1993; Ramakers et al., 1995; Pasinelli et al., 1995), a major model of learning and memory (Fagnou and Tuchek, 1995). Furthermore, these proteins are believed to modulate neuronal plasticity and to be involved in neurotransmitter release (Gispen et al., 1991; Gamby et al., 1996). Although these functions require the activation of molecular cascades resulting in the reorganization of the cytoskeleton, strong evidence for the binding of neuromodulin to actin filaments is still missing, and the function as well as the mode of action of neuromodulin and neurogranin remain to be elucidated (Hens et al., 1993; Biewenga et al., 1996).

MARCKS and MRP

The members of the MARCKS family have a widespread distribution among mammalian species (Albert et al., 1986; Blackshear et al., 1986) and have also been detected in chicken (Graff et al., 1989a) as well as in the electric organ of *Torpedo californica* (Albert et al., 1986). Major cellular events, such as adhesion, mitosis, motility, neurosecretion and phagocytosis, are concomitant with changes in their expression, phosphorylation or subcellular localization (for reviews on MARCKS proteins see: Aderem, 1992; Blackshear, 1993). Several recent gene knockout studies in mice have also demonstrated that MARCKS proteins are essential for brain development and postnatal survival (Stumpo et al., 1995; Wu et al., 1996; Chen et al., 1996). The family comprises two members: MARCKS is an acidic ubiquitous 32 kDa protein, whereas MARCKS-related protein (MRP, also MacMARCKS or F52 in the literature) is a 20 kDa protein which is highly enriched in brain and reproductive tissues. The primary sequences of MARCKS proteins show three conserved domains: 1) The N-terminus is myristoylated (*i.e.*, acylated with a C14 saturated fatty acid) through a cotranslational reaction catalyzed by N-myristoyl transferase and is involved in membrane binding. 2) The MARCKS-homology 2 domain contains 6 residues with sequence identity to the cytoplasmic tail of the cation-independent mannose-6-phosphate receptor. Its function is unknown. 3) The effector domain, a highly basic domain containing 24-25 residues, is central to the function of MARCKS proteins.

MARCKS

MARCKS phosphorylation - In 1982, Wu et al. first found that a calcium/phospholipid kinase regulates the phosphorylation of an "87 kDa" substrate in brain synaptosomes. Since this original discovery, numerous *in vivo* studies have demonstrated that MARCKS can be phosphorylated following activation of cellular PKCs[2]. For example, phosphorylation of MARCKS was demonstrated following activation of fibroblasts with phorbol esters, phospholipase C and growth factors (Rozengurt et al., 1983; Blackshear et al., 1985), during neurosecretion (Kligman and Patel, 1986), following activation of macrophages with bacterial lipopolysaccharides (Aderem et al., 1988), and following activation of neutrophils with N-formyl-Met-Leu-Phe, a chemotactic peptide (Thelen et al., 1991). The phosphorylation induced by this chemotactic peptide is transient and dephosphorylation of MARCKS is inhibited by okadoic acid, suggesting that MARCKS is a substrate for phosphatases (Thelen et al., 1991). Based on similar studies with Swiss 3T3 cells, Clarke et al. (1993) have proposed that protein phosphatases 1 and 2A dephosphorylate MARCKS *in vivo*. *In vitro*, the effector domain of MARCKS is phosphorylated by PKC at three out of four serine residues present in this domain (Rosen et al., 1989; Verghese et al., 1994). These residues are sequentially phosphorylated in the order S156 > S163 > S152 by several PKC isozymes, including PKCβ1, PKCδ and PKCε (Herget et al., 1995). Purified MARCKS is also dephosphorylated by phosphatases 1 and 2A, supporting the conclusions reached *in vivo* (Seki et al., 1995). Interestingly, MARCKS could also be dephosphorylated by calcineurin, suggesting that this enzyme might also regulate the function of MARCKS in brain.

Interactions of MARCKS with CaM - Binding of MARCKS to CaM was originally proposed based on studies with a peptide corresponding to the effector domain (Graff et al., 1989b). These studies were later extended to the intact purified protein (Verghese et al., 1994). In contrast to neuromodulin and neurogranin, MARCKS behaves as a classical CaM ligand since calcium is required for complex formation. A direct demonstration for an *in vivo* interaction between MARCKS and CaM has not been obtained so far. Nonetheless, two studies suggest that such a complex exists in the cell. In the fresh water protozoan *Paramecium,* an action potential results in membrane depolarization and a period of backward swimming; repolarization and a return to forward swimming requires the presence of calmodulin. Microinjection of a peptide, corresponding to the effector domain of MARCKS, caused a 2- to 3-fold increase in the duration of backward swimming, indicating that the peptide can form a complex with CaM *in vivo* (Hinrichsen and Blackshear, 1993). Another piece of evidence

---

[2] The "87 kDa" protein and the calcium/phospholipid-dependent kinase were later named MARCKS and PKC, respectively. Note that MARCKS has a molecular weight of 32 kDa but migrates with an apparent molecular weight of 87 kDa when subjected to SDS polyacrylamide gel electrophoresis.

supporting a role for MARCKS in the regulation of CaM function came from studies in Rat1 cells in which MARCKS was overexpressed: MARCKS significantly decreases the concentrations of CaM antagonists, such as W7 and trifluoperazine, required to inhibit DNA synthesis. Since CaM is essential for cell growth and is required for initiation and progression of DNA synthesis, these results suggest that MARCKS binds to CaM and, consequently, reduces the free concentration of CaM in the cell (Herget et al., 1994).

Subcellular localization of MARCKS - Early phosphorylation experiments with brain extracts have shown that MARCKS is associated with the particulate fraction of cells (Walaas et al., 1983). It was later demonstrated by immunocytochemistry and cell fractionation that MARCKS is associated with the plasma membrane of macrophages (Rosen et al., 1990; Allen and Aderem, 1995a), neurons (Ouimet et al., 1990), and fibroblasts (Swierczynski and Blackshear, 1995; Allen and Aderem, 1995b). Although the association of MARCKS with the plasma membrane is well established, the protein is also found in the cytosol of neuronal and glial cells as well as in association with microtubules in neuronal cells (Ouimet et al., 1990). Furthermore, phosphorylation of MARCKS is concomitant with its translocation to the cytosol of macrophages (Aderem et al., 1988), neutrophils (Thelen et al., 1991) and fibroblasts (Herget and Rozengurt, 1994; Swierczynski and Blackshear, 1995). Translocation to the cytosol might however only be a transient event since phosphorylation induces relocation of MARCKS to the surface of phagosomes in macrophages (Allen and Aderem, 1995a) and lysosomes in fibroblasts (Allen and Aderem, 1995b). These studies also suggest that MARCKS is involved in membrane trafficking.

At the molecular level, several *in vitro* studies with peptides (Kim et al., 1994a) and purified proteins (George and Blackshear, 1992; Nakaoka et al., 1993; Taniguchi and Manenti, 1993; Kim et al., 1994b; Vergères et al., 1995a) as well as studies with mutated proteins expressed in intact cells (Swierczynski and Blackshear, 1995; 1996) show that the binding of MARCKS to membranes is primarily mediated by electrostatic interactions of the basic effector domain with negatively-charged phospholipid membranes, such as the plasma membrane, and by insertion of the myristoyl moiety into the bilayer (Vergères et al., 1995a). Since each of these domains does not provide enough energy to efficiently promote binding of MARCKS to membranes, the myristoyl moiety and the effector domain have been proposed to act synergetically to anchor MARCKS to membranes (McLaughlin and Aderem, 1995). This model provides a rationale for the observations that phosphorylation eventually induces detachment of MARCKS from the surface of acidic lipid vesicles or from the plasma membrane of several cell types; it does not explain, however, the lack of translocation of MARCKS following activation of PKC in myocytes and glioma cells, and the presence of phosphorylated MARCKS on the surface of other cellular organelles such as lysosomes in fibroblasts. The determinants of the interaction of MARCKS with membranes are therefore not completely understood (for reviews on the interactions of MARCKS with membranes see: McLaughlin and Aderem, 1995; Vergères et al., 1995b).

Interactions of MARCKS with the cytoskeleton - The co-localization of MARCKS with microtubules in neurons showed that MARCKS is associated with the cytoskeleton (Ouimet et al., 1990). In macrophages, MARCKS has a punctuate distribution at the cell-substratum interface of pseudopodia and filopodia. At these points, MARCKS co-localizes with vinculin and actin, two components of the membrane cytoskeleton, suggesting an interaction of MARCKS with actin filaments. Such an interaction is supported by an *in vitro* study which demonstrates that MARCKS cross-links actin filaments and that the effector domain is responsible for this activity (Hartwig et al., 1992).

MARCKS in CaM/PKC cross-talk - The multiplicity of interactions taking place at the effector domain results in competition between the various partners of MARCKS, including CaM and PKC. Consequently, phosphorylation regulates the properties of MARCKS by inhibiting its binding to CaM (Graff et al., 1989b; McIlroy et al., 1991; Verghese et al., 1994), and CaM inhibits the phosphorylation of MARCKS (Albert et al., 1984; Sheu et al., 1995). These properties were observed *in vitro* but nonetheless provide the basis for a function of MARCKS in cross-talk between the CaM and PKC signal transduction pathways.

In 1982, Greengard and coworkers already obtained the first indication that MARCKS might mediate a CaM-dependent regulation of the PKC signal transduction pathway: Calmodulin was found to inhibit the Ca^{2+}/phospholipid-dependent phosphorylation of the "87-kDa" protein in crude extracts of rat brain synaptosomal cytosol (Wu et al., 1982).[2] As already mentioned in the previous paragraph, this original observation was later confirmed by several *in vitro* studies. Two recent studies by Chakravarthy and coworkers support this hypothesis *in vivo*. In the first report, incubation of keratinocytes with calcium induced activation of PKC, but failed to stimulate phosphorylation of MARCKS. Since treatment of keratinocytes with cyclosporin A, a calcineurin inhibitor, prior to stimulation with calcium did not increase phosphorylation of

MARCKS, the authors proposed that the apparent failure of the activated PKC to phosphorylate MARCKS was not due to the activation of Ca^{2+}-stimulated phosphatase, but rather to a direct blockage of the phosphorylation site of MARCKS by Ca^{2+}/CaM (Chakravarthy et al., 1995a). In a second report, Chakravarthy et al. (1995b) showed that ionomycin, a calcium ionophore, activates PKC in C6 rat glioma cells without stimulating MARCKS phosphorylation. That this inhibitory effect could result from the binding of CaM to MARCKS is strongly suggested by their observation that pretreating intact glioma cells with a cell-permeable calmodulin antagonist, calmidazolium, prevented ionomycin from blocking MARCKS phosphorylation by PKC.

As already stated above, MARCKS binds to CaM *in vitro* and phosphorylation by PKC disrupts this complex. These results suggest an alternative model for cross-talk, in which MARCKS mediates a PKC-dependent regulation of the CaM pathway. In a model proposed by Blackshear and coworkers, binding of MARCKS to CaM decreases the cytosolic concentration of free CaM, and consequently inhibits the activation of target molecules, such as CaM-dependent kinases; phosphorylation of MARCKS releases CaM, thus allowing activation of its target molecules (Blackshear, 1993). Support for this hypothesis has been found in *Paramecium*: The effects of MARCKS peptides on the swimming behavior of this organism could be completely reversed by activation of PKC with phorbol esters, suggesting that a complex between the effector domain of MARCKS and CaM can be disrupted by PKC *in vivo* (Hinrichsen and Blackshear, 1993). An indirect evidence for this model was also provided in a study in which phorbol ester treatment of PC12 cells increased the cytosolic concentration of CaM. The authors proposed that the phosphorylation of CaM-binding proteins, such as neuromodulin or MARCKS, might increase the availbility of previously bound CaM (MacNicol and Schulman, 1992).

An interesting report by Sawai et al. (1993) indicates that a CaM/MARCKS complex can be actively involved in signal transduction. In cultured mast cells, phorbol esters induce MARCKS phosphorylation and markedly enhance cAMP formation induced by carbacyclin, a stable prostacyclin analogue which activates G-protein-coupled IgE receptors. The enhancing activity of phorbol esters was almost completely suppressed by the CaM inhibitor W-7, in agreement with a role for CaM in adenylate cyclase activation and, consequently, cAMP formation. Importantly, MARCKS had an inhibitory effect on the CaM-induced activation of adenylate cyclase in cells permeabilized with saponin. This inhibitory effect could not be overcome by the addition of excess CaM, suggesting that the sequestration of CaM by MARCKS is not a satisfactory explanation for the inhibitory effect of MARCKS. A model was proposed which implicitly suggests that a CaM/MARCKS complex might actively regulate the cAMP signal transduction pathway in a PKC-dependent manner.

Cross-talk and regulation of the actin cytoskeleton - In 1992 Hartwig et al. showed that modification of the effector domain, either by binding of CaM or by phosphorylation with PKC, disrupts the actin filament cross-linking activity of MARCKS *in vitro*. These observations suggest that CaM and PKC could regulate the interactions of MARCKS with actin filaments. In agreement with this hypothesis, activation of PKC with phorbol esters results in the disappearance of MARCKS from punctuate structures of macrophages together with cell spreading and loss of filopodia (Rosen et al., 1990). Based on these observations, a model was proposed in which the function of MARCKS is to regulate the structure and the plasticity of the actin cytoskeleton (Aderem, 1992). In this model, MARCKS is bound to the plasma membrane of inactivated cells and cross-links actin filaments. Activation of PKC induces phosphorylation of MARCKS and consequently a local release of actin filaments from the membrane. Furthermore, since phosphorylated MARCKS binds to, but does not cross-link, actin filaments, the cytoskeleton becomes more plastic. On the other hand, CaM could also regulate the plasticity of the cytoskeleton in a similar manner: An increase in the intracellular concentration of calcium induces activation of CaM and its binding to MARCKS, again inhibiting the formation of cross-linked actin filaments. In this model of cross-talk, MARCKS integrates signals from either CaM or PKC to modulate its own activity on a target molecule, actin. One should mention that although sufficient data support the co-localization of MARCKS with the actin cytoskeleton *in vivo*, further conclusive data on the regulation of these interactions are still required to demonstrate that the CaM and PKC pathways converge on MARCKS to regulate the plasticity of the cytoskeleton.

MARCKS-related protein

The recent findings that a dominant negative mutant of MRP blocks phagocytosis in macrophages (Zhu et al., 1995) and that MRP is also essential for brain development in mice (Wu et al., 1996), highlight the biological importance of this protein. Since both the myristoylated N-terminus and the effector domain are conserved in MARCKS proteins,

MARCKS and MRP are expected to have similar properties. MRP has however been significantly less characterized than MARCKS and appropriate data are not always available to compare both members of the family. The next paragraphs present a summary of the properties of MRP as well as new data pointing to a function for this protein in cross-talk between the CaM and PKC signal transduction pathways.

Properties of MRP - MRP is phosphorylated by PKC, following activation of macrophages with bacterial lipopolysaccharides (Li and Aderem, 1992). Phosphorylation of MRP has also been reported in lymphocytic leukemia cells treated with phorbol esters (Carballo et al., 1995). *In vitro*, PKC phosphorylates two out of three serine residues in the effector domain of MRP and CaM binds to MRP in a calcium-dependent manner (Blackshear et al., 1992; Li and Aderem, 1992; Verghese et al., 1994). During phagocytosis of zymosan by macrophages, MRP is associated with the plasma membrane on nascent phagosomes (Zhu et al., 1995). Interestingly, and in contrast to MARCKS, MRP was not found on mature phagosomes, suggesting that both proteins have different subcellular localization and, consequently, different functions (Zhu et al., 1995; Allen and Aderem, 1995a,b). A clear picture of the subcellular localization of MRP, in particular of its dynamics, is however still missing and awaits further studies. Finally, although MRP is probably associated with the cytoskeleton, conclusive data on the interaction of MRP with actin have not been documented so far.

Our lab has focused on the characterization of MRP. We have expressed both the unmyristoylated (unmyr) and myristoylated (myr) forms of murine MRP in *E. coli*. A purification procedure was developed which avoids the heat and acid treatment usually employed in purification of MARCKS proteins. In analogy to MARCKS, experiments with phospholipid vesicles have shown that MRP binds to membranes *via* insertion of the myristoyl moiety into the bilayer and *via* electrostatic interactions of the effector domain with the negatively-charged surface of membranes containing acidic phospholipids. The affinity of MRP for acidic lipid membranes is however 20- to 30-fold smaller than reported for MARCKS and a strong synergism between the myristoyl moiety and the effector domain is not observed (Vergères et al., 1995a,b). These results suggest different modes of interactions of MARCKS and MRP with membranes and might account for the observation that these proteins have different subcellular localization in macrophages (see previous paragraph).

Although MRP associates with subcellular membranes, the protein might also be functional in the cytosol. We have therefore characterized MRP in solution. The catalytic subunit of PKC

(PKM) phosphorylates MRP with high affinity ($S_{0.5,\ unmyr}$ = 3.2 μM; $S_{0.5,\ myr}$ = 3.5 μM), positive cooperativity ($n_{H,\ unmyr}$ = 2.1; $n_{H,\ myr}$ = 2.5) and a relatively high turnover number ($k_{p,\ unmyr}$ = 118 min^{-1} ; $k_{p,\ myr}$ = 130 min^{-1}). MRP also binds to CaM with high affinity ($K_{d,\ myr}$ = 4 nM; $K_{d,\ unmyr}$ = 7 nM) in the presence of calcium. Stopped-flow kinetics showed that the mechanism of recognition of unmyr MRP by CaM occurs in two steps: An initial transient complex is rapidly formed ($k_{+1} = 1.6 \times 10^8\ M^{-1}\ s^{-1}$) which relaxes to a final stable complex ($k_{obs2} = 11\ s^{-1}$) (Schleiff et al., 1996). Since an analysis of the hydrodynamic properties of MRP in solution revealed that unmyr MRP has a tendency to form dimers (Schleiff et al., 1996), it is tempting to speculate that the cooperative phosphorylation of MRP by PKM and the biphasic kinetics of complex formation between MRP and CaM reflect the oligomerization state of MRP. Furthermore, oligomerization might well be the mechanism by which MARCKS proteins cross-link actin filaments. In keeping with the theme of this review, it will therefore be interesting to investigate whether CaM and PKC can regulate the oligomerization state of MRP.

MRP in CaM/PKC cross-talk - Apart from the observation that phorbol esters reverse the behavioral effect of a peptide corresponding to the effector domain of MRP in *Paramecium*, no other *in vivo* data are available to support a role for MRP in cross-talk between CaM and PKC signal transduction pathways (Hinrichsen and Blackshear, 1993). Several *in vitro* studies have, however, demonstrated that CaM and PKC act competitively on MRP: Binding of CaM to MRP inhibits the phosphorylation of MRP (Li and Aderem, 1992) and phosphorylation disrupts the MRP/CaM complex (Blackshear et al., 1992; Verghese et al., 1994).

In order to determine the physiological relevance of these competitive experiments, we have taken a quantitative approach and measured the phosphorylation of MRP in the presence of CaM. We found that CaM inhibits the phosphorylation of myr MRP with a half-maximum rate of phosphorylation at a [CaM]/[MRP] ratio of 0.7 (Schleiff et al., 1996). Figure 1 shows that a similar ratio, *i.e.* 1.1, is obtained with unmyr MRP. The result that unmyr and myr MRP behave similarly demonstrates that myristoylation does not modulate the interactions of MRP with CaM and PKC in solution. Since almost complete inhibition of phosphorylation is observed at a [CaM]/[MRP] ratio of 5, which is close to what might be found in brain ([CaM] = 60 μM; [MARCKS] = 12 μM; Blackshear, 1993), these results suggest that CaM might efficiently inhibit the phosphorylation of MARCKS proteins *in vivo*.

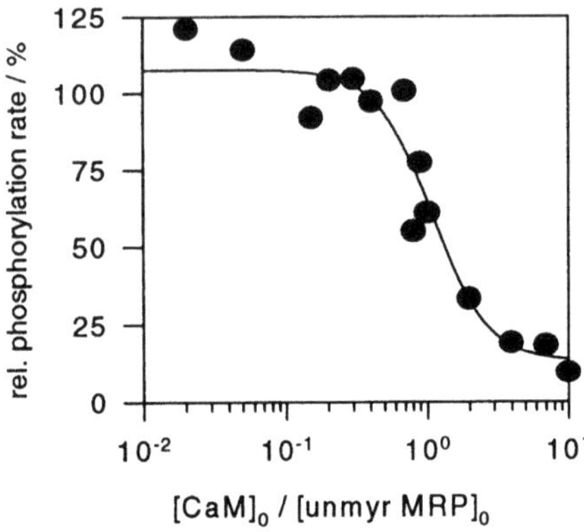

FIG. 1. **CaM inhibits the rate of phosphorylation of MRP.** 1 µM unmyr MRP was phosphorylated by PKM in the presence of 100 µM Ca^{2+} and increasing concentrations of CaM. The data were fitted with a sigmoidal function.

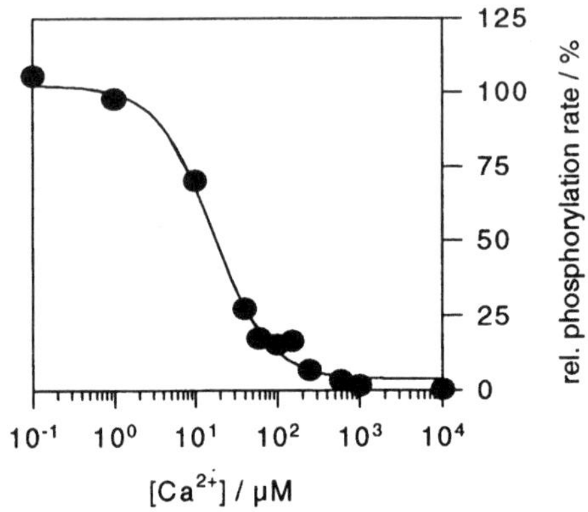

FIG. 2. **Calcium is required for CaM-mediated inhibition of phosphorylation of MRP.** 1 µM unmyr MRP was phosphorylated by PKM in the presence of 5 µM CaM and increasing concentrations of calcium. The data were fitted with a sigmoidal function.

As already mentioned, calcium is an essential modulator of the cellular response and its concentration is tightly regulated. Since both CaM and PKC require calcium for their activities, its concentration could regulate the ability of MARCKS proteins to mediate cross-talk between the CaM and PKC pathways. We have therefore investigated the calcium dependency of the inhibition of the phosphorylation of MRP by PKM. Figure 2 shows preliminary data demonstrating that calcium can efficiently induce the inhibition of phosphorylation of unmyr MRP. Together with the data presented in Figure 1, Figure 2 demonstrates that this inhibition results from the formation of a complex between MRP and CaM. Intriguingly, the concentration of calcium required to obtain half-maximum inhibition is 16 µM. Since the concentration of CaM used in this assay is 5 µM and since the two calcium ions necessary to trigger a conformational change in CaM bind to the carboxy-terminal domain with high affinity ($K_d = 10^{-7}$ M) (James et al., 1995), we would expect that *ca.* 5 µM calcium should be sufficient to activate half of the CaM molecules. Our data thus indicate that calcium is not only required to activate CaM but might also be necessary to activate MRP and promote its binding to CaM. Although a direct proof of the binding of calcium to MRP still remains to be demonstrated, it is not surprising that the putative binding of calcium to MRP could have escaped detection so far. First, MARCKS proteins do not possess EF-hand domains which would clearly indicate a calcium-binding property; since MARCKS proteins are unusually acidic (pIs of *ca.* 4.5), other non-classical calcium-binding motifs might however be present in these proteins. Second, since both CaM and PKC need calcium for their activities, binding of calcium to MRP could be masked by the requirement of those proteins for calcium. Finally, as an alternative to the hypothesis that MRP binds calcium, MRP could well significantly decrease the affinity of CaM for calcium. We are currently performing experiments to test these models.

Outlook

As a component of a signaling cascade, a CaM substrate could potentially transfer a signal from CaM to downstream elements in the form of a complex with CaM. In analogy, a PKC substrate would propagate a signal in the form of a phospho-protein. Cross-talk between these proteins can be established by allowing CaM and PKC to act on the same substrate. Since a relatively large number of proteins are recognized by both CaM and PKC, cross-talk is not limited to the action of few specialized molecules (see Table I). However, the substrates listed in Table I could be classified in several groups, depending on their properties. This observation suggests that a limited number of cellular events, such as the reorganization of the cytoskeleton, requires cross-talk between the CaM and PKC pathways.

The opportunities of cross-talk are increased by allowing molecular interactions at different levels of these signaling cascades. For example, several PKC substrates interact with downstream targets of CaM (see Table I). Many proteins in these cascades have also evolved into protein families containing numerous isozymes, in order to increase the specificity of the cellular response. Dramatic examples of such molecular diversity are the proteins of the PKC family as well as upstream elements such as cell surface receptors and heterotrimeric G proteins. In this context, MARCKS and MRP have very similar sequences but almost certainly different functions.

The flow of information along a signaling cascade can be regulated by proteins which act upstream on their own pathway, allowing a feedback regulation. In this respect, a recent study demonstrating that MARCKS can inhibit the activity of PLC by sequestering phosphatidylinositol 4,5-biphosphate in lateral domains indicates that MARCKS could be involved in feedback regulation of the PKC pathway (Glaser et al., 1996). Furthermore, the ability of several CaM-dependent enzymes, such as phosphorylase kinase, calcineurin, and NO synthase, to regulate the activity of neuromodulin or neurogranin demonstrates that, in addition to feedback regulation, complex connections are established between the CaM and PKC pathways, as well as with other pathways.

The multiplicity of these mechanisms indicates that a specific cell type could potentially initiate an almost infinite number of molecular connections. However, it is clear that only specific subsets of these connections are activated in response to particular stimuli. In order to solve this challenging problem several aspects must be considered which are central to signal transduction.

First, two proteins can only interact provided they are located in the same subcellular compartment. The question of the subcellular localization is particularly well illustrated by the isozymes of the PKC family. The subcellular localization is also critical for the interactions of CaM, a cytosolic protein, with MARCKS. Since MARCKS is a membrane-bound protein in its unphosphorylated form and since phosphorylated MARCKS cannot interact with CaM, it is rather challenging to build up a model in which both proteins can interact.

Also, several substrates might compete for the same effector molecule. In this case, it is often not possible to predict which one of these interactions will take place in the cell. A determination of the cellular concentrations of these molecules and of their affinities for each other would clarify this point. For example, the *in vitro* reconstitution experiments presented in this article (see section "MARCKS-related protein") suggest that the cellular concentration of CaM should be sufficient to effectively inhibit phosphorylation of MRP by PKC, provided all three proteins are located in the same subcellular compartment.

Finally, the interactions of proteins involved in signal transduction require an appropriate timing. On a long time scale, this phenomenon is controlled by regulating the expression of these proteins. For example, bacterial lipopolysaccharides dramatically increase the expression of MARCKS proteins in macrophages (Li and Aderem, 1992). On a short time scale, protein-protein interactions can be regulated by altering the levels of second messengers, such as calcium and DAG.

With respect to the points mentioned above, the different models of cross-talk proposed for MARCKS are not mutually exclusive (see section "MARCKS"). Each of these models could well be relevant in different cellular situations. For example, CaM can only inhibit the phosphorylation of MARCKS by PKC, if a CaM/MARCKS complex forms before PKC is activated. Another interesting example is brought by a comparison of neuromodulin and neurogranin with MARCKS proteins. Whereas the proteins of the MARCKS family require calcium for their binding to CaM, neuromodulin and neurogranin bind strongly to CaM only in the absence of calcium. This "simple" difference in the mechanisms of recognition is sufficient to allow activation of these proteins under different cellular conditions. These examples further demonstrate the importance of calcium as a second messenger in the regulation of spatio-temporal events in signaling cascades (Berridge and Dupont, 1994).

Acknowledgments

This work was supported by Swiss National Foundation Grant 3100-042045.94 (to G. Schwarz). The participation of A.S. in the *Spetses summer school* was made possible by a grant from German Academic Exchange Service (DAAD), Bonn-Bad Godesberg. The authors would like to thank Gerhard Schwarz for his support and Christoph Stürzinger for purifying MRP.

References

Aderem AA, Albert KA, Keum MM, Wang JKT, Greengard P and Cohn ZA (1988) Nature 332:362-64

Aderem AA (1992) Cell 71:713-16

Albert KA, Wu WCS, Nairn AC and Greengard P (1984) Proc. Natl. Acad. Sci. USA 81:3622-25

Albert KA, Walaas SI, Wang JKT and Greengard P (1986) Proc. Natl. Acad. Sci. USA 83:2822-26

Allen LAH and Aderem AA (1995a) J. Exp. Med. 182:829-40

Allen LAH and Aderem AA (1995b) EMBO J. 14:1109-21

Apel ED, Byford MF, Au D, Walsh KA and Storm DR (1990) Biochemistry 29:2330-35

Apel ED and Storm DR (1992) Persp. Dev. Neurobiol. 1:3-11

Asaoka Y, Nakamura S, Yoshida K and Nishizuka Y (1992) Trends Biochem. Sci. 17:414-17

Azzi A, Boscoboinik D and Hensey C (1992) Eur. J. Biochem. 208:547-57
Baudier J, Deloulme JC, van Dorsselaer A, Black D and Matthes HWD (1991) J. Biol. Chem. 266:229-37
Bell RM (1986) Cell 45:631-32
Beltman J, Sonnenburg WK and Beavo JA (1993) Mol. Cell. Biochem. 127-128:239-53
Berridge MJ and Dupont G (1994) Curr. Opin. Cell Biol. 6:267-74
Biewenga JE, Schrama LH and Gispen WH (1996) Acta Biochim. Pol. 43:327-38
Blackshear PJ, Witters LE, Girard PR, Kuo JF and Quamo SN (1985) J. Biol. Chem. 260:13304-15
Blackshear PJ, Wen L, Glynn BP and Witters LE (1986) J. Biol. Chem. 261:1459-69
Blackshear PJ, Verghese GM, Johnson JD, Haupt DM and Stumpo DJ (1992) J. Biol. Chem. 267:13540-46
Blackshear PJ (1993) J. Biol. Chem. 268:1501-04
Bredt DS, Ferris CD and Snyder SH (1992) J. Biol. Chem. 267:10976-81
Bredt DS and Snyder SH (1994) Annu. Rev. Biochem. 63:175-95
Carballo E, Colomer D, Vives Corrons JL, Blackshear PJ and Gil J (1995) Leukemia 9:834-39
Chakravarthy BR, Isaacs RJ, Morley P, Durkin JP and Whitfield JF (1995a) J. Biol. Chem. 270:1362-68
Chakravarthy BR, Isaacs RJ, Morley P and Whitfield JF (1995b) J. Biol. Chem. 270:24911-16
Chapman ER, Au D, Alexander KA, Nicolson TA and Storm DR (1991) J. Biol. Chem. 266:207-13
Chen J, Chang S, Duncan SA, Okano HJ, Fishell G and Aderem AA (1996) Proc. Natl. Acad. Sci. USA 93: 6275-79
Clarke PR, Siddhanti SR, Cohen P and Blackshear PJ (1993) FEBS Letters 336:37-42
Crivici A and Ikura M (1995) Annu. Rev. Biophys. Biomol. Struct. 24:85-116
Dekker LV and Parker PJ (1994) Trends Biochem. Sci. 19:73-77
Estep RP, Alexander KA and Storm DR (1990) Curr. Top. Cell. Reg. 31:161-80
Fagnou DD and Tuchek JM (1995) Mol. Cell. Biochem. 149/150:279-86
Faux MC and Scott JD (1996) Trends Biochem. Sci. 21:312-15
Gamby C, Waage MC, Allen RG and Baizer L (1996) J. Biol. Chem. 271:26698-705
George DJ and Blackshear PJ (1992) J. Biol. Chem. 267:24879-85
Gerendasy DD, Herron SR, Watson JB and Sutcliffe JG (1994) J. Biol. Chem. 269:22420-26
Gerendasy DD, Herron SR, Jennings PA and Sutcliffe JG (1995) J. Biol. Chem. 270:6741-50
Gispen WH, Nielander HB, De Graan PNE, Oestreicher AB, Schrama LH and Schotman P (1991) Mol. Neurobiol. 5:61-85
Glaser M, Wanaski S, Buser CA, Boguslavsky V, Rashidzada W, Morris A, Rebecchi M, Scarlata SF, Runnels LW, Prestwich GD, Chen J, Aderem A, Ahn J and McLaughlin S (1996) J. Biol. Chem. 271: 26187-93
Graff JM, Stumpo DJ and Blackshear PJ (1989a) Mol. Endocrinol. 3:1903-06
Graff JM, Young TN, Johnson JD and Blackshear PJ (1989b) J. Biol. Chem. 264:21818-23
Hartwig JH, Thelen M, Rosen A, Janmey PA, Nairn AC and Aderem AA (1992) Nature 356:618-22
Hens JJH, Benfenati F, Nielander HB, Valtorta F, Gispen WH and De Graan PNE (1993) J. Neurochem. 61:1530-33
Herget T and Rozengurt E (1994) Eur. J. Biochem. 225:539-48
Herget T, Broad S and Rozengurt E (1994) Eur. J. Biochem. 225:549-56
Herget T, Oehrlein SA, Pappin DJ, Rozengurt E and Parker PJ (1995) Eur. J. Biochem. 233:448-57
Hess DT, Patterson SI, Smith DS and Skene JHP (1993) Nature 366:562-65

Hinrichsen RD and Blackshear PJ (1993) Proc. Natl. Acad. Sci. USA 90:1585-89
Huang K-P (1989) Trends Neurosci. 12:425-32
Huang K-P and Huang FL (1993) Neurochem. Int. 22:417-33
Hug H and Sarre TF (1993) Biochem. J. 291:329-43
Jaken S (1996) Curr. Opin. Cell Biol. 8:168-73
James P, Vorherr H and Carafoli E (1995) Trends Biochem. Sci. 20:38-42
Kemp BE and Pearson RB (1990) Trends Biochem. Sci. 15:342-46
Kennelly PJ and Krebs EG (1991) J. Biol. Chem. 266:15555-58
Kim J, Blackshear PJ, Johnson JD and McLaughlin S (1994a) Biophys. J. 67:227-37
Kim J, Shishido T, Jiang X, Aderem AA and McLaughlin S (1994b) J. Biol. Chem. 269: 28214-19
Klee CB (1991) Neurochem. Res. 16:1059-65
Kligman D and Patel J (1986) J. Neurochem. 47:298-303
Li J and Aderem AA (1992) Cell 70:791-801
Liscovitch M (1992) Trends Biochem. Sci. 17:393-99
Liu Y and Storm DR (1989) J. Biol. Chem. 264:12800-904
MacNicol M and Schulman H (1992) J. Biol. Chem. 267:12197-201
Mahoney CW, Pak JH and Huang K-P (1996) J. Biol. Chem. 271:28798-804
Martzen MR and Slemmon JR (1995) J. Neurochem. 64:92-100
McIlroy BK, Walters JD, Blackshear PJ and Johnson JD (1991) J. Biol. Chem. 266:4959-64
McLaughlin S and Aderem AA (1995) Trends Biochem. Sci. 20:272-76
Means AR (1994) FEBS Letters 347:1-4
Meberg PJ, Barnes CA, McNaughton BL and Routtenberg A (1993) Proc. Natl. Acad. Sci. USA 90:12050-54
Nakaoka T, Kojima N, Hamamoto T, Kurosawa N, Lee YC, Kawasaki H, Suzuki K and Tsuji S (1993) J. Biochem. 114:449-52
Nishizuka Y (1986) Science 233:305-12
Nishizuka Y (1988) Nature 334:661-65
Nishizuka Y (1992a) Trends Biochem. Sci. 17:367
Nishizuka Y (1992b) Science 258:607-14
Nishizuka Y (1995) FASEB J. 9:484-96
O'Neil KT and DeGrado WF (1990) Trends Biochem. Sci. 15:59-64
Ouimet CC, Wang JKT, Walaas SI, Albert KA and Greengard P (1990) J. Neurosci. 10:1683-98
Pasinelli P, Ramakers GMJ, Urban IJA, Hens JJH, Oestreicher AB, de Graan PNE and Gispen WH (1995) Beh. Brain Res. 66:53-59
Paudel HK, Zwiers H and Wang JH (1993) J. Biol. Chem. 268:6207-13
Pickett-Gies CA and Walsh DA (1986) In: The Enzymes, Vol. XVII (Boyer PD and Krebs EG, eds). Academic Press, Inc., Orlando, FL, USA, p. 395-459
Ramakers GMJ, De Graan PNE, Urban IJA, Kraay D, Tang T, Pasinelli P, Oestreicher AB and Gispen WH (1995) J. Biol. Chem. 270:13892-98
Rasmussen CD and Means AR (1989) Trends Neurosci. 12:433-38
Rosen A, Nairn AC, Greengard P, Cohn ZA and Aderem AA (1989) J. Biol. Chem. 264:9118-21
Rosen A, Keenan KF, Thelen M, Nairn AC and Aderem AA (1990) J. Exp. Med. 172:1211-15
Rozengurt E, Rodriguez-Pena M and Smith KA (1983) Proc. Natl. Acad. Sci. USA 80:7244-48

Sawai T, Manabu N, Nishigaki N, Ohno T and Ichikawa A (1993) J. Biol. Chem. 268:1995-2000
Schleiff E, Schmitz A, McIlhinney RAJ, Manenti S and Vergères G (1996) J. Biol. Chem. 271:26794-802
Schulman H (1993) Curr. Opin. Cell Biol. 5:247-53
Seki K, Chen H-C and Huang K-P (1995) Arch. Biochem. Biophys. 316:673-79
Sheu F-S, Huang FL and Huang K-P (1995) Arch. Biochem. Biophys. 316:335-42
Sheu F-S, Mahoney CW, Seki K, and Huang K-P (1996) J. Biol. Chem. 271:22407-13
Slemmon JR and Martzen MR (1994) Biochemistry 33:5653-60
Stumpo DJ, Bock BC, Tuttle JS, and Blackshear PJ (1995) Proc. Natl. Acad. Sci. USA 92:944-48
Swierczynski SL and Blackshear PJ (1995) J. Biol. Chem. 270:13436-45
Swierczynski SL and Blackshear PJ (1996) J. Biol. Chem. 271:23424-30
Taniguchi H and Manenti S (1993) J. Biol. Chem. 268:9960-63
Taniguchi H, Suzuki M, Manenti S, and Titani K (1994) J. Biol. Chem. 269:22481-84
Taylor CW and Marshall ICB (1992) Trends Biochem. Sci. 17:403-07
Thelen M, Rosen A, Nairn AC, and Aderem AA (1991) Nature 351:320-22
Török K and Whitaker M (1994) BioEssays 16:221-24
Urbauer JL, Short JH, Dow LK, and Wand AJ (1995) Biochemistry 34:8099-109
Vergères G, Manenti S, Weber T, and Stürzinger C (1995a) J. Biol. Chem. 270:19879-87
Vergères G, Manenti S, and Weber T (1995b) In: Signalling Mechanisms - from Transcription Factors to Oxidative Stress (Packer L, Wirtz KWA, eds). Springer Verlag Berlin Heidelberg NATO ASI H 92:125-37
Verghese GM, Johnson JD, Vasulka C, Haupt DM, Stumpo DJ, and Blackshear PJ (1994) J. Biol. Chem. 269:9361-67
Walaas SI, Nairn AC, and Greengard P (1983) J. Neurosci. 3:291-301
Williams RJ (1992) Cell Calcium 13:355-62
Wu WCS, Walaas SI, Nairn AC, and Greengard P (1982) Proc. Natl. Acad. Sci. USA 79:5249-53
Wu M, Chen DF, Sasaoka T, and Tonegawa S (1996) Proc. Natl. Acad. Sci. USA 93:2110-15
Zhu Z, Bao Z, and Li J (1995) J. Biol. Chem. 270:17652-55

A function for Eps15 in EGF-receptor endocytosis?

Sanne van Delft, Arie J. Verkleij and Paul M.P. van Bergen en Henegouwen.

Department of Molecular Cell Biology
Institute of Biomembranes
Utrecht University
Padualaan 8
3584 CH Utrecht
The Netherlands

Introduction

Binding of growth factors, such as the epidermal growth factor (EGF), to their specific receptors on the cell surface causes the initiation of a signal transduction cascade which leads to changes in gene expression and finally to cell division. Inactivation of the EGF-receptor can occur via several mechanisms, such as receptor transmodulation (Northwood and Davis, 1990), receptor dephosphorylation (Faure *et al.*, 1992) and receptor down-regulation (for review see Sorkin and Waters, 1993). The importance of down-regulation as a negative regulatory mechanism of receptor tyrosine kinase signaling is stressed by the observation that defects in this regulation can facilitate cellular transformation (Wells *et al.*, 1990) and tumor formation (Masui *et al.*, 1991). Receptor down-regulation results in the loss of EGF binding sites from the plasma membrane by internalization of the receptors. EGF-receptors enter the cell via receptor mediated endocytosis, a process involving clathrin coated pits and clathrin coated vesicles. The coat is composed of a number of proteins, such as the adaptor proteins (APs), the heavy and light chain of clathrin, forming the clathrin lattice (for review see

NATO ASI Series, Vol. H 101
Molecular Mechanisms of Signalling
and Membrane Transport
Edited by Karel W. A. Wirtz

Schmid, 1992) and, as recently has been demonstrated, Eps15 (Tebar *et al.,* 1996; van Delft *et al.,* 1997). Two classes of APs have been described: AP-1, which is found in the trans-Golgi network and AP-2, which is found at the plasma membrane (Robinson, 1987; Kirchhausen *et al.*, 1993). AP-1 is a heterotetramer composed of two large subunits, γ-adaptin and β-adaptin (100-115 kD), a medium subunit μ1 of 47 kD and a small subunit δ1 of 20 kD, whereas AP-2 consists of α-adaptin, β-adaptin and two polypeptides of 50 kD and 17 kD (μ2 and δ2)(Pearse and Robinson, 1984; Keen, 1987). AP-2 has been shown to bind to the regulatory domain in the C-terminal tail of the EGF-receptor, a region containing the sequence YRAL (Nesterov *et al.*, 1995). This sequence has recently been shown to bind directly to the μ2 subunit of the AP-2 complex (Boll *et al.*,1996). The interaction of AP-2 with the EGF-receptor has led to the hypothesis that recruitment of the EGF-receptor into the clathrin coated pits is mediated by AP-2 (Sorkin and Waters 1993).

Eps15 (EGF-receptor pathway substrate clone #15) has been described as one of the EGF-receptor kinase substrates (Fazioli *et al.*, 1993). Furthermore it has been shown that phosphorylation of Eps15 occurs specifically by activated EGF-receptors but not by activated PDGF- or insulin-receptors (van Delft *et al.*, 1997). Eps15 has an apparent molecular weight of 142 kD and consists of three structural domains. Domain I is the putative regulatory domain, containing a candidate tyrosine phosphorylation site, two EF-hand-type calcium-binding domains (Fazioli *et al.*, 1993) and three protein binding domains, designated as Eps15 homology (EH) domain (Wong *et al.*, 1995). Domain II has the features of a coiled-coil structure and domain III exhibits repeated DPF motifs, a motif which is conserved in several methyl-transferases, and a proline-rich-motif which can bind *in vitro* to the SH3 domain of c-Crk and v-Crk (Schumacher *et al.*, 1995). Additionally, this domain is involved in the constitutive association of Eps15 with the AP-2 complex via the COOH-terminal appendage (ear) of α-adaptin (Benmerah *et al.*, 1995, 1996). Furthermore, Eps15 has been shown to be a component of clathrin coated pits and vesicles (Tebar *et al.*, 1996; van Delft *et al.* 1997) and has homology with the yeast protein End3 which is involved in receptor-mediated endocytosis of the α-factor in *S. cerevisiae* (Bénédetti *et al.*, 1994). These data suggest a function for Eps15 in the down-regulation of EGF-receptors.

Tyrosine kinase activity of the EGF-receptor is required for the post-translational modification of Eps15.

In order to study the function of Eps15 in EGF-receptor down-regulation, we used NIH 3T3 fibroblasts stably transfected with cDNA of the human EGF-receptor. These cells, designated as HER14 cells, express approximately 400,000 receptors per cell (Rotin *et al.*, 1992). Stimulation of HER14 cells with EGF results in the activation of the EGF-receptor tyrosine kinase. As a result, Eps15 becomes phosphorylated on tyrosine residues which is shown in figure 1. In addition, EGF stimulation induced a transient mobility shift of phosphorylated Eps15 from 142 kD to 150 kD (Fig. 1). Eps15 has one putative tyrosine phosphorylation site (Fazioli *et al.*, 1993) and it seems unlikely that phosphorylation of this single site can cause an 8 kD mobility shift. Moreover, the 142 kD form of Eps15 is already phosphorylated. This indicates that Eps15 is not only modified by tyrosine phosphorylation but also by another, as yet unknown post-translational modification process. To determine

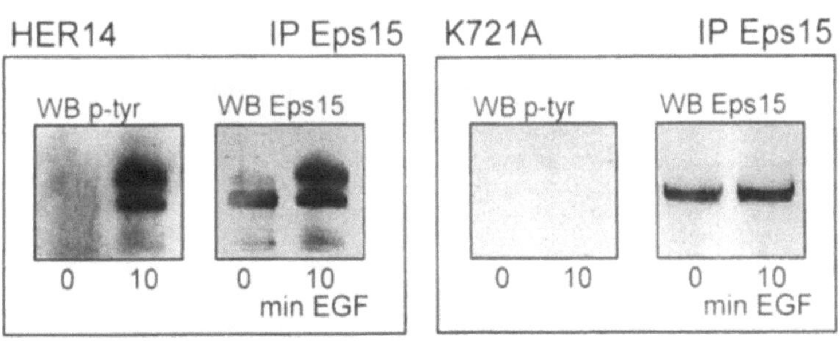

Figure 1 **Post-translational modifications of Eps15.**
HER14 and K721A fibroblasts were left unstimulated or were treated for 10 min with 50 ng/ml EGF. Eps15 was immunoprecipitated from cell lysates and samples were separated on 8% SDS PAGE and the Western blots were probed with anti-phospho-tyrosine or anti-Eps15 antibodies. Immunoprecipitations were performed as previously described (van Delft *et al.*, 1995).

whether the tyrosine kinase activity of the EGF-receptor is required for this post-translational modification, we investigated Eps15 modification in NIH 3T3 fibroblasts stably transfected with kinase inactive EGF-receptor cDNA (K721A). The cells were either left unstimulated or stimulated for 10 min with EGF. Eps15 was immunoprecipitated from the lysates and the proteins on Western blot were detected with anti-phospho-tyrosine antibodies. As expected, phosphorylation of Eps15 as found in HER14 cells did not occur in K721A cells (Fig. 1). Subsequently, the Western blot was stripped and reprobed with anti-Eps15 antibodies. Eps15 could be detected in unstimulated and stimulated samples of both cell types, but the 8 kD mobility shift was only visible in EGF-treated HER14 cells indicating that no post-translational modification had taken place in the K721A cells (Fig. 1). This result suggests that the tyrosine kinase activity of the EGF-receptor is required for the EGF-induced post-translational modifications of Eps15.

Association of Eps15 with the EGF-receptor, AP-2 and clathrin LC

By an *in vitro* kinase assay, Eps15 has been shown to be a direct substrate of the EGF-receptor (Fazioli *et al.*, 1993). To study the possible associations of Eps15 with the EGF-

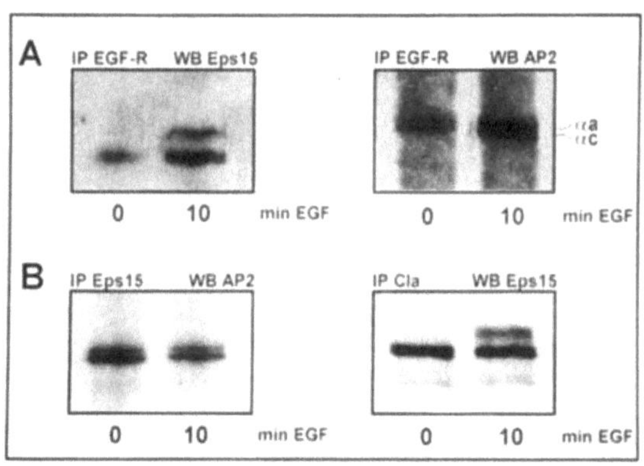

Figure 2 **Association of Eps15 with EGF-receptor, AP-2 and clathrin LC.** (A) EGF-receptor immunoprecipitates from HER14 cells which were either left unstimulated or treated for 10 min with 50 ng/ml EGF. The proteins were separated on 8% SDS-PAGE and the Western blots were probed with anti-Eps15 or anti-α-adaptin (AP-2) antibodies. (B) Eps15 and clathrin LC (Cla) immunoprecipitates from HER14 cells which were left unstimulated or stimulated for 10 min with 50 ng/ml EGF. Western blots were probed with anti-α-adaptin or anti-Eps15 antibodies. Immunoprecipitations were performed as previously described (van Delft *et al.*, 1995).

receptor *in vivo*, EGF-receptors were immunoprecipitated from cell lysates. The precipitated proteins were separated by SDS-PAGE, blotted onto PVDF-membrane and Western blots were analyzed for the presence of Eps15. A modest association between the EGF-receptor and Eps15 was observed in unstimulated cells (Fig. 2A, lane 1). However, stimulation of the cells with EGF resulted in a dramatic increase in the binding of Eps15 to the EGF-receptor. Similarly, an EGF-receptor immunoprecipitate probed with anti-α-adaptin antibodies showed an association of the EGF-receptor with AP-2. This association was increased upon EGF stimulation (Fig 2A, lanes 3 and 4) which is in agreement with data of Boll and coworkers (1995). The double band visible on the Western blot probed with anti-α-adaptin represents the αa and the αc form of this protein (Boll *et al.*, 1995). The effect of EGF on the association of Eps15 and AP-2 was investigated by co-immunoprecipitations. Eps15 was immunoprecipitated from unstimulated and EGF-stimulated cells and Western blots were probed with anti-α-adaptin antibodies (Fig. 2B, lanes 1 and 2). The bands were analyzed by densitometry. EGF treatment did not affect the interaction between Eps15 and AP-2 which indicates that the binding of Eps15 to AP-2 is constitutive. The AP-complexes have been shown to bind directly with the β-adaptin subunit to the clathrin heavy chain (Pearse and Crowther, 1987). In order to check the association of Eps15 with clathrin we investigated whether clathrin could be co-immunoprecipitated with Eps15 in an EGF dependent manner. HER14 cells were treated with EGF or left untreated and clathrin was immunoprecipitated using specific anti-clathrin light chain (LC) antibodies. The Western blot was subsequently probed with anti-Eps15 antibodies and revealed an association of Eps15 with the clathrin light chain. This association was not altered by EGF stimulation (Fig. 2B, lanes 3 and 4). In conclusion, these experiments show that Eps15 is associated with both AP-2 and clathrin light chain and that these interactions are not altered by EGF. In contrast, EGF induces a strong increase in the binding of the EGF-receptor to both Eps15 and AP-2.

Colocalization of Eps15 with AP-2 and clathrin

We next analyzed the subcellular localization of Eps15, AP-2 and clathrin heavy chain (HC) using immunofluorescence microscopy. HER14 fibroblasts were fixed, permeabilized and stained with the relevant antibodies and subsequently analyzed in an immunofluorescence light microscope. Staining of the cells with anti-Eps15 antibodies gave a clear punctuated staining pattern throughout the whole cell (Fig. 3A,C). Staining of the same cells with anti-α-adaptin antibodies revealed a similar punctuated staining pattern representing coated pits and coated vesicles (Fig. 3B; Guagliardi *et al*, 1990). Staining with anti-clathrin heavy chain (HC) antibodies led to a punctuated pattern throughout the cytoplasm, representing endocytotic vesicles and, in addition (Fig. 3C) a peri-nuclear staining, representing exocytotic vesicles

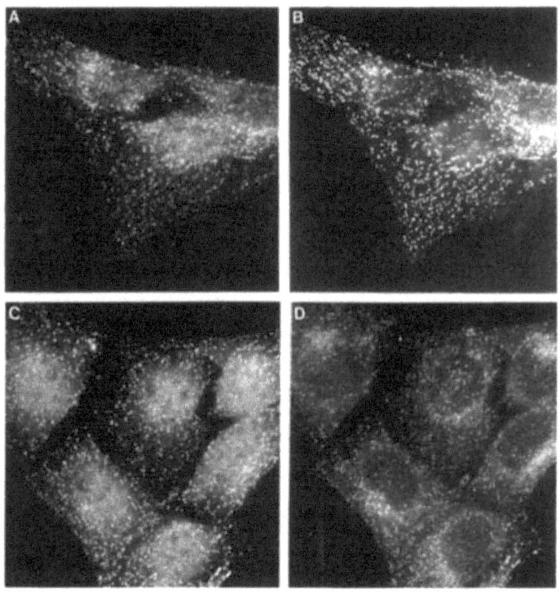

Figure 3 **Colocalization of Eps15 with AP-2 and clathrin LC.**
HER14 cells were grown in DMEM supplemented with 7.5% FCS, fixed in 3% formaldehyde and permeabilized in 0.2% Triton X-100 in PBS. Cells were stained with anti-Eps15 (A,C), anti-α-adaptin (B) or anti-clathrin HC antibodies (D).

from the trans-Golgi-network. Comparison of these images strongly suggests a colocalization of Eps15 with AP-2. The colocalization of Eps15 and clathrin HC is only apparent in endocytotic vesicles but absent in the clathrin-coated vesicles in the peri-nuclear region (Fig. 3A-D). In contrast to Eps15, clathrin showed additional staining around the nucleus. Additional experiments revealed no colocalization of Eps15 with markers for early endosomes, rab4 and rab5 (van Delft *et al.*, 1997). These data demonstrate that Eps15 is present in coated pits and coated vesicles, suggesting an involvement of Eps15 in the endocytosis of EGF-receptors from the plasma membrane. This localization excludes an involvement of Eps15 in exocytosis occurring from the trans-Golgi network. Moreover, the absence of Eps15 in early endosomes suggests that the function of Eps15 is restricted to coated pits and coated vesicles.

Discussion

The down-regulation of EGF receptors is started by stimulation of the receptor by EGF. The AP-2 complex has been suggested to recruit the EGF-receptor into coated pits thereby initiating the EGF mediated endocytosis (Sorkin *et al.*, 1995). Binding of the AP-2 complex to the EGF receptor has been suggested to be mediated by the 50 kD component μ2. Using a peptide library, it has recently been shown that μ2 binds directly to the non-

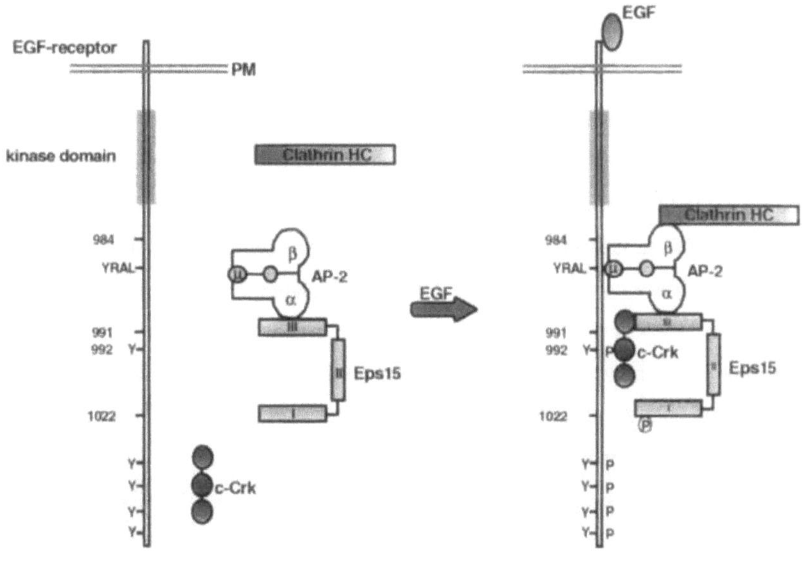

Figure 4 **Schematic representation of the binding partners of Eps15.**
In unstimulated cells Eps15 is present in the cytoplasm associated to AP-2. After stimulation a complex is formed consisting of Eps15, AP-2 and clathrin which binds to the EGF-receptor possibly mediated by c-Crk. For further details see text.

phosphorylated YRAL sequence which is present in the EGF-receptor at amino acid 974-977 (Boll *et al.,* 1996). In agreement with this is the observation that receptor autophosphorylation is not required for AP-2 binding to the EGF-receptor (Sorkin and Carpenter, 1993; Boll et al 1995; Sorkin et al, 1995). However, in this paper we show that binding of AP-2 to the cytoplasmic tail of the EGF-receptor increases after receptor activation. Moreover, deletion of the AP-2 binding site does not abolish EGF-induced receptor down-regulation (Nesterov *et al.*, 1995). These observations suggest that processes other than AP-2 mediated retention are involved in EGF-receptor endocytosis. It has recently been shown that kinase-deficient receptors fail to undergo ligand-induced sequestration into coated pits (Lamaze and Schmid, 1995) and that, as a result, kinase-deficient receptors are not internalized via coated vesicles. Recruitment into coated pits could be restored by the addition of a soluble EGF-receptor tyrosine kinase. Therefore, it has been proposed that the phosphorylation of another protein, an as yet unknown EGF-receptor substrate, is required for the efficient recruitment of EGF-receptors into coated pits (Lamaze and Schmid, 1995).

Based on the studies presented in this paper, we suggest that Eps15 might be involved in EGF-receptor internalization. Eps15 is a EGF-receptor tyrosine kinase substrate, it associates with the EGF receptor and binds directly to AP-2. The interaction between Eps15 and AP-2 is EGF-independent whereas the association of the EGF-receptor to both AP-2 and Eps15 is EGF-dependent. This could indicate that AP-2 and Eps15 can bind the EGF-receptor as a pre-existing complex (Fig. 4). Alternatively, some receptors may bind to AP-2 while others may associate with Eps15. Considering the fact that the interaction of AP-2 with the receptor is phosphorylation independent suggests that Eps15 is responsible for the increased binding of both AP-2 and Eps15 to the receptor. The binding of Eps15 and AP-2 to the activated EGF receptor as a complex is therefore our favorite model.

At this moment, the domain of Eps15 that interacts with the EGF receptor is unknown. Domain III contains two proline-rich motifs and one of these motifs has been shown to bind *in vitro* to the SH3 domain of the adaptor molecule Crk (Schumacher *et al.,* 1996). Interestingly, the SH2 domain of Crk has been shown to bind to the phosphorylated YxxP motif, a sequence that is present at residue 992-995 in the intracellular domain of the EGF receptor. The Crk-mediated binding of Eps15 to the EGF receptor would explain the increase in binding as a result of autophosphorylation of Y992 in response to EGF (Fig. 4). In addition, this idea is supported by the observation that the EGF receptor mutant lacking residues 991-

1021 is no longer able to phosphorylate Eps15 (Alvarez *et al.,* 1995). Preliminary experiments, however, do not reveal an *in vivo* binding of c-Crk or v-Crk with neither the EGF receptor or Eps15. Clearly, more experiments are required to completely understand the interactions of Eps15 at the molecular level.

Conclusions

Our observations all point to the idea that Eps15 is involved in the internalization of EGF-receptors via the coated vesicle pathway. In this model depicted in figure 4, we suggest that the Eps15/AP-2 complex mediates the interaction between activated EGF-receptors and the clathrin heavy chain in the coated pit. The binding site for Eps15 on the EGF-receptor may be located between residues 991-1021 and is possibly mediated by Crk. The increased binding of the Eps15/AP-2 complex to activated EGF-receptors may be regulated by increased binding of an Eps15/Crk complex to the tyrosine phosphorylated EGF receptor. This suggests that Eps15 selectively recruits activated EGF-receptors into the coated pit region. Inactive receptors, however, will not bind Eps15 and as a consequence will either stay at the plasma membrane, or will be endocytosed via the constitutive pathway. Recent studies have shown that EGF-receptor residues 1022-1123 are involved in the lysosomal targeting of the EGF-receptor (Kornilova *et al.,* 1996). This would indicate that the EGF-receptor domain between residues 970-1123 is a binding domain for different proteins, each with a different function in the process of EGF-induced EGF-receptor down-regulation.

References

Alvarez, C.V., Shon K.J., Miloso, M. and Bequinot, L. 1995. Structural requirements of the epidermal growth factor receptor for tyrosine phosphorylation of Eps8 and Eps15, substrates lacking Src SH2 homology domains. J. Biol. Chem. 270: 16271-16276

Bénédetti, H., Raths, S., Crausaz, F. and Riezman, H. 1994. The END3 gene encodes a protein that is required for the internalization step of endocytosis and for actin cytoskeleton organization in yeast. Mol. Biol. Cell 5: 1023-1037

Benmerah, A., Gagnon, J., Begue, B., Megarbane , B., Dautry-Varsat, A. and Cerf-Bensussan, N. 1995. The tyrosine kinase substrate eps15 is constitutively associated with the plasma membrane adaptor AP-2. J. Cell Biol. 131: 1831-1838

Benmerah, A., Beque, J., Dautry-Varsat, A. and Cerf-Bensussan, N. 1996. The ear of α-adaptin interacts with the COOH-terminal domain of the Eps15 protein. J. Biol.Chem. 271: 12111-12116

Boll, W., Gallusser and Kirchhausen T. 1995. Role of the regulatory domain of the EGF-receptor cytoplasmic tail in selective binding of the clathrin-associated complex AP-2. Current Biology 5: 1168-1178

Boll, W., Ohno, H., Songyang, Z., Rapoport, I., Cantley, L.C., Bonifacino, J.S. and Kirchhausen, T. 1996. Sequence requirements for the recognition of tyrosine-based endocytic signals by clathrin AP-2 complexes. EMBO J. 15: 5789-5795

Faure, R., Baquiran, G., Bergeron, J.J.M. and Posner, B.I. 1992. Dephosphorylation of insulin and epidermal growth factor receptors. J. Biol. Chem. 267: 11215-11221

Fazioli, F., Minichiello, L., Matoskova, B., Wong, W.T. and Di Fiore, P.P. 1993. Eps15, a novel tyrosine kinase substrate, exhibits transforming activity. Mol. Cell. Biol. 13: 5814-5828

Guagliardi, L, Koppelman, B., Blum, J.S., Marks, M.S., Cresswell, P. and Brodsky, F. 1990. Colocalization of molecules involved in antigen processing and presentation in an early endocytic compartment. Nature 343: 133-139

Keen, J.H. 1987. Clathrin assembly proteins: affinity purification and a model for coat assembly. J. Cell Biol. 105: 1989-1998

Kirchhausen, T. 1993. Coated pits and coated vesicles-sorting it all out. Curr. Opin. Struc. Biol. 3: 182-188

Kornilova, E., Sorkina, T., Bequinot, L. and Sorkin, A. 1996. Lysosomal targetting of epidermal growth factor recepotrs via a kinase-dependent pathway is mediated by the receptor carboxyl-terminal residues 1022-1123. J. Biol. Chem. 271: 30340-30346

Lamaze, C. and Schmid, S.L. 1995. Recruitment of epidermal growth factor receptors into coated pits requires their activated tyrosine kinase. J. Cell Biol. 129, 47-54

Masui, H., Wells, A., Lazar, C.S., Rosenfeld, M.G. and Gill, G.N. 1991. Enhanced tumorigenesis of NR6 cells which express non-down-regulating epidermal growth factor receptors. Cancer Res. 51: 6170-6175

Nesterov, A., Wiley, H.S. and Gill, G.N. 1995. Ligand-induced endocytosis of epidermal growth factor receptors that are defective in binding adaptor proteins. Proc. Natl. Acad. Sci. USA 92: 8719-8723

Northwood, I.C. and Davis, R.J. 1990. Signal transduction by the epidermal growth factor receptor after functional desensitization of the receptor tyrosine kinase activity. Proc. Natl. Acad. Sci. USA 87: 6107-6111

Pearse, B.M.F. and Robinson, M.S. 1984. Purification and properties of 100-kd proteins from coated vesicles and their reconstitution with clathrin. EMBO J. 3: 1951-1957

Pearse, B.M.F. and Crowther, R.A. 1987. Structure and assembly of coated vesicles. Ann. Rev. Bioph. Biochem. Comm. 16: 49-68

Robinson, M.S. 1987. 100 kD coated vesicle proteins: molecular heterogeneity and intracellular distribution studied with monoclonal antibodies. J. Cell Biol. 104: 887-895

Rotin, D., Margolis, B., Mohammadi, M., Daly, R.J., Daum, G., Li, N., Fischer, E.H., Burgess, W.H.M Ullrich, A. and Schlessinger, J. 1992. SH2 domains prevent tyrosine dephosphorylation of the EGF receptor at Tyr992 as the high-affinity binding site for SH2 domains of phospholipase C. EMBO J. 11: 559-567.

Schmid, S.L. 1992. The mechanism of receptor-mediated-endocytosis: more questions than answers. BioEssays 14: 589-596

Sorkin, A. and Carpenter, G. 1993. Interaction of activated EGF receptors with coated pit adaptins. Science 261: 612-615

Sorkin, A. and Waters, C.M. 1993. Endocytosis of growth factor receptors. BioEssays 15: 375-382

Sorkin, A., McKinsey, T., Shih, W., Kirchhausen, T. and Carpenter, G. 1995. Stoichiometric interaction of the epidermal growth factor receptor with the clathrin-associated protein complex AP-2. J. Biol. Chem. 270: 619-625

Schumacher, C., Knudsen, B.S., Ohuchi, T., Di Fiori, P.P., Glassman, R.H. and Hanafusa, H. 1995. The SH3 domain of Crk binds specifically to a conserved proline-rich motif in Eps15 and Eps15R. J. Biol. Chem. 270: 15341-15347

Tebar, F., Sorkina, T., Sorkin, A., Ericsson, M. and Kirchhausen, T. 1996. Eps15 is a component of clathrin-coated pits and vesicles and is located at the rim of coated pits. J. Biol. Chem. 271, 28727-28732

van Delft, S., Verkleij, A.J., Boonstra, J. and van Bergen en Henegouwen P.M.P. 1995. Epidermal growth factor induces serine phosphorylation of actin. FEBS letters 357: 251-254

van Delft, S., Schumacher, C., Hage, W., Verkleij, A.J. and van Bergen en Henegouwen, P.M.P. 1997. Association and colocalization of Eps15 with AP-2 and clathrin. J. Cell Biol. In press

Wells, A., Welsh, J.B., Lazar, C.S., Wiley, H.S., Gill, G.N. and Rosenfeld, M.G. 1990. Ligand-induced transformation by a non-internalizing epidermal growth factor receptor. Science 247: 962-964

Wong, W.T., Schumacher, C., Salcini, A.E., Romano, A., Castagnino, P., Pelicci, P.G. and Di Fiore, P.P. 1995. A protein-binding domain, EH, identified in the receptor tyrosine kinase substrate Eps15 and conserved in evolution. Proc. Natl. Acad. Sci. USA 92: 9530-9534

Acknowledgments: We wish to thank Christoph Schumacher (Ciba-Geigy Pharmaceutical Division, Summit NJ) for his reagents and support during the coarse of this work, Willem Stoorvogel and Peter van der Sluijs (Department of Cell biology, Utrecht University) for stimulating discussions and antibodies, and Lisette Verspui and Theo van der Krift for photographic reproductions.

This work was supported by the Life Sciences Foundation (SLW, grant 17.182), which is subsidized by the Netherlands Organization for Scientific Research (NWO).

Proteins, Sorted. The Secretory Pathway from the Endoplasmic Reticulum to the Golgi, and Beyond.

Sean Munro
MRC Laboratory of Molecular Biology
Hills Road
Cambridge CB2 2QH
UK

All eukaryotic cells are both defined and protected by an impermeable plasma membrane. However eukaryotic cells need to export proteins to alter their environments, or those of the organisms they form part of, and also to send signals to other cells within such a multicellular organism. To accomplish this, eukaryotic cells have a secretory pathway which is made up of a set of membranous compartments. The pathway starts in the endoplasmic reticulum (ER) where membrane proteins, and proteins destined for secretion, are co-translationally translocated from the cytosol into the lumen of the ER via specific channel structures (Walter and Johnson, 1994). Once in the ER, the proteins are modified by the addition of N-linked sugars, and then they fold and assemble.

Proteins Folding and Assembly in the ER

Protein folding in the ER is aided by chaperone proteins such as protein-disulphide isomerase (PDI) and members of the hsp70 family. The set of chaperones in the ER overlaps those involved in protein folding in the cytosol (Hartl et al., 1994). Thus both compartments contain members of the hsp70 family and the prolylisomerase (or cyclophilin) family. Interestingly the ER lumen has no homologue of the hsp60/GroEL chaperonin family even though this system is conserved in the cytosol of prokaryotes and chloroplasts. The ER of vertebrates contains a homologue of hsp90 (grp94) but the recently completed sequence of the genome of the budding yeast Saccharomyces cerevisiae reveals that this protein is missing in this unicellular eukaryote, although cytosolic forms are present. This suggests that hsp70 is perhaps the fundamental chaperone that is uniquely

NATO ASI Series, Vol. H 101
Molecular Mechanisms of Signalling
and Membrane Transport
Edited by Karel W. A. Wirtz

conserved between all protein folding environments. The ER also contains a family of chaperones that are unique to this environment. These are the protein disulphide isomerases involved in the formation and rearrangement of disulphide bonds in the oxidising environment of the ER. The cytosol is maintained in a highly reduced state, obviating disulphide bonds and hence the need for cytosolic homologues of the PDI family. The time taken for protein folding in the ER can take from seconds to many minutes depending on the size and complexity of the protein.

Folding chaperones are not the only proteins which interact with the newly synthesised proteins in the ER. There are also enzymes that carry out various post translational modifications of the nascent peptides, often very soon after their synthesis. The best known and characterised of these is the addition of N-linked carbohydrate structures which is thought to occur co-translationally. A core structure is assembled on the giant lipid dolichol and then transferred to specific sites on the proteins during translocation by the enzyme oligosaccharyltransferase (Kleene and Berger, 1993). The purpose of this modification has been the topic of much debate over many years (Gahmberg and Tolvanen, 1996). In some cases it may simply provide a framework for sugar modifications which the cell needs to display at its surface, regardless of the protein that is carrying them. In other cases it may aid the folding and solubility of the glycosylated protein. Finally the N-linked carbohydrate can also provide a frame work for carrying signals used to sort specific proteins in the secretory pathway. The best characterised example of the this is the addition of mannose-6-phosphate in the Golgi which serves as a signal for sorting of proteins to the lysosome as they exit the Golgi (Trowbridge et al., 1993).

There is also a more recently characterised system that recognises N-linked sugars in the ER itself. This is the calnexin/calreticulin quality control system (Bergeron et al., 1994). The N-linked carbohydrate that is initially added to proteins contains nine mannoses with an additional three glucose residues present at the end of one branch. These glucoses are rapidly removed by ER glucosidases. However these can be replaced by a glucosyltransferase that appears to act only if the protein is still unfolded. These glucoses are recognised by two ER resident proteins: calnexin, a membrane protein, and calreticulin, a structurally related protein which is apparently free in the lumen of the ER. It is believed that these proteins hold the glucosylated proteins until they are completely folded and hence are no longer substrates for the glucosyltransferase and therefore lack glucose residues. It is believed that this mechanism serves to hold partially folded proteins in the chaperone rich environment of the ER, and so prevent them from leaving until they are completely and correctly folded and assembled. N-linked glycosylation is not the only modification that occurs in the ER. Proline residues in collagens are hydroxylated, O-linked sugar residues are added and palmitylation of cysteines in membrane proteins occurs in the ER.

Vesicular Transport from ER to Golgi

Once a secreted or membrane protein has been folded, assembled and modified it is now ready and free to exit the ER and be transported to it final destination. Movement of proteins from the ER occurs by the same a mechanism found at many steps within the secretory pathway - inclusion in transport vesicles which bud of the ER and fuse with the next compartment in the secretory pathway. Transport vesicles from the ER were originally observed in EM studies and some of the proteins required for their formation have recently been identified by use of both biochemical and genetic means, primarily in the yeast S. cerevisiae (Schekman and Orci, 1996). These studies have revealed that a set of cytosolic proteins is assembled on the ER membrane to form a vesicle coat. This coat is called COPII (for reasons explained below) and its recruitment is regulated by a small GTP-binding protein called Sar1p. The binding of GTP to this GTPase is promoted by an ER nucleotide exchange factor called Sec12p. Although originally identified in yeast, homologues of the COPII system have since been found in mammalian cells and it appears that they mediate the pinching-off of cargo-laden vesicles from the ER, and hence initiate the process of exocytosis (Aridor et al., 1995).

Once a transport vesicle carrying newly made proteins has left the ER it must accurately deliver its contents to the next compartment in the pathway by specifically fusing with it. Transport vesicles derived from the ER fuse with the *cis*-Golgi network, (also called the ER-Golgi intermediate compartment). Although there are many different organelles in the cell, and many different vesicles moving between them, secretory vesicles are believed to be able to target the correct target membrane by virtue of specific address molecules on both the vesicle and the target organelle. These address molecules are called SNARES and there are specific classes on the vesicle and the target membranes (v- and t-SNARES; Sollner et al., 1993). The existence of this family of proteins was originally noted in yeast, and evidence for their central role in vesicle docking was obtained in yeast, biochemical and synaptic systems. The v-SNARES on the vesicle are thought to interact directly with the organelle specific t-SNARE, hence providing the specificity in vesicular transport.

Once the vesicle has docked accurately a fusion event occurs. It was originally proposed that a protein factor, NSF (NEM-sensitive fusion protein) was directly involved in fusion, but recent evidence has suggested that this protein may play a role at some other point in the cycle of docking and fusion (Morgan and Burgoyne, 1995; Mayer et al., 1996). Thus it is not currently understood how the the two membranes brought into close apposition by the SNARES go on to fuse.

After Vesicle Fusion, What Next?

Once the ER-derived transport vesicle has fused with the *cis*-Golgi a round of transport has been completed. However, the mechanism described above does not actually account for any specific sorting. When a vesicle pinches off a section of ER it will contain a mix of newly-made proteins and ER residents, and so simply transferring them to another compartment could simply move the ER along to another place in the cell. This is clearly not what happens in vivo as the ER residents are able to remain in place whilst the proteins they help fold exit the cell, or are incorporated into post-ER organelles. How this specificity in transport occurs - the protein trafficking question - has been a central problem in studies of the secretory pathway. An early clue to solving this problem came from examining the sequence of several of the seemingly soluble proteins that are able to maintain residence in the ER lumen. This revealed a conserved tetrapeptide KDEL which found at the C-terminus of these proteins (Munro and Pelham, 1987). If this sequence was transferred to a protein that would normally be secreted, then the protein instead stayed in the cell, accumulating in the ER. A series of experiments in both mammalian cells and yeast indicated that the KDEL sequence (HDEL in yeast) exerts its affect by it being used to recognise ER residents when they arrive at the *cis*-Golgi and collect them into vesicles for return to the ER (Dean and Pelham, 1990; Pelham, 1995). The receptor that binds -KDEL sequences is the product of the ERD2 gene.

These studies raised the question of what proteins are required for forming the vesicles for this "retrograde" transport back from the Golgi to the ER. The unexpected answer to this came from studies on a second Golgi to ER retrieval signal. This is the KKXX motif that is found on at the end the cytoplasmic tail of some membrane proteins that reside in the ER (such as calnexin mentioned above). Like KDEL, the KKXX sequence was shown to mediate the return of escaped ER residents back from the *cis*-Golgi to the ER (Jackson et al., 1993). A genetic screen in yeast for mutations in which this retrieval was defective produced a surprising result - mutations in a set of vesicle coat proteins called COPI are defective in this retrograde transport step (Letourneur et al., 1994). Furthermore, COPI can bind directly to KKXX signals in vitro. The COPI coat was originally identified in an in vitro biochemical system designed to study vesicular transport through the compartments of the Golgi (Orci et al., 1993; Rothman and Wieland, 1996). It had been assumed that it was involved in anterograde transport. However it now appears that COPI is the coat that forms the vesicles that pinch of the *cis*-Golgi to carry escaped ER residents back to the ER. It has been postulated that COPI may also be involved in forward transport but there is no definitive evidence for this and as discussed below the question of forward transport in the Golgi remains open.

Thus the sample of ER pinched off by the COPII coat and delivered to the *cis*-Golgi is cleaned of ER residents by signal-mediated retrograde transport. This contributes to the enormous enrichment of newly made proteins relative to ER residents seen between the ER and the Golgi apparatus. However there is recent evidence that a second mechanism is also involved in this sorting process - the selective removal of secreted proteins from the ER. Until recently it was widely believed that there was a rapid and nonspecific movement of proteins forward in the secretory pathway which obviated the need for exocytic signals - the "Bulk Flow" hypothesis (Pfeffer and Rothman, 1987). This view survived for so long because it was observed that removal of specific retrieval signals such as KDEL resulted in the secretion of proteins that were never designed for this, albeit at often rather slow rates. However various pieces of evidence are now accumulating which are inconsistent with the bulk-flow model (Balch and Farquar, 1995). Perhaps the most striking is an experiment in which the two different coat systems, COPI and COPII were used to form vesicles in vitro from the ER of yeast (Bednarek et al., 1995). It is not known if COPI normally ever binds to the ER but because its recruitment is regulated by a Sar1p-like small GTPase (called ARF), it can be forced to form vesicles on many membranes by the addition of non-hydrolysable analogs of GTP. In these conditions COPI and COPII will form vesicles from ER membranes with equal efficiency. However analysis of the content of the vesicles showed that only the vesicles produced by COPII were enriched in a newly made soluble secreted protein. This clearly suggests the existence of cargo-receptors which specifically recruit secreted proteins into vesicles leaving the ER.

Of course, the size and shape of the vesicle (a sphere c70nM in diameter) means that it is probably impossible to completely exclude all ER residents, but the retrieval sequences carried by these proteins will allow their efficient return from the *cis*-Golgi as described above. Thus an enormous enrichment is achieved between ER and Golgi by this two vesicle system. A first step going forwards, and then a switch at the Golgi when all that was specifically carried is released, and that which was non-specific contamination is now specifically retrieved. This switch in specificity is thought to be achieved, at least in part, by differing pH levels between the two compartments (Wilson and Pelham, 1994). Many of aspects of this model have yet to be fully confirmed - in particular the identity of the forward moving receptors and the signals they recognise on the cargo. However candidate molecules for these receptors have already been found such as ERGIC53 and the emp24p family and the detailed characterisation of these proteins may well confirm the demise of the bulk flow hypothesis (Schimmoller et al., 1995; Itin et al., 1996).

Transport Through the Golgi

Once newly-made secreted proteins arrive at the *cis*-Golgi and any ER escapees have been rounded up and removed, the proteins are ready for the next step of the secretory pathway: traversing the Golgi complex. The Golgi complex consists of a set of discrete compartments, or cisternae, often arranged in a stack. These cisternae are named in the order that proteins are believed to move through them - i.e. the first compartment is the *cis*-Golgi, followed by the *medial*-, the *trans*-, and then the *trans*-Golgi network or TGN which is last compartment. The Golgi contains many enzymes which carry out post-translation modification of secreted proteins and it is also where glycolipids and sphingomyelin are synthesised (Kleene and Berger, 1993; Van Meer, 1989). These modifications are mainly concerned with the building of complex carbohydrate structures on N-linked sugars, but also include sulphation and proteolysis of the secreted proteins. Finally, the last compartment of the Golgi, the TGN, is also an important junction point in the secretory pathway with vesicles leaving to either the lysosome or the plasma membrane and, in some polarised cells, specific vesicles even leaving to each of the two domains of the plasma membrane. In cells with a regulated secretory pathway, the TGN is also the point at which secretory granules are formed.

The exact structure of the Golgi is somewhat heterogeneous with the number of cisternae varying both between species, and between different cell types in the same species. The enzymes of the Golgi are usually located within the structure in the order in which they act, spread out over two or more cisterna (Roth, 1987; Rabouille et al., 1995). Even this is variable with the same enzyme being found in different locations in different tissues (Roth et al., 1986; Velasco et al., 1993).

However the biggest mystery about protein trafficking in the Golgi is how proteins traverse this structure. For many years it has been assumed that secreted proteins move from *cis* to *medial* to *trans* by the same process of vesicle budding and fusion that occurs elsewhere in the pathway. The isolation of COPI coated vesicles from Golgi membranes, and the ability of conditional mutations in yeast of the gene encoding NSF (SEC18) to block protein trafficking in different Golgi compartments, gave strong support to this view (Graham and Emr, 1991; Orci et al., 1993) . However the recently discovered role of COPI in retrograde transport, and the current uncertainty over the exact function of NSF, has once again opened the door to alternative views. Moreover, if COPI is not involved in retrograde transport, then genetics and biochemistry have so far failed to identify any candidates for a coat structure for forward moving vesicles in the Golgi itself. Of course such coats may yet appear but at present there is little to exclude a second older model for intra-Golgi transport - cisternal progression or maturation. In this model the cisternae form at the *cis* face of the Golgi by coalescence of ER-derived vesicles. These cisternae then gradually mature with new cisternae forming behind them until they reach the *trans* face where they either vesiculate completely, or fuse with preexisting TGN structures. Some vesicles would have to be moving back through the stack to return escaped ER-residents and to reposition Golgi enzymes. The advantage of this model is that accounts for the trafficking of very large structures such as algal scales, viruses and lipoproteins which can be seen in Golgis by electron microscopy but which are too large to enter transport vesicles (Becker et al., 1995). In the mammary gland casein particles can seen forming be in the *cis* Golgi, and then are present throught the stack, but are never seen in Golgi-associated vesicles (Clermont et al., 1993).

Whilst the question of intra-Golgi transport remains controversial, it's final product is clear: the arrival of newly made, and now modified, secretory proteins at the final compartment of the Golgi, the TGN. This is the end of the biosynthetic phase of protein trafficking and, as mentioned above, proteins now travel to their final destinations in the cell: the cell surface; the lysosome; or regulated secretory structures. Transport to the lysosome and to the plasma membrane is clearly mediated by vesicles. For traffic to the lysosome, clathrin-coated structures form similar to those used for endocytosis at the plasma membrane, although containing a different subset of coat proteins or "adapters" (Robinson, 1994; Hunziker and Geuze, 1996). Again this coat recognises a specific signal in the cytoplasmic tail of membrane proteins destined for the lysosome, including the mannose-6-phosphate receptors which recognise soluble lysosomal proteins marked with this modification which is added in the Golgi (Trowbridge et al., 1993). The coat for transport to the plasma membrane has not been unequivocally identified, although a candidate has recently been identified which is related to the clathrin coat adapters, although these constitutive vesicles do not use clathrin itself (Simpson et al., 1996).

Protein Sorting in the Golgi.

The existence of Golgi-specific enzymes presents the cell with a similar problem to that it faces with ER-residents - how to maintain the location of these enzymes in their compartment whilst the proteins and lipids that they modify are rapidly leaving. All Golgi enzymes examined so far are membrane proteins and it appears that two mechanisms are involved in specifying their intracellular location. For at least two proteins of the TGN, the protease furin and TGN38 (a protein of unknown function), there are retrieval signals in the cytoplasmic tails of these proteins which specify their return from the plasma membrane (Humphrey et al., 1993; Chapman and Munro, 1994). Thus it appears that, like ER proteins, they have signals that allow them to be retrieved to their correct location if they escape (Luzio and Banting 1993). However, the vast majority of Golgi enzymes have short cytoplasmic tails with no obvious signals and moreover they do not appear to be recycling through the surface (Chapman and Munro, 1994; Teasdale et al., 1994). Thus there must be a mechanism which prevents them from getting into the transport vesicles which are leaving the Golgi. Investigation of what part of the protein is required for this retention revealed for several proteins that their single membrane-spanning transmembrane domains (TMDs) are both necessary and sufficient for this effect (Munro, 1991, Nilsson et al., 1991; Machamer, 1993).

The mechanism by which TMDs are able to prevent Golgi enzymes exiting the Golgi has been the subject of debate, and two main models have been proposed. In the first the Golgi proteins form large aggregates or "kin hetero-oligomers" which are too large to enter transport vesicles (Pfeffer and Rothman, 1987; Weisz et al., 1993; Nilsson et al., 1994). In the second it is proposed that proteins segregate between lipid microdomains in the Golgi and that only the lipid domains which exclude the Golgi enzymes are used for formation of transport vesicles (Bretscher and Munro, 1993). We proposed the latter model because of the apparent lack of a specific sequence requirement for retention and because of a difference we observed in the average length of TMDs of Golgi and plasma membrane proteins (Munro, 1991; Munro, 1995a). This difference is observed in both yeast and mammalian cells. Moreover it appears that altering TMD length is sufficient to determine if a protein is localised to the Golgi or the plasma membrane of mammalian cells (Munro, 1995b). Thus the lipid sorting model would require that the domains that form in the Golgi are of a different thickness, and that the thicker lipid moves forward. This is consistent with the increase in cholesterol and sphingolipid content of the membrane that occurs in the TGN with these lipids being scarce in the ER and highly enriched in the plasma membrane (Orci et al., 1981; Van Meer, 1989). It is also known from in vitro studies that these lipids thicken the lipid bilayer and it is thought that they are enriched in the plasma membrane to ensure that this bilayer is a more effective barrier (Grover et al., 1968; Sankaram and Thompson, 1990). Although there is no direct evidence to confirm the existence of these lipid domains in the TGN, a recent study using photobleaching has shown that Golgi enzymes are extremely mobile within the bilayer which is more consistent with them being excluded from vesicles by a means which does not involve them forming large aggregates (Cole et al., 1996).

Summary

Thus the secretory pathway represents a specialised site of synthesis of secreted proteins, topologically equivalent to the cell exterior but allowing a distinct contained environment for folding (the ER), followed by a sorting cycle to extract the newly made proteins from their chaperones (ER to Golgi transport). This is then followed by a bilayer transition in the Golgi to allow the ER and plasma membrane to adopt and maintain very different compositions, followed by arrival at a final sorting compartment, the TGN, from which matured proteins, and plasma membrane compatible lipid bilayers, are sorted to their final destinations.

References

Aridor M, Bannykh SI, Rowe T, Balch WE (1995) Sequential coupling between COPII and COPI vesicle coats in endoplasmic-reticulum to Golgi transport. J Cell Biol 131: 875-893

Balch WE, Farquhar MG (1995) Beyond bulk flow. Trends Cell Biol 5 :16-19

Becker B, Bolinger B, Melkonian M (1995) Anterograde transport of algal scales through the Goli complex is not mediated by vesicles. Trends Cell Biol 5: 305-306

Bednarek SY, Ravazzola M, Hosobuchi M, Amherdt M, Perrelet A, Schekman R, Orci L (1995) COPI-coated and COPII-coated vesicles bud directly from the endoplasmic-reticulum in yeast. Cell 83: 1183-1196

Bergeron JJM, Brenner MB, Thomas DY, Williams DB (1994) Calnexin - a membrane-bound chaperone of the endoplasmic reticulum. Trends Biochem Sci 19: 124-128

Bioessays 18: 379-389

Bretscher MSB, Munro S (1993) Cholesterol and the Golgi apparatus. Science 261: 1280-1281

Chapman RE, Munro S (1994) Retrieval of TGN proteins from the cell surface requires endosomal acidification. EMBO J 13 2305-2312

Clermont Y, Xia L, Rambourg A, Turner JD, Hermo L (1993) Transport of caesin submicelles and formation of secretion granules in the Golgi appartus of epithelial cells of the lactating mammary gland of the rat. Anat Record 235: 363-373

Cole NB, Smith CL, Sciaky N, Terasaki M, Edidin M, Lippincott-Schwartz J (1996) Diffusional mobility of Golgi proteins in membranes of living cells. Science 273 797-800

Dean N, Pelham HRB (1990) Recycling of proteins from the Golgi compartment to the ER in yeast. J Cell Biol 111 369-377

Gahmberg CG, Tolvanen M (1996) Why mammalian cell surface proteins are glycosylated. Trends Biochem Sci 21: 308-311

Graham TR, Emr SD (1991) Compartmental organisation of Golgi-specific protein modification and vacuolar sorting events defined in a yeast sec18 (NSF) mutant. J Cell Biol 114: 207-218

Grove SN, Bracker CE, Morre DJ (1968) Cytomembrane differentiation in the endoplasmic reticulum-Golgi apparatus-vesicle complex. Science 101: 171-173

Hartl FU, Hlodan R, Langer T (1994) Molecular Chaperones in Protein-Folding. Trends Biochem Sci 19: 20-25

Humphrey JS, Peters PJ, Yuan LC, Bonifacino JS (1993) Localization of TGN38 to the trans-Golgi network: involvement of a cytoplasmic tyrosine-containing sequence. J Cell Biol 120: 1123-1135

Hunziker W, Geuze HJ (1996) Intracellular trafficking of lysosomal membrane-proteins.

Itin C, Roche AC, Monsigny M, Hauri HP (1996) ERGIC-53 is a functional mannose-selective and calcium-dependent human homolog of leguminous lectins. Mol Biol Cell 7: 483-493

Jackson MR, Nilsson T, Peterson PA (1993) Retrieval of transmembrane proteins to the endoplasmic reticulum. J Cell Biol 121: 317-333

Kleene R, Berger EG (1993) The molecular and cell biology of glycosyltransferases. Biochim Biophys Acta 1154: 283-325

Letourneur F, Gaynor EC, Hennecke S, Demolliere C, Duden R, Emr SD, Riezman H, Cosson P (1994) Coatomer is essential for retrieval of dilysine-tagged proteins the endoplasmic-reticulum. Cell 79: 1199-1207

Luzio JP, Banting G (1993) Eukaryotic membrane traffic - retrieval and retention mechanisms to achieve organelle residence. Trends Biochem Sci 18: 395-398

Machamer CE (1993) Targeting and retention of Golgi membrane proteins. Current Opin Cell Biol 5: 606-612

Mayer A, Wickner W, Haas A (1996) Sec18p (NSF)-driven release of Sec17p (alpha-SNAP) can precede docking and fusion of yeast vacuoles. Cell 85: 83-94

Morgan A, Burgoyne RD (1995) Is NSF a fusion protein? Trends Cell Biol 5: 335-339

Munro S (1991) Sequences within and adjacent to the transmembrane segment of a-2,6-sialyltransferase specify Golgi retention. EMBO J 10: 3577-3588

Munro S (1995a) A comparison of the transmembrane domains of Golgi and plasma membrane proteins. Biochem Soc Trans 23: 527-530

Munro S (1995b) An investigation of the role of transmembrane domains in Golgi protein retention. EMBO J 14: 4695-4704

Munro S, Pelham HRB (1987) A C-terminal signal prevents secretion of luminal ER proteins. Cell 48: 899-907

Nilsson T, Hoe MH, Slusarewicz P, Rabouille C, Watson R, Hunte F, Watzele G, Berger EG Warren G (1994) Kin recognition between *medial* Golgi enzymes in HeLa cells. EMBO J 13: 562-574

Nilsson T, Lucocq JM, Mackay D, Warren G (1991) The membrane spanning domain of b-1,4-galactosyltransferase specifies *trans*Golgi localization. EMBO J 10: 3567-3575

Nilsson T, Slusarewicz P, Hoe ME, Warren G (1993) Kin recognition: a model for the retention of Golgi enzymes. FEBS Letts 330: 1-4

Orci L, Montesano R, Meda P, Malaisse-Lagae F, Brown D, Perrelet A, Vassalli P (1981) Heterogeneous distribution of filipin-cholesterol complexes across the cisternae of the Golgi apparatus. Proc Natl Acad Sci USA 78: 293-297

Orci L, Palmer DJ, Amherdt M, Rothman JE (1993) Coated vesicle assembly in the Golgi requires only coatomer and ARF proteins from the cytosol. Nature 364: 732-734

Pelham HRB (1995) Sorting and retrieval between the endoplasmic-reticulum and Golgi-apparatus. Curr Opin Cell Biol 7: 530-535

Pelham HRB, Munro S (1993) Sorting of membrane proteins in the secretory pathway Cell 75: 603-605

Pfeffer S, Rothman JE (1987) Biosynthetic protein transport and sorting by the endoplasmic reticulum and Golgi. Ann Rev Biochem 56: 829-852

Rabouille C, Hui N, Hunte F, Kieckbusch R, Berger EG, Warren G, Nilsson T (1995) Mapping the distribution of Golgi enzymes involved in the construction of complex oligosaccharides. J Cell Sci 108: 1617-1627

Robinson MS (1994) The role of clathrin, adapters and dynamin in endocytosis. Curr Opin Cell Biol 6: 538-544

Roth J (1987) Subcellular organization of glycosylation in mammalian cells. Biochim Biophys Acta 906: 405-436

Roth J, Taatjes DJ, Weinstein J, Paulson JC, Greenwell P, Watkins WM (1986) Differential subcompartmentation of terminal glycosylation in the Golgi apparatus of intestinal absorptive and goblet cells. J Biol Chem 261: 14307-14312

Rothman JE, Wieland FT (1996) Protein sorting by transport vesicles. Science 272: 227-234

Sankaram MB, Thompson TE (1990) Modulation of phospholipid acyl chain order by cholesterol. A solid state 2H nuclear magnetic resonance study. Biochemistry 29: 10676-10684

Schekman R, Orci L (1996) Coat proteins and vesicle budding. Science 271: 1526-1533

Schimmöller F, Singer-Krüger B, Schröder S, Krüger U, Barlowe C, Riezman H (1995) The absence of Emp24p, a component of ER-derived COPII-coated vesicles, causes a defect in transport of selected proteins to the Golgi. EMBO J 14: 1329-1339

Simpson F, Bright NA, West MA, Newman LS, Darnell RB, Robinson MS (1996) A Novel Adapter-Related Protein Complex. J Cell Biol 133: 749-760

Sollner T, Whiteheart SW, Brunner M, Erdjument-Bromage H, Geromanos S, Tempst P, Rothman JE (1993) SNAP receptors implicated in vesicle targeting and fusion. Nature 362: 318-324

Teasdale RD, Matheson F, Gleeson PA (1994) Post-translational modifications distinguish cell surface from Golgi-retained b1,4 galactosyltransferase molecules. Golgi localization involves active retention. Glycobiology 4: 917-928

Trowbridge IS, Collawn JF, Hopkins CR (1993) Signal-dependent membrane-protein trafficking in the endocytic pathway. Ann Rev Cell Biol 9: 129-161

Van Meer G (1989) Lipid traffic in mammalian cells. Ann Rev Cell Biol 5: 247-275

Velasco A, Hendricks L, Moremen KW, Tulsiani DRP, Touster O, Farquhar MG (1993) Cell-type dependent variations in the subcellular-distribution of a-mannosidase-I and II. J Cell Biol 122: 39-51

Walter P, Johnson AE (1994) Signal sequence recognition and protein targeting to the endoplasmic reticulum membrane. Ann Rev Cell Biol 10: 87-119

Weisz OA, Swift AM, Machamer CE (1993) Oligomerization of a membrane protein correlates with its retention in the Golgi complex. J Cell Biol 122: 1185-1196

Wilson DW, Lewis MJ, Pelham HRB (1993) PH-dependent binding of KDEL to its receptor in vitro. J Biol Chem 268: 7465-7468

The roles of PI3Ks in cellular regulation

A. Eguinoa, S. Krugmann, J. Coadwell, L. Stephens and P. Hawkins
Signalling Department
The Babraham Institute
Babraham
Cambridge CB2 4AT

Definition of PI3Ks

The term phosphoinositide 3OH kinase (PI3K) is given to a family of enzymes which can phosphorylate one or more of the conventional inositol phospholipids found in cells in the 3-position of their inositol headgroup (Fig. 1). It is now clear that these lipids act as regulators of intracellular metabolism and at least one of them, PtdIns(3,4,5)P_3, shows all the credentials of being a major 'second-messenger' in signalling pathways used by cell-surface receptors for growth factors, inflammatory stimuli and antigens (Stephens *et al.*, 1993; Cantley *et al.*, 1991).

(A)	PtdIns	→	PtdIns3P
	PtdIns4P	→	PtdIns(3,4)P_2
	PtdIns(4,5)P_2	→	PtdIns(3,4,5)P_3

(B)

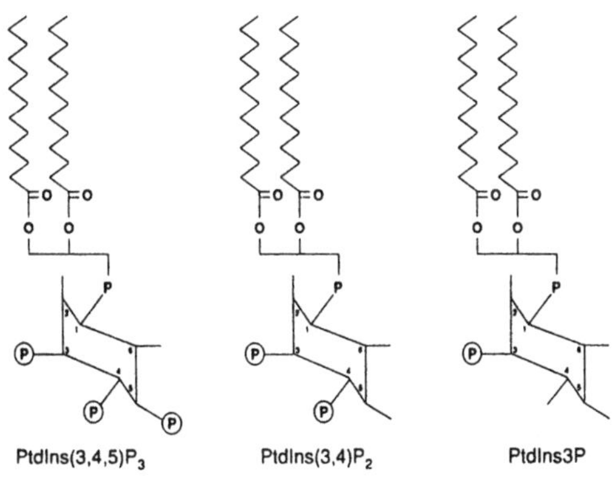

Fig. 1 (A) Reactions catalysed by PI3Ks. (B) Structures of 3-phosphorylated inositol lipids.

NATO ASI Series, Vol. H 101
Molecular Mechanisms of Signalling
and Membrane Transport
Edited by Karel W. A. Wirtz

The family of PI3Ks

Since the original discovery of a PI3K activity which could be pulled out of cells tightly associated with oncogenic protein tyrosine kinases (Whitman *et al.*, 1988) there has been ever increasing interest in characterising PI3K activities in cells and deducing their involvement in cellular regulation. Thus we now have a substantial body of information about the substrate specificities of these enzymes and how some of them may be controlled by cell-surface receptors. The current situation however is far from complete and, in particular, the current speed with which DNA homology-based cloning strategies are adding to the list of PI3K-like sequences is greater than the accumulation of biochemical and physiological information about the roles they may play in cells.

PI3Ks can be subdivided on the basis of their substrate-specificity *in vitro* and *in vivo* and the type of regulation they receive (Table 1). Agonist-sensitive PI3Ks appear to use all three potential substrates *in vitro* (with a small preference for PtdIns(4,5)P_2) but are relatively specific for PtdIns(4,5)P_2 *in vivo* (see next section). PI3Ks apparently unlinked to receptor activation are relatively specific for PtdIns *in vitro*. This functional classification is strengthened by a comparison of available cDNA sequences for putative PI3Ks. A simple analysis of similarity between their catalytic domains shows they segregate across a very wide range of species according to their proposed differences in function (Fig. 2).

Table 1 Classification of PI3Ks (functional)

Substrate specificity	Regulation	Identity	Basis of information	References
In vitro: PtdIns PtdIns4P PtdIns(4,5)P_2 (best) *In vivo*: PtdIns(4,5)P_2 (PtdIns4P?)	Activation by receptor-driven protein tyrosine kinases	Heterodimers of p85/p110 - p85α/β isoforms - p50-55 splice variants and isoforms of p85 - p110α/β isoforms	Purified protein cDNA cloning and expression	Otsu *et al* (1991) Skolnik *et al* (1991) Escobedo *et al* (1991) Hiles *et al* (1992) Pons *et al* (1995) Hu *et al* (1993) Antonetti *et al* (1996) Inukai *et al* 1996)
	Activation by receptor-driven formation of Gβγ	Heterodimers of p101/p110γ	Purified protein cDNA cloning and expression	Stephens *et al* (1994) Stephens *et al* (1996) Stoyanov *et al* (1995)
In vitro: PtdIns *In vivo*: PtdIns	Upstream protein kinases?	Approx. 100 kD PI3K catalytic subunits (homologues of VPS34) - probably in complex with approx 150 kD protein kinases (VPS15)	Partial protein purification cDNA cloning and expression Gene knockout in yeast	Shu *et al* (1993) Stephens *et al* (1994) Volinia *et al* (1995)
In vitro: PtdIns (best) PtdIns4P *In vivo*:?	?	Approx 200 kD PI3K catalytic subunits with characteristic C-terminal C2 domain	cDNA cloning and expression	MacDougall *et al* (1995) Virbasius *et al* (1996) Molz *et al* (1996)

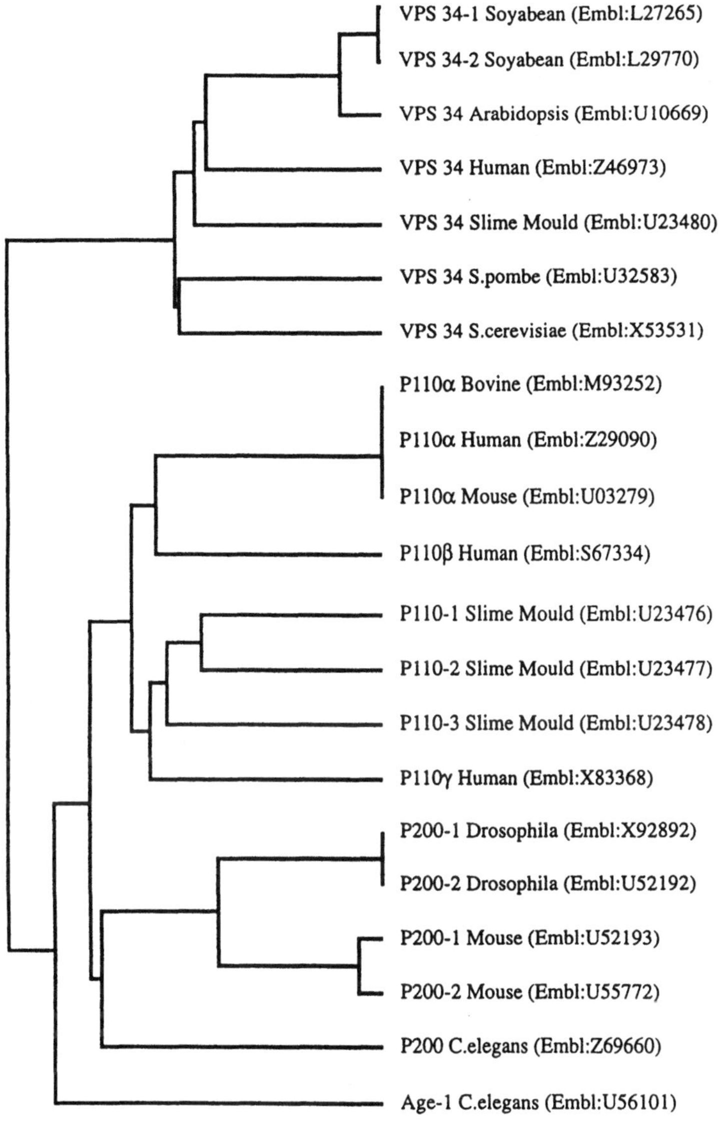

Fig. 2 Comparison of amino acid sequences of PI3K catalytic domains (programme Pileup; accession nos. shown in parentheses). The entries are named after the prototypic member of each group and a species abbreviation. A 'loop' of 24 amino acids was removed from p110 *C.Elegans* sequence for this comparison.

The most intensively studied group of PI3Ks are those regulated by protein tyrosine kinases. The prototypic member of this group was purified and cloned on the basis of its translocation to activated growth factor receptors. This PI3K is a heterodimer of p85 regulatory and p110 catalytic subunits (Hiles *et al.*, 1992). The cDNA encoding the p110 subunit showed unexpected homology with a yeast vacuolar sorting mutant, VPS34. This led to the discovery that VPS34 is indeed a PI3K but that it is PtdIns-specific *in vitro* and responsible for the synthesis of the high steady-state levels of PtdIns3P found in *S.cerevisiae* (Schu *et al.*, 1993). The cDNAs encoding p110 and VPS34 then represented the start point for discovering homologous cDNAs in a variety of species (Fig. 2). In parallel, a mammalian PI3K which is activated by heterotrimeric G-protein subunits has also been characterised (Stephens *et al.*, 1994).

An interesting finding to emerge from the purification and cloning of the original p110 PI3K is that this enzyme can also use its lipid catalytic site to phosphorylate a serine residue on the p85 regulatory subunit (Dhand *et al.*, 1994), an event which appears to inhibit the enzyme. Whether this protein kinase-property of PI3K will have roles outside of the internal regulation of these enzymes, or indeed to what extent it is shared by other PI3Ks, has yet to be investigated.

Agonist-stimulated synthesis of 3-phosphorylated inositol lipids in mammalian cells

PtdIns(3,4,5)P_3 was discovered independently, and in parallel, to the discovery of PI3K as a novel, highly polar lipid which was synthesised in neutrophils in response to stimulation with the chemotactic peptide FMLP (Traynor-Kaplan *et al.*, 1989). Since then, changes in the levels of 3-phosphorylated inositol lipids have been documented in response to a number of agonists which activate cell surface receptors (some examples are given in Table 2). The basic pattern of changes seen is remarkably similar across different combinations of both tissues and agonists (an example is shown in Fig. 3). Unstimulated cells contain significant quantities of PtdIns3P (in the range 2-10% of the levels of PtdIns4P) whose levels do not change substantially on appropriate stimulation. In contrast, the levels of PtdIns(3,4)P_2 and PtdIns(3,4,5)P_3 are usually close to undetectable in unstimulated cells but their levels rise dramatically on stimulation (Auger *et al.*, 1989; Stephens *et al.*, 1991).

Table 2 Examples of agonists which activate PI3K (see Stephens *et al.*, 1993 for refs.)

Activation via protein tyrosine kinase domains present within the 1° structure of the receptor	
PDGF	FGF
EGF	CSF-1
Insulin	SL/CSF
IGF-1	HGF/Scatter factor
Activation via non-receptor protein tyrosine kinases (e.g. members of the src-family)	
GM-CSF	CD3/TcR
IL-3	Growth Hormone
IL-4	Erythropoetin
mIgM	Thrombin
CD4	FMLP
CD2	(Oncogenes v-src, v-yes, v-fyn)
	(Middle T-transformation)
Activation via heterotrimeric G-proteins	
FMLP	PAF
ATP	LTB4
Histamine	LTD4
Thrombin	LTE4

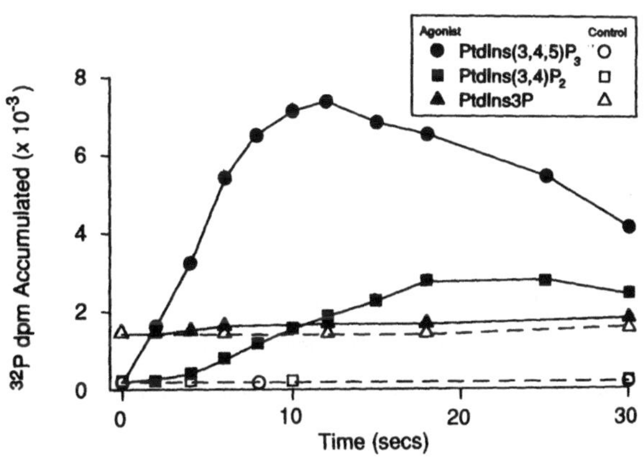

Fig. 3 Time course of changes in the levels of 3-phosphorylated inositol lipids in human neutrophils stimulated with FMLP (see Stephens *et al.*, 1993b).

A combination of techniques have been used to decipher the pathway by which these changes occur, including measuring changes in the levels of the 3-phosphorylated inositol lipids over short periods of stimulation, a [^{32}P]Pi-radiotracer analysis of the order in which individual phosphate groups are added to the inositol ring and an investigation into the nature of phosphatase activities in cells capable of degrading 3-phosphorylated inositol lipids (Stephens *et al.*, 1991; Hawkins *et al.*, 1992). These studies suggested the pathways outlined in Fig. 4. The rapid rise in the levels of PtdIns(3,4,5)P_3 is driven by an agonist-activated PtdIns(4,5)P_2-specific PI3K and the lagged production of PtdIns(3,4)P_2 is a consequence of active 5-phosphomonesterase activity on the emerging pool of PtdIns(3,4,5)P_3 (though some synthesis of PtdIns(3,4)P_2 via direct action of a PI3K on PtdIns4P cannot be excluded). The pool of PtdIns3P found in unstimulated cells is envisaged to be synthesised via the action of independent, agonist-insensitive PtdIns-specific PI3Ks (Stephens *et al.*, 1994b). These conclusions fit very nicely with what is now known about the properties of PI3K activities (see Table 1 and Fig. 2).

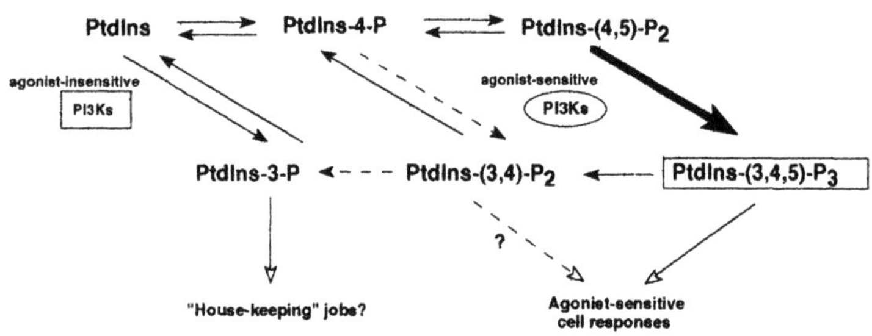

Fig. 4 Pattern of changes in 3-phosphorylated inositol lipids on agonist stimulation (see Stephens *et al.*, 1993).

The low levels, rapid turnover, and the precedent set by the established 'second messenger' Ins(1,4,5)P_3 all suggested that 3-phosphorylated inositol lipids were likely to act as allosteric regulators of metabolism and, in particular, PtdIns(3,4,5)P_3 appeared to be a good candidate for a new membrane-captive 'second messenger' used by a wide variety of growth factors and other stimuli (Auger *et al.*, 1989; Traynor-Kaplan *et al.*, 1989; Stephens *et al.*, 1991).

Mechanisms by which agonists regulate PI3Ks

A substantial body of work has now provided a foundation for understanding how the p85/p110 family of PI3Ks are activated by protein tyrosine kinases (e.g. Schlessinger & Ullrich

1992; see Fig. 5). Activation of protein tyrosine kinases by agonists leads to the phosphorylation of specific tyrosines in receptor tails or associated proteins. These phosphotyrosines, by virtue of their local amino acid sequence, act as specific targets for SH2-domain containing signalling proteins. The SH2-domains within the p85 regulatory subunit of PI3K have been shown to have preference for phosphotyrosines within YXXM motifs (Songyang *et al.*, 1993; Cantley *et al.*, 1991). The docking of PI3K onto activating phosphotyrosines is thought to activate the lipid catalytic activity of PI3K by a combination of translocation to the membrane interface (where its substrate is present) and an allosteric effect of SH2 domain docking (Carpenter *et al.*, 1993; Backer *et al.*, 1992). Recently it has also been shown that the small G-protein ras is able to activate this form of PI3K (Rodriguez-Viciana *et al.*, 1994). Ras is able to bind directly to the N-terminal half of p110 and this binding has been shown to activate PI3K catalytic activity, particularly in the context of SH2-domain docking to phosphotyrosines (Rodriguez-Viciana *et al.*, 1996). Given the large number of agonists able to activate ras and the importance of ras in controlling cell growth and transformation, ras-modulation of the PI3K signalling system may turn out to be an important and widespread phenomenon.

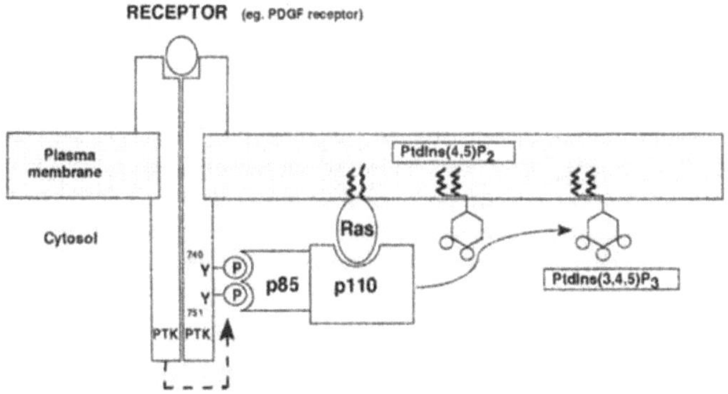

Fig. 5 Activation of PI3K by protein tyrosine kinase driven mechanisms. The example given is based on activation by a receptor with an intrinsic protein tyrosine kinase which autophosphorylates to provide PI3K-docking sites. There are many examples however where the protein tyrosine kinase(s) is activated indirectly by receptors (e.g. members of the src-family and cytokine receptors) and the docking site is not the receptor but a separate protein (e.g. IRS-1 in the case of insulin signalling). Ras is drawn in the model as a component of the activation mechanism that is required for full activation of PI3K - ras in turn is under the control of cell surface receptors which in principle may or may not be the same as those driving the tyrosine-kinase arm of the activation mechanism.

Although activation of PI3Ks by protein tyrosine kinases is a more widespread and well studied pathway, the most intense bursts of agonist-stimulated PtdIns(3,4,5)P_3-synthesis are seen in cells of myeloid origin, in response to agonists which use heterotrimeric G-proteins to access signalling pathways (e.g. Stephens *et al.*, 1993b). Recently a new form of PI3K has been purified which can be activated directly by G-protein βγ subunits (Stephens *et al.*, 1994; Thomason *et al.*, 1994). This isoform is comprised of a p110-like catalytic subunit (110γ; Stoyanov *et al.*, 1995) and a novel regulatory subunit, p101, with no homology to p85 (Stephens *et al.*, 1996). It seems likely that this isoform is structurally adapted to receive regulatory inputs from heterotrimeric G-proteins (see Fig. 6).

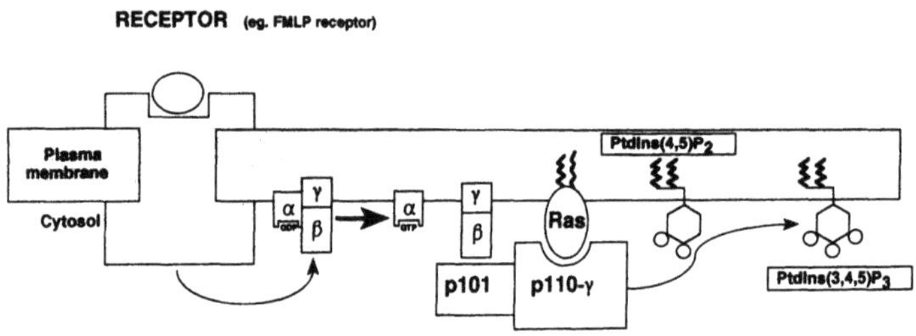

Fig. 6 Activation of PI3K by heterotrimeric G-proteins. Gβγ subunits are known to activate a p101/p110γ PI3K directly, but it is not known which part of the complex actually binds the βγ-subunits. Ras is drawn in the model performing a similar role to that described for the p85/p110 PI3Ks on the basis of a region of homology between the N-terminus of p110γ and the ras-binding site on p110α (this interaction has yet to be shown to operate in cells).

The function of agonist-stimulated synthesis of PtdIns(3,4,5)P_3

Within the last two years a number of strategies have become available to evaluate the role of agonist-stimulated PI3Ks in cellular regulation. Activation of PI3Ks has been inhibited by use of the relatively specific lipid catalytic site inhibitors wortmannin (Arcaro & Wymann, 1993; Yano *et al.*, 1993) and LY294002 (Vlahos *et al.*, 1994). Further, activation of PI3Ks by protein tyrosine kinase-driven mechanisms has been inhibited by the use of site-directed receptor mutants (Fantl *et al.*, 1992; Kashishian *et al.*, 1992), phosphopeptides mimicking the docking sites on receptors (Kotani *et al.*, 1994) and a mutant PI3K regulatory subunit (Δp85) that can dock with activating phosphotyrosines but cannot bind the p110 catalytic subunit (Wennström *et al.*, 1994); these strategies are summarised in Fig. 7.

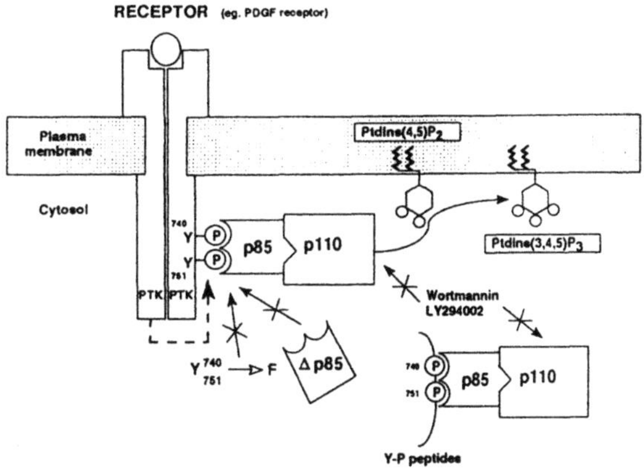

Fig. 7 Schematic diagram of the mechanisms which have been used to inhibit agonist-stimulation of PI3K.

In the last few months a number of groups have also provided strategies for creating highly active p110 PI3Ks. This has been achieved by targeting PI3Ks to the membrane or by mutation of either the ras-binding domain or the p85-p110 interaction domain (M. Thelen, S. Brasselmann, personal communications; Klippel *et al.*, 1996; Rodriguez-Viciana *et al.*, 1996): these are summarised in Fig. 8.

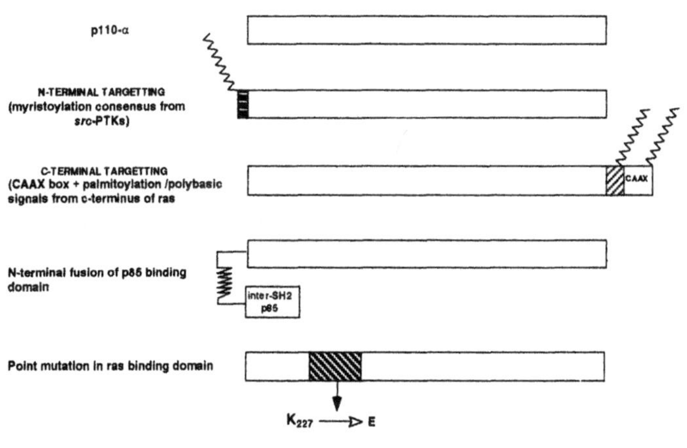

Fig. 8 Strategies used to construct cDNAs encoding PI3Ks of high basal lipid catalytic activity.

Using these methods a number of important cellular responses have been identified as depending upon receptor driven activation of PI3Ks (Table 3). The hierarchy of the PI3K signalling web and the molecular details by which PtdIns(3,4,5)P_3 (and perhaps PtdIns(3,4)P_2) sets it in motion are still far from understood. There is good evidence that the small G-protein rac and the protein kinase PKB/AKT are key intermediaries relatively high up in these signalling pathways since they have been shown to be important for co-ordinating a number of PI3K-dependent responses. Rac is thought to be involved in cytoskeletal changes (Ridley *et al.*, 1992), superoxide formation (Segal & Abo, 1993), activation of $p70^{s6k}$ (Chou & Blenis, 1996), mast cell secretion (Price *et al.*, 1995), and cell growth (Qiu *et al.*, 1995). PKB is thought to control the activation of $p70^{s6k}$ (Burgering & Coffer, 1995) and glycogen synthase kinase-3 (Cross *et al.*, 1995). Clearly it will be important to establish the relationship between rac and PKB/AKT.

Table 3 Cellular responses downstream of PI3K-activation by cell-surface receptors. Where links between components of this signalling pathway have been shown, they are indicated by arrows

PI3K →

GTP-rac	Hawkins *et al* (1995)
PKB-AKT	Burgering & Coffer (1995) Franke *et al* (1995)
PKC isoforms (novel, atypical)	Nakanishi *et al* (1993) Akimoto *et al* (1996) Toker *et al* (1994

membrane ruffling/cell movement	Wennström *et al* (1994) Kotani *et al* (1994) Klundra *et al* (1994)
mast cell secretion	Yano *et al* (1993)
superoxide formation	Arcaro & Wymann (1993)
cell division	Fantl *et al* (1992) Valius & Kazlauskas (1993)
$p70^{s6k}$	Cheatham *et al* (1994) Chung *et al* (1994)
GSK-3	Cross *et al* (1994) Shepherd *et al* (1995) Yamamoto-Honda *et al* (1995)
ras/MAPK/c-fos	Cross *et al* (1994) Yamauchi *et al* (1993) Hu *et al* (1995)
pleckstrin phosphorylation	Zhang *et al* (1995)
activation of integrins	Zhang *et al* (1996)
neurite outgrowth	Kimura *et al* (1994) Jackson *et al* (1996)
receptor endocytosis	Joly *et al* (1994)
glucose transport	Hara *et al* (1994) Okada *et al* (1994)
phospholipase D	Bonser *et al* (1991)

A number of proteins have been suggested to bind PtdIns(3,4,5)P_3 (and usually PtdIns(3,4)P_2) directly, such as certain PKC isoforms (Nakanishi *et al.*, 1993; Palmer *et al.*, 1995;

Toker *et al.*, 1994), PKB/AKT (Franke *et al.*, 1995; James *et al.*, 1996) and some SH2-domain containing proteins (Rameh *et al.*, 1995). In the case of PKB these lipids are predicted to bind via a PH homology domain (a domain known to bind inositol phospholipids in other proteins (Shaw, 1996)) and given the number of relevant signalling proteins which possess PH homology domains (e.g. exchangers for small G-proteins) it has been suggested that interaction with PH domains may represent the mechanism by which the lipid products of PI3Ks regulate the activity of other proteins (Franke *et al.*, 1995). However in each of these cases, the demonstrated specificity of lipid binding (i.e. discrimination between inositol lipids) and the magnitude of their effects on the target protein's activity, together with the very plausible argument that PH domains may be present simply as a mechanism for membrane localisation, mean that there is no really compelling case yet for the direct signalling targets of PtdIns(3,4,5)P_3.

The VPS34 phenotype in *S.cerevisiae* is a defect in vacuolar protein sorting. Analogies can be made between movement of specific vesicle populations in yeast and some of the agonist-sensitive responses listed in Table 3 (e.g. movement of glucose-transporter vesicles to the plasma membrane), suggesting that perhaps the mechanism of action of different PI3K lipid products may share some common features (related perhaps to the idea that as membrane-captive molecules they may be involved in processes which are membrane-domain relevant - in a spatial sense - such as vesicle budding and targeting). Given the widely accepted importance of small G-proteins in all aspects of vesicle management this idea can be extended to suggest that PI3K lipid products are common regulators of the function of certain small G-proteins, responsible for targeting their activation to membrane domains in which they are formed (Stephens, 1995).

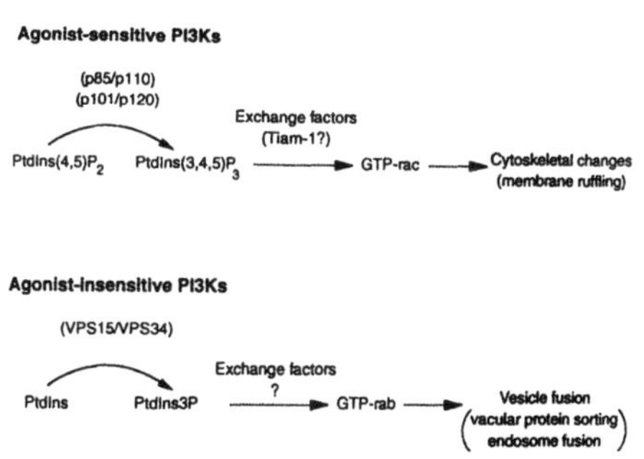

Fig. 9 PI3Ks may be upstream regulators of small GTPases. This idea is based on data suggesting that agonist-stimulated PI3Ks regulate the GTP-loading of rac (Hawkins *et al.*, 1995) and that a PI3K activity can regulate endosome fusion via putative GTP-loading of a rab (Li *et al.*, 1995), together with the discovery that a PtdIns-specific PI3K (VPS34) is on a vacuolar-sorting pathway together with a rab protein (VPS21) in yeast (e.g. Shu *et al.*, 1993).

References

Arcaro A, Wymann MP (1993) Biochem J 296:297-301
Akimoto K, Takahashi , Moriya S, Nishioka N *et al.* (1996) EMBO J 15:788-798
Antonetti, Algenstaedt P, Kahn CR (1996) Mol Cell Biol 16:2195-2203
Auger KR, Serunian LA, Soltoff SP, Libby P, Cantley LC (1989) Cell 57:167-175
Backer JM, Myers MG, Shelson SE, Chin DJ, Sun X-J, Miralpeix M, Hu P, Margolis B, Skolnik Y, Schlessinger J, White MF (1992) EMBO 11:3469-3479
Bonser RW, Thompson NT, Randall RW, Tateson JE, Spacey GD, Hodson HF,Garland LG (1991) Br. J. Pharmacol. 103:1237-1241
Burgering BMT, Coffer PJ (1995) Nature 376:599-602
Cantley LC, Auger KR, Carpenter C, Duckworth B, Graziani A, Kapeller R, Soltoff S (1991) Cell 64:281-302
Carpenter CL, Auger KR, Chaudhuri M, Yoakim M, Schaffhausen B, Schoelson S, Cantley LC (1993) J Biol Chem 268:9478-9483
Cheatham B, Vlahos CJ, Cheatham L, Wang L, Blenis J, Kahn CR (1994) Mol Cell Biol 14:4902-4911
Chou MM, Blenis J (1996) Cell 85:573-583
Chung J, Grammar TC, Lemon KP, Kazlanskas A, Blenis J (1994) Nature 370:71-75
Cross DAE, Alessi DR, Vandenheede JR, McDowell HE, Hundal HS, Cohen P (1994) Biochem J 303:21-26
Cross DAE, Alessi DR, Cohen P, Andjelkonich M, Hemmings BGA (1995) Nature 378:785-789
Dhand R, Hiles I, Panayotou G, Roche S, Fry MJ, Gout I *et al.* (1994) EMBO J 13:522-533
Escobedo J, Navankasattusas S, Kavanaugh WN, Milfay D, Fried VA, Williams LT (1991) Cell 65:75-82
Fantl W, Escobedo JA, Martin GA, Turck CW, del Rosario M, McCormick F *et al.* (1992) Cell 69:413-423
Franke TF, Yang S-I, Chan TO, Datta K, Kazlauskas A, Morrison DK, Kaplan DR, Tsichlis PN (1995) Cell 81:727-736
Hara K, Yonezawa K, Sakane H, Ando A, Kotanyk *et al.* (1994) Proc Natl Acad Sci 91:7415-7419
Hawkins PT, Jackson TR, Stephens LR (1992) Nature 358:157-159
Hawkins PT, Eguinoa A, Qiu R-G, Stokoe D, Cooke FT, Walters R, Wennström S, Claesson-Welsh L, Evans T, Symons M, Stephens L (1995) Current Biology 5:393-403
Hiles I, Otsu M, Volinia S, Fry MJ, Gout I, Dhand R *et al.* (1992) Cell 70:419-429
Hu P, Mondino A, Skolnik EY , Schlessinger J (1993) Mol Cell Biol 13:7677-7688
Inukai K, Anai M, Van Breda EV, Hosaka T, Katagiri H, Funaki M *et al* (1996) J Biol chem 271:5317-5320
Jackson TR, Blader IJ, Hammonds-Odie LP, Bunga CR, Cooke F *et al* (1996) J Cell Science 109:289-300
James SR, Downes CP, Gigg R, Grove SJA, Holmes AB, Alessi DR (1996) Biochim J 315:709-713
Joly M, Kazlauskas A, Fay FS, Corvera S (1994) Science 263:684-687
Kashishian A, Kazlauskas A, Cooper JA (1992) EMBO J 11:1373-1382
Kimura K, Hattori S, Kakuyama Y, Shizawa Y, Takayanugi J *et al* (1994) J Biol Chem 269:18961-18967.
Klippel A, Reinhard CW, Kavanaugh M, Apell G, Escobedo MA, Williams LT (1996) Mol Cell Biol 16:4117-4127

Kotani K, Yonezawa K, Hara K, Ueda H, Kitamura Y, Sakaune H *et al.* (1994) EMBO J 13:2313-2321
Kundra V, Escobedo J, Kazlauskas H, Lin K, Phee SG, Williams L, Zetter BR (1994) Nature 367:474-476
Li G, D'Souza-Schorey, Barbieri MA, Roberts RL, Klippel A, Williams L, Stahl PD (1995) Proc Natl Acad Sci 92:10207-10211
MacDougall LK, Domin J, Waterfield MD (1995) Current Biology 5:1405-1414
Molz L, Chen Y-W, Hirano M, Williams LT (1996) J Biol Chem 271: 13892-13899
Nakanishi H, Brewer KA, Exton JH (1993) J Biol Chem 268:13-16
Okada T, Kawano Y, Sakakibara T, Hazeki O, Ui M (1994) J Biol Chem 269:3568-3573
Otsu M, Hiles I, Gout I, Fry MJ, Ruiz-Larrea F, Panayotou G *et al.* (1991) Cell 65:91-104
Palmer RH, Dekker LV, Woscholski R, LeGood A, Gigg R, Parker PJ (1995) J Biol Chem 270:22412-22416
Pons S, Asano T, Glasheen E, Miralpeix M, Zhang Y, Fisher TL *et al.* (1995) Mol Cell Biol 15:4453-4465
Price LS, Norman JC, Ridley AJ, Koffer A (1995) Current Biology 5:68-73
Qiu RG, Chen J, Kirn D, McCormick F, Symons M (1995) Nature 374:457-459
Rameh LE, Chen CS, Cantley LC (1995) Cell 83:821-830
Ridley A, Paterson HF, Johnston CL, Diekman D, Hall A (1992) Cell 70:401-410
Rodriguez-Viciana P, Warne PH, Dhand R, Vanhaesebroeck B, Gout I, Fry MJ, Waterfield M, Downward J (1994) Nature 370:527-532
Rodriguez-Viciana P, Warne PH, Vanhaesebroeck B, Waterfield MD, Downward J (1996) EMBO J 15:2442-2451
Schlessinger J, Ullrich A (1992) Neuron 9:383-391
Schu PV, Takegawa K, Fry M, Stack JH, Waterfield MD, Emr SD (1993) Science 260:88-91
Segal AW, Abo A (1993) Trend Biochem Sci 18:43-47
Shaw (1996) BioEssays 18:35-46
Shepherd PR, Navé BT, Siddle K (1995) Biochem J 305:25-28
Skolnik EY, Margolis B, Mohammadi M, Lowenstein E, Fischer R, Drepps A *et al.* (1991) Cell 65:83-90
Songyang Z, Shoelson SE, Chaudhuri M, Gish G, Pawson T, Haswer WG *et al.* (1993) Cell 72:767-778
Stephens LR, Hughes KT, Irvine RF (1991) Nature 351:33-39
Stephens L (1995) Biochem Soc Trans 23:207-221
Stephens LR, Eguinoa A, Erdjument-Bromage H, Lui M, Cooke F, Coadwell J, Smrcka A, Thelen M, Cadwallader K, Tempst P, Hawkins PT (1996) Submitted
Stephens L, Smrcka A, Cooke FT, Jackson TR, Sternweis PC, Hawkins PT (1994) Cell 77:83-93
Stephens L, Cooke FT, Walters R, Jackson T, Volinia S, Gout I, Waterfield MD, Hawkins PT (1994b) Curr Biol 4:203-214
Stephens L, Jackson TJ, Hawkins PT (1993) Biochim Biophys Acta 1179:27-75
Stephens L, Eguinoa A, Corey S, Jackson TJ, Hawkins PT (1993b) EMBO J 12:2265-2273
Stoyanov B, Volinia S, Hanck T, Rubio I, Loubtchenkov M, Malek D *et al.* (1995) Science 269:690-693
Thomason PA, James SR, Casey PJ, Downes CP (1994) J Biol Chem 269:16525-16528
Toker A, Meyer M, Reddy KK, Falck JR, Aneja R, Aneja S, Parra A, Burns DJ, Ballas LM, Cantley LC (1994) J Biol Chem 269:32358-32367

Traynor-Kaplan AE, Thompson BL, Harris AL, Taylor P, Omann GM, Sklar LA (1989) J Biol Chem. 264:15668-157673
Valius M, Kazlaukas A (1993) Cell 73:321-334
Virbasius JV, Guilherme A, Czech MP (1996) J Biol Chem 271:13304-13307
Vlahos CJ, Matter WF, Hui KY, Brown RF (1994) J Biol Chem 269:5241-5248
Volinia S, Dhand R, Vanhaesebroeck B, MacDougall LK, Stein R, Zvelebil MJ *et al.* (1995) EMBO J 14:3339-3348
Wennström S, Hawkins P, Cooke F, Hara K, Yonezawa K, Kasuga M, Jackson T, Claesson-Welsh L, Stephens L (1994) Current Biology 4:385-393
Whitman M, Downes CP, Keeler M, Keller T, Cantley L (1988) Nature 332:644-646
Yamamoto-Honda R, Tobe K, Kaburagi Y, Ueki K *et al* (1995) J Biol Chem 270:2729-2734
Yamauchi K, Holt K, Pessin JE (1993) J Biol Chem 268:14597-14600
Yan H, Nakanishi S, Kimura K, Hanni N, Saitoh Y, Fukui Y, Novomura Y, Matsuda Y (1993) J Biol Chem 268:25846-25856
Zhang J, Falck JR, Reddy KK, Abrams CS, Zhao W, Rittenhouse SE (1995) J Biol Chem 270:22807-22810
Zhang J, Zhang J, Shattil SJ, Cunningham MC, Rittenhouse SE (1996) J Biol Chem 271:6265-6272

Phosphatidylinositol Transfer Protein, Phosphoinositides and Cell Function

Karel W.A. Wirtz, Jan Westerman and Gerry T. Snoek
Centre for Biomembranes and Lipid Enzymology
Institute of Biomembranes
Utrecht University
P.O. Box 80.054
3508 TB Utrecht
The Netherlands

Phosphorylated derivatives of phosphatidylinositol are among the most studied of lipids because of their roles in intracellular signaling. Agonist-induced receptor activation may lead to phospholipase C (PLC)-mediated breakdown of phosphatidylinositol 4,5-bisphosphate (PIP_2) yielding the second messengers diacylglycerol (DAG) and inositol 1,4,5-P_3 (IP_3). Cells stimulated by growth factors (PDGF, EGF) may give rise to a rapid accumulation of PI-3,4-P_2 and PI-3,4,5-P_3 suggesting that their stimulated synthesis is part of a signaling event (Majerus *et al.*, 1990; Stephens *et al.*, 1991; Jackson *et al.*, 1992). These phosphoinositides may also be involved in membrane vesicle flow, specifically in the process of vesicle budding and fusion, possibly by mediating the recruitment of specific cytosolic proteins to the donor membrane (Liscovitch and Cantley, 1994). Subcellular fractionation studies have indicated that in mammalian cells the bulk of phosphoinositides (PI-4-P, PIP_2) is present in the plasma membrane. However, these phosphoinositides have also been found in the micosomal/Golgi fraction of Chinese hamster ovary (CHO) cells and in the nuclei of different cell types (Helms *et al.*, 1991; Divecha and Irvine, 1995). Isotope-labeling studies have clearly indicated that in the cell we are dealing with a hormone-sensitive and hormone-insensitive PIP_2 pool (reviewed in Monaco and Gershengorn, 1992). Recently, evidence was provided that in A431 cells the hormone-sensitive PIP_2 pool is present in caveolae, distinct areas of the plasma membrane that account for 1% or less of this structure (Pike and Casey, 1996). It is very likely that the remaining PIP_2 within the plasma membrane has a structural role (interactions with proteins) and, hence, should be classified as being part of the hormone-insensitive pool.

The intracellular localization of phosphoinositide (*e.g.* PIP_2) synthesis and degradation is still a matter of conjecture. It is clear that PIP_2 synthesis may occur in or on the plasma membrane (Lundberg and Jergil, 1988). This synthesis has also been demonstrated in the microsomal/Golgi fraction of CHO cells (Helms *et al.*, 1991). Moreover, an independent phosphoinositide cycle

NATO ASI Series, Vol. H 101
Molecular Mechanisms of Signalling
and Membrane Transport
Edited by Karel W. A. Wirtz

exists in the nucleus which may be under control of receptors in the plasma membrane (Divecha and Irvine, 1995; Michell, 1992; Cocco *et al.*, 1994). At these sites of metabolism (*i.e.* plasma membrane, nucleus) one has failed to identify the activities required for *de novo* synthesis of PI. This implies that PIP_2 degradation as part of the phosphoinositide cycle at these subcellular sites requires a continuous supply of PI, phosphorylation of which maintains the PIP_2 pool. In most mammalian cells investigated to date, PI synthesis is restricted to the endoplasmic reticulum (ER) (Morris *et al.*, 1990). This separation of enzymatic activities requires transport of PI from the ER to the sites of PIP_2 formation and degradation, *cq.* the site of IP_3-production. Recent evidence indicates that the phosphatidylinositol-transfer protein (PI-TP) could play an essential role in this process.

Different forms of PI-TP

PI-TP belongs to the class of intracellular proteins that *in vitro* carry phospholipids between membranes (Helmkamp, 1990; Wirtz, 1991). This protein occurs in all eukaryotic species investigated to date including a great variety of mammalian, avian, yeast and *Xenopus laevis*. cells. Because of its abundance in bovine brain, PI-TP (35 kDa) was first purified from this tissue (15). With the availability of the appropriate cDNAs, mouse and rat PI-TP are now routinely purified from *Escherichia coli* (Geijtenbeek *et al.*, 1994; Tremblay *et al.*, 1996). *In vitro* PI-TP mediates the transfer of PI and PC (Helmkamp, 1990). In line with this dual specificity, PI-TP from bovine brain occurs in a PI-containing form (pI 5.5) and a PC-containing form (pI 5.3) (van Paridon *et al.*, 1987) The high affinity of PI-TP for phospholipids was borne out by the fact that PI-TP purified from *E. coli* was found to contain 1 mol of PG (Geijtenbeek *et al.*, 1994; Tremblay *et al.*, 1996). The counterpart of PI-TP in yeast is SEC14p, a cytosolic protein known to be essential for secretory vesicle flow (Aitken *et al.*, 1990; Bankaitis *et al.*, 1990).
Given the proposed role of PI-TP in the phosphoinositide cycle we considered the possibility that a similar protein may exist being involved in the sphingomyelin (SM)-ceramide cycle (Hannun, 1994). By assaying for SM-transfer actvity. an active protein was identified and purified, initially from chicken liver (Westerman *et al.*, 1995) and subsequently from bovine brain (de Vries *et al.*, 1995). Interestingly, in addition to SM, this protein (36 kD) was able to transfer PI and PC suggesting a close relationship with the 35 kD PI-TP (PI-TPα). Determination of the N-terminal amino acid sequence showed that both the chicken liver and bovine brain protein were identical to rat brain PI-TPβ, the cDNA of which was identified by its ability to rescue SEC14 yeast mutants (Tanaka and Hosaka, 1994). It is to be noted that the amino acid sequence of PI-TPβ is 77%

identical to that of PI-TPα the latter sequence being highly conserved (99% identity) between mammalian species (Dickeson *et al.*, 1994). So far it is the only protein known to transfer SM. It remains to be established whether it has a role in the SM-cycle.

In addition to these soluble PI-TPs, a membrane-associated form of PI-TP (160 kD) denoted as Drosophila retinal degeneration B (rdgB) protein is present in photoreceptor cells and the brain of *Drosophila melanogaster*. (Vihtelic *et al.*, 1993). Specifically, the amino terminal peptide segment of the rdgB protein is >40% identical to rat brain PI-TP. Expression in *E. coli* showed that indeed this peptide segment has PI-transfer activity. Evidence was obtained that rdgB may be involved in photoreceptive membrane transport.

Role in the phosphoinositide cycle

Given that nerve ending particles from rat brain were highly enriched in PI-transfer activity (Wirtz *et al.*, 1976), it was suggested that PI-TP may play a role in neuronal processes the progress of which is linked to PI-metabolism (Michell, 1975; Wirtz *et al.*, 1978). Recently, it was shown in permeabilized, cytosol-depleted HL60 cells that PI-TPα is able to reconstitute the GTPγS-dependent PLCβ activity by stimulating the synthesis of PIP_2 (Thomas *et al.*, 1993; Cunningham *et al.*, 1995). In a comparable study with cytosol-depleted A431 cells it was shown that EGF greatly stimulated IP_n production when both PI-TP and PLCγ were added back to the medium (Kauffmann-Zeh *et al.*, 1995). This indicated that the expression of the EGF-receptor controlled PLCγ activity was dependent on the presence of PI-TP. This protein was also required for the reconstitution of GTPγS-mediated PLCβ- and of tyrosine kinase-controlled PLCγ activity in cytosol-depleted rat basophilic leukemia (RBL-2H3) cells (Cunningham *et al.*, 1996). In the latter study both PI-TPα, PI-TPβ and SEC14p were equally effective, strongly suggesting that the ability shared by all three proteins to bind and transfer PI was sufficient to restore inositol lipid signaling.

In these reconstitution studies PI-TP presumably delivers PI to sites of PI-4-kinase activity the product of which being phosphorylated to PIP_2. In fact, it was postulated that during phosphorylation the products remain bound to PI-TP for PIP_2 then to be degraded by PLC (Cunningham *et al.*, 1995). However, this is hard to reconcile with the recent observation that stimulation of A431 cells with EGF or bradykinin resulted in a specific hydrolysis of caveolar PIP_2 (Pike and Casey, 1996). Moreover, PIP_2 is known to effectively inhibit the transfer activity of PI-TP (van Paridon *et al.*, 1988). It is not yet known whether this inhibition is due to a strong interaction of PI-TP with the interface containing PIP_2 or due to monomer PIP_2 occupying the lipid binding site of PI-TP. Because of its distinct affinity for PC, it was previously proposed that *in vivo* PI-TP may catalyze a net transfer of PI to sites of metabolism in exchange for PC (van Paridon *et al.*, 1987). If caveolae are the primary site of receptor-controlled PIP_2-hydrolysis one should consider the possibility that PI-TP delivers PI directly at these sites. On the other hand, PI-TP is an essential factor in vesicle budding and fission suggesting the possibility that PI is delivered to the caveolae by PI-TP-controlled vesicle flow.

Association with the Golgi system

By using antibodies raised against peptides representing predicted epitope regions of PI-TPα, this protein and a cross-reactive 36 kD-protein (PI-TPβ) were identified in Swiss mouse 3T3 cells (Snoek *et al.*, 1992). Permeabilization of these cells by streptolysin O indicated that PI-TPβ was preferentially retained in the cells, a major part of which being associated with the perinuclear Golgi system (de Vries *et al.*, 1994). By indirect immunofluoresence using antibodies that discriminated between PI-TPβ and PI-TPα, we were able to show that indeed in exponentially growing 3T3 cells PI-TPβ is mainly associated with perinuclear membrane structures (de Vries *et al.*, 1995). As for PI-TPα, this isoform was primarily localized in the nucleus and in the cytosol.
The intracellular localization was also investigated by microinjection of fluorescently labeled PI-TPs into living cells (de Vries *et al.*, 1996). For this purpose, purified bovine brain PI-TPα and PI-TPβ were covalently labeled with the fluorescent dyes Cy3 and Cy5, which are structurally highly similar molecules but which have different excitation and emission wavelenghts. A 1:1 mixture of

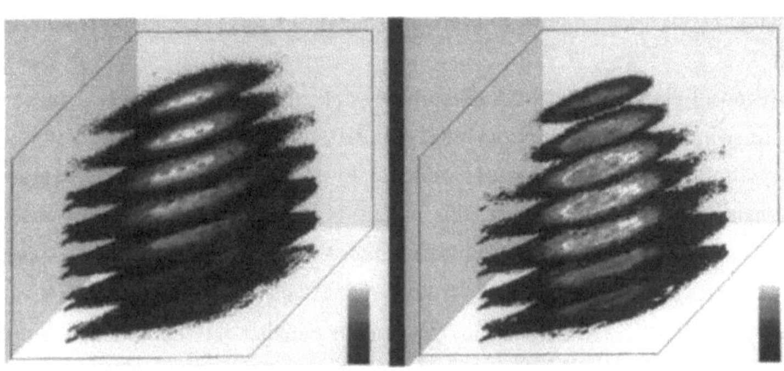

Fig.1. Localization of Cy3-PI-TPα (left panel) and Cy5-PI-TPβ (right panel) in confocal optical sections of fetal bovine heart endothelial cells (adapted from ref. 38).

CY3-PI-TPα and Cy5-PI-TPβ was microinjected into living fetal bovine heart epithelial cells. After 30 minutes the cells were fixed and the fluorescent signals were analyzed by confocal laser scanning microscopy (Fig. 1). As shown in the left panel, the fluorescence signal of CY3-PI-TPα has the highest intensity in the nucleus. Labeled protein was also found in the cytosol, possibly associated with some structures around the nucleus. When in the same optical slices the fluorescence signal of Cy5-PI-TPβ was analyzed (right panel) the highest intensity was observed with the perinuclear structures. In none of the studies carried out did we observe an association of PI-TP with the plasma membrane.

Role in vesicle flow

That the PI-TPs have an important role in Golgi function, was borne out by some recent studies on the identification of the cytosolic factors that control protein secretion. In semi-intact PC12 cells it was shown that PI-TPα is one of the three cytosolic proteins which are necessary for the reconstitution of the Ca^{2+}-, ATP-dependent secretion of noradrenalin (Hay and Martin, 1993). In addition, GTPγS-stimulated protein secretion in permeabilized HL60 cells was restored by addition of both PI-TPα and β (Fensome *et al.*, 1996). In the latter study, PI-TPβ was shown to be distinctly more potent in restoring protein secretion as compared to PI-TPα which may reflect the preferential association of PI-TPβ with the Golgi complex. In both cell systems restoration of secretion was linked to the ablity of PI-TP to promote PIP_2 -synthesis (Fensome *et al.*, 1996; Hay *et al.*, 1995). Recently we have shown that in a cell-free system derived from PC12 cells both PI-TPα and β were also able to stimulate the budding of secretory vesicles from *trans*-Golgi membranes (Ohashi *et al.*, 42). These pronounced effects of PI-TP on vesicle fusion and fission strongly suggest that these processes are critically dependent on PI-TP-controlled steps in phosphoinositide metabolism. In line with these studies, yeast mutants lacking the *sec14* gene were defective in vesicular transport (Bankaitis *et al.*, 1990). In yeast PI-TP (SEC14p) may control the PI/PC ratio of the Golgi complex which parameter is proposed to be linked to those processes that regulate the budding of secretory vesicles from the *trans*-Golgi network (McGee *et al.*, 1994; Skinner *et al.*, 1995). SEC14p exerts its effect on this PI/PC ratio by direct regulation of the CDP-choline pathway of PC-biosynthesis (Skinner *et al.*, 1995). On the other hand, SEC14p also stimulates the budding of secretory vesicles in a mammalian cell-free system (Ohashi *et al.*, 1995). At this stage it is not known how the stimulatory effect of Sec14p in this cell-free system is related to its function in yeast.

Concluding remarks

The above studies indicate an apparent discrepancy between the distinct cellular localization of PI-TPα and PI-TPβ and the fact that both isoforms are effective in restoring GTPγS-dependent PLC activity and secretory vesicle flow. This strongly suggests that additional factors control the intracellular localization of these PI-TPs thereby regulating their function One of these factors may be the protein kinase C-dependent phosphorylation of PI-TPα (Snoek *et al.*, 1993). Other factors may include Golgi-associated docking proteins interacting with one and not with the other PI-TP isomer . Another point to be noted is that, in contrast to PI-TPα, PI-TPβ is able to transfer SM, a lipid which is prominently present in Golgi membranes. Although many questions remain to be solved, it is evident that the PI-TPs form an interface between phospholipid metabolism and crucial cellular processes.

References

Aitken JF, van Heusden GPH, Temkin M, Dowhan W (1990) J. Biol. Chem. 265:4711-4717
Bankaitis VA, Aitken JR, Cleves AE, Dowhan W (1990) Nature 347:561-562
Cocco L, Martelli AM, Gilmour RS (1994) Cell Signal. 6:481-485
Cunningham E, Tan SK, Swigart P, Hsuan J, Bankaitis V, Cockcroft, S (1996) Proc. Natl. Acad. Sci. USA 93:6589-6593
Cunningham E, Thomas GMH, Ball A, Hiles I, Cockcroft S (1995) Curr. Biol. 5:775-783
De Vries KJ, Heinrichs AJA, Cunningham E, Brunink F, Westerman J, Somerharju PJ, Cockcroft S, Wirtz KWA, Snoek GT (1995) Biochem. J. 310:643-649
De Vries KJ, Momchilova-Pankova A, Snoek GT, Wirtz K W A (1994) Exp. Cell Research. 215:109-113
De Vries KJ, Westerman J, Bastiaens PIH, Jovin TM, Wirtz KWA, Snoek GT (1996) Exp. Cell Res. 227:33-39
Dickeson SK, Helmkamp GM, Yarbrough LR (1994) Gene 142:301-305
Divecha N, Irvine RF (1995) Cell 80:269
Fensome A, Cunningham E, Prosser S, Khoon Tan S, Swigart P, Thomas G, Hsuan J, Cockcroft S (1996) Curr. Biol.. 6:731-738
Geijtenbeek TBH, de Groot E, van Baal J, Brunink F, Westerman J, Snoek GT, Wirtz KWA (1994) Biochim. Biophys. Acta 1213:309-318
Hannun YA (1994) J. Biol. Chem. 269:3125-3128
Hay JC, Martin TFJ (1993) Nature 366:572-575
Hay JC, Fisette PL, Jenkins GH, Fukami K, Takenawa T, Anderson RE, Martin TFJ (1995) Nature 374:173-177
Helmkamp GM (1990) in Subcellular Biochemistry (ed. H.J. Hilderson) p. 129-174 (Plenum, New York)
Helmkamp GM, Harvey MS, Wirtz KWA, Van Deenen LLM (1974) J. Biol. Chem. 249:6382-6389
Helms JB, De Vries KJ, Wirtz KWA (1991) J. Biol. Chem. 266:21368-21374
Jackson TR, Stephens LR, Hawkins PT (1992) J. Biol. Chem. 267:16627-16636

Kauffmann-Zeh A, Thomas GMH, Ball A, Prosser S, Cunningham E, Cockcroft S, Hsuan JJ (1995) Science 268:1188-1190
Liscovitch M, Cantley LC (1994) Cell 77:329
Lundberg GA, Jergil B (1988) FEBS Lett 240:171-176
Majerus PW, Ross TS, Cunningham TW, Caldwell KK, Jefferson AB, Bansal VS (1990) Cell 63:459-465
McGee TP, Skinner HB, Whitters EA, Henry SA, Bankaitis VA (1994) J. Cell Biol. 124:273-287
Michell RH (1975) Biochim. Biophys. Acta 415:81-147
Michell RH (1992) Current Biol. 2:200
Monaco ME, Gershengorn MC (1992) Endocr. Rev. 13:707-718
Morris SJ, Cook HW, Byers DM, Spence MW, Palmer FB (1990) Biochim. Biophys. Acta 1022:339-347
Ohashi M, De Vries KJ, Frank R, Snoek GT, Bankaitis VA, Wirtz KWA, Huttner WB (1995) Nature 377:544-547
Pike LJ, Casey L (1996) J. Biol. Chem. 271:26453-26456
Skinner HB, McGee TP, Mc Master CR, Fry MR, Bell RM, Bankaitis VA (1995) Proc. Natl. Acad. Sci. USA 92:112-116
Snoek GT, De Wit ISC, Van Mourik JHG, Wirtz KWA (1992) J. Cell. Biochem. 49:339-348
Snoek GT, Westerman J, Wouters FS, Wirtz KWA (1993) Biochem. J. 291:649-656
Stephens LR, Hughes KT, Irvine RF (1991) Nature 351:33-39
Tanaka S, Hosaka K (1994) J. Biochem. 115:981-984
Thomas GMH, Cunningham E, Fensome A, Ball A, Totty NF, Truong O, Hsuan JJ, Cockcroft S (1993) Cell 74:919-928
Tremblay JM, Helmkamp GM, Yarbrough LR (1996) J. Biol. Chem. 271:21075-21080
Van Paridon PA, Gadella TWJ, Wirtz KWA (1988) Biochim. Biophys. Acta 943:76-86
Van Paridon PA, Gadella TWJ, Somerharju PJ, Wirtz KWA (1987) Biochim. Biophys. Acta 903:68-77
Van Paridon PA, Visser AJWG, Wirtz, KWA (1987) Biochim. Biophys. Acta 898:172-180
Vihtelic TS, Goebl M, Milligan S, O'Tousa JE, Hyde DR (1993) J. Cell Biol. 122:1013-1022
Westerman J, De Vries KJ, Somerharju P, Timmermans-Hereijgers JLPM, Snoek GT, Wirtz KWA (1995) J. Biol. Chem. 270:14263-14266
Wirtz KWA (1991) Annu. Rev. Biochem. 60:73-99
Wirtz KWA, Helmkamp GM, Demel RA (1978) in Protides of the Biological Fluids (ed. H. Peeters) p. 25-32 (Pergamon Press Oxford)
Wirtz KWA, Jolles J, Westerman J, Neys F (1976) Nature 260:354-355

Receptor Regulation of Phospholipases C and D

Martina Schmidt, Ulrich Rümenapp, Chunyi Zhang, Jutta Keller, Barbara Lohmann and Karl H. Jakobs
Institut für Pharmakologie
Universität GH Essen
D-45122 Essen
Germany

Receptor Regulation of Phospholipase C

A large variety of hormones, neurotransmitters and growth factors regulate cellular functions by stimulating phospholipase C (PLC) enzymes (Berridge, 1993; Divecha & Irvine, 1995; Exton, 1996). These phospholipases hydrolyze phosphatidylinositol 4,5-bisphosphate [PtdIns(4,5)P_2], a rare and uniquely polar plasma membrane phospholipid, generating the two second messengers, inositol 1,4,5-trisphosphate ($InsP_3$) and diacylglycerol. $InsP_3$ exerts its effect by binding to specific receptors located on components of the endoplasmatic reticulum, thereby leading to a quantal release of calcium (Clapham, 1995). The accumulation of diacylglycerol in the plasma membrane due to the hydrolysis of PtdIns(4,5)P_2 induces the translocation of certain protein kinase C (PKC) isozymes from the cytosol to the membrane and their concurrent activation (Nishizuka, 1995). The classical or conventional PKC isozymes (α, β1, β2, and γ) possess C1 and C2 domains, binding diacylglycerol or phorbol ester and calcium, respectively, and are thus calcium-dependent, whereas the other isoforms (δ, ε, η, θ) lack the C2 domain and are therefore calcium-independent. The atypical PKC isozymes ζ and λ are not only calcium-independent, but also diacylglycerol- or phorbol ester-independent, due to deletions or differences in the C1 domain. However, these enzymes are still dependent upon phosphatidylserine and are activated by other lipids (Nishizuka, 1995). The increase in cytoplasmic calcium and activation of different PKC isozymes initiated by PLC-catalyzed hydrolysis of PtdIns(4,5)P_2 apparently participate in the transduction of mitogenic signals across the plasma membrane into the nucleus leading to cell growth and cell transformation (Berridge, 1993; Nishizuka, 1995).

NATO ASI Series, Vol. H 101
Molecular Mechanisms of Signalling
and Membrane Transport
Edited by Karel W. A. Wirtz

PLC Isozymes

At least 10 mammalian PLC isozymes, divided into the three families, β (130 - 155 kDa), γ (145 kDa) and δ (~85 kDa), have been identified with biochemical and molecular biology approaches (Rhee & Choi, 1992; Lee & Rhee, 1995). The different PLC isozymes are significantly homologous in two amino acid sequence regions, one of ~170 amino acids and the other one of ~260 amino acids, which are designed as the X and Y domains, respectively, and probably are the catalytic domains. There are four mammalian PLC-β isozymes (β1 - β4), all of them are characterized by a long C-terminal sequence beyond the Y domain, which is responsible for the specific binding and activation of PLC-β isozymes by G protein α_q subunits (see below). Two γ isozymes (γ1 and γ2) have been identified, which differ from the other subtypes by the presence of Src homology (SH2 and SH3) and a pleckstrin homology (PH) domain between the X and Y domain. The four δ isozymes are missing the C-terminal amino acid extension and do not contain SH domains, which are involved in the regulation of PLC-γ isozymes by tyrosine kinase-linked receptors (Rhee & Choi, 1992; Lee & Rhee, 1995). Tyrosine kinase receptors and heptahelical receptors coupled to heterotrimeric guanine nucleotide-binding proteins (G proteins) generally activate distinct PLC isoenzymes. In the following the receptor regulation of the different PLC isozymes will be discussed.

PLC-β Isozymes

G protein-coupled receptors activate PLC isozymes of the β family by two distinct mechanisms. The pertussis toxin (PTX)-insensitive activation of PLC-β isozymes is apparently mediated by activated α subunits of the G_q class of G proteins (α_q, α_{11}, α_{14}, α_{15} and α_{16}). All PLC-β isozymes are activated by these Gα proteins, with a hierarchy of PLC-β4 ≥ PLC-β1 ≥ PLC β3 > PLC-β2 (Lee & Rhee, 1995). The PTX-sensitive stimulation of PLC-β isozymes appears to be due to free βγ dimers of G_i type proteins, with a hierarchy of PLC-β3 ≥ PLC-β2 > PLC-β1 >> PLC-β4 (Lee & Rhee, 1995). The concentration of βγ dimers required for half-maximal activation of PLC-β3 or PLC-β2 is much higher than that required for activation of PLC-β1 by α_q (Camps *et al.*, 1992). The G proteins releasing the activating βγ dimers are not identified with certainty, however, there is ample evidence that they belong to the G_i and/or G_o proteins. At least five different G protein β subtypes and twelve different G protein γ subtypes exhibiting distinct specificity in their interaction with each other and in their tissue distribution have been identified (Ueda *et al.*, 1194; Hamm & Gilchrist, 1996). Studies with different antisense oligonucleotides to specific β and γ subunits have demonstrated that the specificity of the interaction of G proteins with receptors is determined not only by their complement of α subunits, but also of β and γ subunits (Kleuss *et al.*, 1992; 1993). Furthermore, it has been already shown that certain βγ complexes stimulate PLC-β isozymes more efficiently than others (Katz *et al.*, 1992; Watson *et al.*, 1994). Finally, it has recently been reported that G protein-coupled receptors, *e.g.*, the angiotensin II receptor, can induce tyrosine phosphorylation of PLC-γ1, most likely *via* a cytosolic tyrosine kinase, resulting in stimulated $InsP_3$ production which was abolished by tyrosine kinase inhibitors (Marrero *et al.*, 1996). Thus, there is apparently a major intracellular cross-talk in the cellular integration of tyrosine kinase receptor- and G protein-coupled receptor-mediated activation of PLC isozymes.

PLC-γ Isozymes

Tyrosine kinase receptors, *e.g.*, those for platelet-derived growth factor (PDGF), epidermal growth factor (EGF), fibroblast growth factor and nerve growth factor, cause stimulation of PLC–γ isozymes. Tyrosine kinase receptors dimerize after binding of their appropriate ligands followed by activation of their intrinsic tyrosine kinase activity. The stimulation of the intrinsic receptor tyrosine kinase activity induces phosphorylation of distinct tyrosine residues in the cytoplasmic receptor domain, to which various cytoplasmic signal transduction components including PLC-γ isozymes become associated (Claesson-Welsh, 1994). PLC-γ isozymes bind with high affinity to different autophosphorylated sites *via* their SH2 domains. Tyrosine kinase receptor binding of PLC-γl isozyme induces tyrosine phosphorylation of the phospholipase on the amino acid residues 771, 783 and 1254, correlating with the stimulation of PLC-γ isozyme activity probably due to the conformational changes in the enzyme structure. Interestingly, at least in rat hepatocytes it has been demonstrated that activation of PLC isozymes by tyrosine kinase receptors can involve a heterotrimeric G protein (Yang *et al.*, 1991).

PLC-δ Isozymes

Recently, the crystal structure of a mammalian PLC-δ1 has been identified (Essen *et al.*, 1996). PLC-δ1 is a multi-domain protein consisting of 4 distinct domains, which are from the N- to the C-terminus: 1) a PH domain, which probably confers the membrane association of PLC-δ1; 2) a helical-loop-helix unit (EF hand), probably responsible for the calcium sensitivity of PLC-δ isozymes; 3) a catalytic TIM barrel (triosephosphate isomerase-like) domain; 4) a C-terminal β-sandwich domain bearing a C2 domain, a potential phospholipid-binding domain (Essen *et al.*, 1996). Thus, PLC-δ1 contains at least two motifs, PH and C2 domains, which are important for membrane association and are also found in other proteins, *e.g.*, PKC and cytosolic PLA_2 enzymes. The C2 domain is also involved in the regulation of

PLC-δ1 by calcium, thus is a putative calcium-dependent lipid-binding site. Interestingly, deletions of this C2 domain in PLC-δ1 results in a complete loss of enzyme activity. The regulatory mechanisms being involved in receptor activation of PLC-δ isozymes are much less characterized than the processes leading to stimulation of PLC-β and PLC-γ isozymes. Studies in CHO cells indicated that activation of the thrombin receptor stimulates PLC-δ1, a process apparently involving both G protein(s) and calcium (Banno *et al.*, 1994). Furthermore, PLC-δ1 has recently been demonstrated to be the effector of the G_h (transglutaminase II)-mediated signaling pathway (Feng *et al.*, 1996). Activation of PLC-δ isozymes by tyrosine kinase receptors has not yet been demonstrated, probably due to the lack of SH domains.

PLC Substrate Supply

Reconstitution studies with purified protein components supports the general assumption that only three factors are required for efficient PLC activation by G protein-coupled receptors, *i.e.*, the receptor, the relevant heterotrimeric G protein and a PLC-β isozyme (Berstein *et al.*, 1992). However, in intact cells, the regulatory mechanisms involved in PLC stimulation are apparently by far more complex, particularly the supply of the PLC substrate, PtdIns(4,5)P_2, seems to be crucial to maintain receptor signaling to PLC. Indeed, it has been demonstrated recently that the phosphatidylinositol (PtdIns) transfer protein, involved in the transport of PtdIns from intracellular compartments to the plasma membrane for conversion to PtdIns(4,5)P_2, dictates the production of $InsP_3$ by formyl peptide receptor- and G protein-mediated stimulation of PLC-β enzymes in HL-60 cells and by EGF receptor-mediated activation of PLC-γ isozymes in A431 human epidermoid carcinoma cells through the effect on PtdIns(4,5)P_2 synthesis (Thomas *et al.*, 1993; Kauffmann-Zeh *et al.*, 1995; Cunningham *et al.*, 1995). Furthermore, CDP-diacylglycerol synthase catalyzing formation of PtdIns from CDP-diacylglycerol has been demonstrated to be indispensable for receptor-induced PLC signaling (Wu *et al.*, 1995). The lipid kinase, PtdIns4P 5-kinase(s), which converts PtdIns4P to PtdIns(4,5)P_2, apparently also plays a critical role in PLC stimulation (Carpenter & Cantley, 1996). Recently, evidence has been provided that PtdIns4P 5-kinase is stimulated by small molecular weight G proteins of the Rho family. For example, specific inactivation of RhoA by the *Clostridium botulinum* C3 exoenzyme in mouse fibroblasts has demonstrated that this small molecular weight G protein is involved in the regulation of PtdIns(4,5)P_2 synthesis and PDGF-induced calcium signaling (Chong *et al.*, 1994). Likewise, studies with C3 exoenzyme and *Clostridium difficile* toxin B, which also inactivates Rho family G proteins, have demonstrated that Rho proteins regulate the supply of the PLC substrate and receptor-mediated inositol phosphate production in HEK-293 cells and N1E-115 neuroblastoma cells (Schmidt *et al.*, 1996a; Zhang *et al.*, 1996). Furthermore, Rac1 has recently been reported to be involved in the synthesis of PtdIns(4,5)P_2 in human platelets

(Hartwig *et al.*, 1995). However, the mechanisms by which Rho proteins stimulate synthesis of PtdIns(4,5)P_2 by the phosphoinositide 5-kinase(s) are not yet clear (Tolias *et al.*, 1995). PtdIns(4,5)P_2 synthesis can apparently also be stimulated by G protein-coupled receptors as shown in human neutrophils and rat pancreatic acinar cells (Stephens *et al.*, 1993; Lods *et al.*, 1995). Most interestingly, recent studies in HEK-293 cells have shown that short-term agonist treatment can increase the level of PtdIns(4,5)P_2, resulting in long-lasting potentiation of receptor signaling by the phosphoinositide pathway (Schmidt *et al.*, 1995; 1996b).

Receptor Regulation of Phospholipase D

Receptor-mediated activation of phospholipase D (PLD) has been reported in a wide range of cell types in response to various hormones, neurotransmitters and growth factors (Billah, 1993; Boarder, 1994; Exton, 1994; Morris *et al.*, 1996). Stimulation of PLD activity has been implicated in various physiological processes, including metabolic regulation, inflammation, immune responses, secretion, and cell growth and differentiation. In addition, PLD is apparently also involved in the regulation of membrane trafficking in eukaryotic cells (Liscovitch & Cantley, 1995). The hydrolysis of the major membrane phospholipid, phosphatidylcholine, by PLD results in the formation of phosphatidic acid (PA). This lipid may act by itself as an intracellular signaling molecule (Cockcroft, 1992; Billah, 1993). For example, PA has recently been demonstrated to regulate the membrane translocation of Raf-1 kinase in Madin-Darby canine kidney cells (Ghosh *et al.*, 1996). Furthermore, PA has been shown to stimulate PtdIns4P 5-kinase, generating PtdIns(4,5)P_2 (Moritz *et al.*, 1992; Jenkins *et al.*, 1994), which not only acts as PLC substrate but is apparently also an essential cofactor for PLD stimulation. Furthermore, PA can regulate the activity of a GTPase specific for ADP-ribosylation factors (ARFs) (Randazzo *et al.*, 1994), which have been implicated in the regulation of PLD activity in many cellular systems (see below). The complexation of Rho proteins with the guanine nucleotide dissociation inhibitor (RhoGDI) is also regulated by PA (Chuang *et al.*, 1993). The formation of PA apparently also modulates the organization of the actin cytoskeleton, as has been shown in IIC9 fibroblasts and porcine aortic endothelial cells (Ha & Exton, 1993b; Cross *et al.*, 1996). As Rho proteins are likewise involved in the regulation of the actin cytoskeleton and, as mentioned above, the PLD product PA regulates the interaction between Rho proteins and RhoGDI, it is tempting to propose a positive feedback loop between PLD and Rho proteins resulting in rapid changes of the cytoskeleton. In addition to its direct effects, PA can be converted upon deacylation and dephosphorylation to the extra- and intracellular signaling molecules, lysophosphatidic acid and diacylglycerol, respectively, the latter may activate specific PKC isozymes when derived from phosphatidylcholine breakdown (Ha & Exton, 1993a).

Mechanisms Involved in PLD Activation

Receptor-mediated PLD activation has been reported in a wide range of cell types, including neutrophils, blood platelets, hepatocytes, lymphocytes, fibroblasts, neuronal cells, muscle cells and endothelial cells. Among the many receptors mediating PLD stimulation are both tyrosine kinase receptors and G protein-coupled receptors (Billah, 1993; Exton, 1994). Stimulation of PLD activity by PDGF has been demonstrated in TRMP canine kidney epithelial cells and in NIH 3T3 fibroblasts (Yeo *et al.*, 1994; Lee *et al.*, 1994). In both cell lines, the PDGF receptor-mediated stimulation of PLD is dependent on the level of PLC-γ1. EGF receptor-mediated PLD stimulation has been demonstrated in Swiss 3T3 cells, apparently occurring in the absence of phosphoinositide hydrolysis (Cook & Wakelam, 1992). Different subtypes of G protein-coupled receptors have been shown to stimulate PLD activity, *e.g.*, muscarinic acetylcholine receptors (mAChR), bradykinin receptors, thrombin receptors and formyl peptide receptors. The type of heterotrimeric G protein, PTX-insensitive and/or -sensitive, involved in receptor-mediated PLD stimulation seems to be different from one cell type to another. For example, in neutrophils receptor activation of PLD is mediated by PTX-sensitive G_i proteins (Cockcroft, 1992). In contrast, m3 mAChR-mediated stimulation of PLD activity in HEK-293 cells is mediated *via* PTX-insensitive $G_{q/11}$ proteins (Offermanns *et al.*, 1994; Schmidt *et al.*, 1994). Many agonists that cause activation of PLD also evoke stimulation of PLC activity. And indeed, in many cellular systems PLD can be activated by raising the intracellular calcium concentration or by direct activation of PKC with phorbol esters (Billah, 1993; Exton, 1994). Furthermore, pervanadate, an inhibitor of protein tyrosine phosphatases inducing accumulation of tyrosine phosphorylated proteins, has been shown to increase PLD activity in many cell types, indicating that tyrosine phosphorylation-dependent mechanisms are also crucial for regulation of PLD activity (see below). In general, however, although the stimulation of PLD by membrane receptors is now well established, the exact mechanisms linking cell surface receptors to PLD yet remain unknown.

Activation of PLD by Small Molecular Weight G Proteins

Besides the receptors and heterotrimeric G proteins, there is increasing evidence that small molecular weight G proteins of different families are involved in the regulation of PLD activity. First, participation of ARF proteins in the stimulation of PLD activity by the stable GTP analog GTPγS, initially reported in HL-60 cells and human neutrophils (Brown *et al.*, 1993; Cockcroft *et al.*, 1994), has now been demonstrated for various cell types, *e.g.*, CHO cells and rat and porcine brain membranes (Ktistakis *et al.*, 1995; Massenburg *et al.*, 1994; Brown *et al.*, 1995). Based on these observations, and taking into account that the PLD product PA can stimulate PtdIns4P 5-kinase and ARF-specific GTPase, the group of Liscovitch proposed a positive feedback loop between PLD activation, PtdIns(4,5)P_2

synthesis and ARF-regulated vesicular fusion (Liscovitch *et al.*, 1994). Studies on ARF cytosol-membrane translocation in HEK-293 cells and on effects of the macrolide antibiotic brefeldin A, a specific inhibitor of the ARF guanine nucleotide exchange factor, provided the first evidence that ARF proteins and their guanine nucleotide exchange factor are involved in receptor (m3 mAChR)-mediated stimulation of PLD activity (Rümenapp *et al.*, 1995). Otherwise, the involvement of ARF proteins in receptor signaling to PLD in different cell types has not been studied in great detail. Likewise, the mechanisms by which ARF proteins are activated and then stimulate PLD activity are not yet clear. In human neutrophils, ARF proteins act in concert with a 50-kDa factor, probably a ARF-specific regulatory protein, to stimulate PLD activity (Lambeth *et al.*, 1995). In addition to and apparently synergistic with ARF, PLD activity is stimulated by GTPγS-activated Rho proteins, as demonstrated *e.g.* in human neutrophils, HL-60 cells, Madin-Darby canine kidney cells, rat liver and rat and porcine brain (Bowman *et al.*, 1993; Malcolm *et al.*, 1994; Siddiqi *et al.*, 1995; Singer *et al.*, 1995; Kwak *et al.*, 1995; Kuribara *et al.*, 1995; Balboa *et al.*, 1995). Studies with the Rho-inactivating agents, C3 exoenzyme and toxin B, in HEK-293 cells demonstrated that Rho proteins, most likely RhoA, are involved in receptor-mediated PLD stimulation (Schmidt *et al.*, 1996a). It is presently unclear whether Rho proteins regulate PLD activity directly and/or indirectly, *e.g.*, *via* supply of the PLD cofactor PtdIns(4,5)P_2 (Schmidt *et al.*, 1996b). Finally, in addition to ARF and Rho proteins, Ral proteins are apparently also involved in PLD stimulation, as demonstrated in Balb/c 3T3 and NIH 3T3 fibroblasts (Jiang *et al.*, 1995). Stimulation of PLD activity induced by v-Src in these cells was mediated by Ras and RalA, the latter acting downstream of Ras. Thus, GTPase cascades are apparently not only involved in cell proliferation and organization of the cytoskeleton (Nobes & Hall, 1995; Chant & Stowers, 1995), but may also participate in stimulation of PLD activity. The exact mechanisms, however, which couple the various small molecular weight G proteins to PLD and to cell surface receptors have not yet been established.

Regulation of PLD by Phosphorylation Reactions

Studies with various specific and nonspecific protein kinase inhibitors (staurosporine, calphostin C, genistein, tyrphostins) in different cellular systems have demonstrated that PKC and tyrosine kinase(s) are involved in receptor- and G protein-mediated stimulation of PLD activity (Billah, 1993; Uings *et al.*, 1992; Schmidt *et al.*, 1994). In cell-free systems, addition of MgATP often potentiates stimulation of PLD activity by GTPγS (Dubyak *et al.*, 1993; Ward *et al.*, 1995), supporting a role for phosphorylation reactions in PLD activation. Specifically, PKC activation has been shown to mediate EGF-induced PLD stimulation in Swiss 3T3 cells (Yeo & Exton, 1995). Likewise, involvement of PKC has been reported for the formyl peptide receptor-mediated PLD stimulation in human neutrophils (Lopez *et al.*,

1995). However, phosphorylation-independent PLD activation by phorbol esters, nevertheless involving PKC, has been described in some cellular systems (Conricode *et al.*, 1992; Singer *et al.*, 1996). In addition to protein kinases, increasing evidence has been provided that lipid kinases, specifically the PtdIns(4,5)P_2-synthesizing PtdIns4P 5-kinase(s), are involved in the regulation of PLD activity. As demonstrated by Liscovitch and its coworkers in various cellular systems, PtdIns(4,5)P_2 is an essential cofactor for PLD activity, and its synthesis is required for PLD stimulation by stable GTP analogs (Liscovitch *et al.*, 1994; Pertile *et al.*, 1995). As synthesis of PtdIns(4,5)P_2 is regulated by Rho proteins, and as PLD activity inhibited by inactivation of Rho proteins is restored by PtdIns(4,5)P_2 (Schmidt *et al.*, 1996b), it is tempting to speculate that Rho regulation of PLD activity is due to supply of PtdIns(4,5)P_2. The mechanism involved in PLD activation by PtdIns(4,5)P_2 is not yet clear, but recent data suggest that PtdIns(4,5)P_2 may serve as an anchor to localize PLD at the plasma membrane (Yokozeki *et al.*, 1996).

PLD Isozymes

Like for PLC enzymes, there is increasing evidence for the existence of different PLD isozymes (Massenburg *et al.*, 1994). Recently, cloning and expression of the first yeast and human PLD isozymes have been reported (Hammond *et al.*, 1995; Waksman *et al.*, 1996). The yeast PLD enzyme (195 kDa) catalyzes the PLD-specific transphosphatidylation reaction, is PtdIns(4,5)P_2-sensitive and oleate-independent (Waksman *et al.*, 1996). The first human PLD isozyme cloned (120 kDa) is apparently membrane-associated, selective for phosphatidylcholine as enzyme substrate, stimulated by ARF and PtdIns(4,5)P_2 and inhibited by oleate. In contrast to PLC isozymes, the human PLD, which is highly homologous to the yeast enzyme, contains no domain structures, *e.g.*, SH2, SH3 and/or PH domains (Hammond *et al.*, 1996; Morris *et al.*, 1996). Thus, identification of the PLD enzymes will most likely rapidly advance our understanding of the cellular functions and apparently very complex regulation of this signaling pathway.

Acknowledgements

The authors' studies reported herein were supported by the Deutsche Forschungsgemeinschaft, the Fonds der Chemischen Industrie and an Alexander von Humboldt-Foundation fellowship to C. Z.

REFERENCES

Phospholipase C

Banno Y, Okano Y, Nozawa Y (1994) Thrombin-mediated phosphoinositide hydrolysis in Chinese Hamster Ovary cells overexpressing phospholipase C-δ1. J Biol Chem 269:15846-15852

Berridge MJ (1993) Inositol trisphosphate and calcium signalling. Nature 361:315-325

Berstein G, Blank JL, Smrcka AV, Higashijima T, Sternweis PC, Exton JH, Ross EM (1992) Reconstitution of agonist-stimulated phosphatidylinositol 4,5-bisphosphate hydrolysis using purified m1 muscarinic receptor, $G_{q/11}$, and phospholipase C-β1. J Biol Chem 267:8081-8088

Camps M, Carozzi A, Schnabel P, Scheer P, Parker PJ, Gierschik P (1992) Isozyme-selective stimulation of phospholipase C-β2 by G protein βγ subunits. Nature 360:684-686

Carpenter CL, Cantley LC (1996) Phosphoinositide kinases. Curr Opinion Cell Biol 8:153-158

Chong LD, Traynor-Kaplan A, Bokoch GM, Schwartz MA (1994) The small GTP-binding protein Rho regulates a phosphatidylinositol 4-phosphate 5-kinase in mammalian cells. Cell 79:507-513

Claesson-Welsh L (1994) Platelet-derived growth factor receptor signals. J Biol Chem 269:32023-32026

Clapham DE (1995) Calcium signaling. Cell 80: 259-268

Cunningham E, Thomas GMH, Ball A, Hiles I, Cockcroft S (1995) Phosphatidylinositol transfer protein dictates the rate of inositol trisphosphate production by promoting the synthesis of PIP_2. Curr Biol 5:775-783

Divecha N, Irvine RF (1995) Phospholipid signaling. Cell 80:269-278

Essen LO, Perisic O, Cheung R, Katan M, Williams RL (1996) Crystal structure of a mammalian phosphoinositide-specific phospholipase Cδ. Nature 380:595-602

Exton JH (1996) Regulation of phosphoinositide phospholipases by hormones, neurotransmitters, and other agonists linked to G proteins. Annu Rev Pharmacol Toxicol 36:481-509

Feng JF, Rhee SG, Im, MJ (1996) Evidence that phospholipase δ1 is the effector in the G_h (transglutaminase II)-mediated signaling. J Biol Chem 271:16451-16454

Hamm HE, Gilchrist A (1996) Heterotrimeric G proteins. Curr Opinion Cell Biol 8:189-196

Hartwig JH, Bokoch GM, Carpenter CL, Janmey PA, Taylor LA, Toker A, Stossel TP (1995) Thrombin receptor ligation and activated Rac uncap actin filament barbed ends through phosphoinositide synthesis in permeabilized human platelets. Cell 82:643-653

Katz A, Wu D, Simon MI (1992) Subunits βγ of heterotrimeric G protein activate β2 isoform of phospholipase C. Nature 360:686-689

Kauffmann-Zeh A, Thomas GMH, Ball A, Prosser S, Cunningham E, Cockcroft S, Hsuan, JJ (1995) Requirement for phosphatidylinositol transfer protein in epidermal growth factor signaling. Science 268:1188-1190

Kleuss C, Scherübl H, Hescheler J, Schultz G, Wittig B (1992) Different β-subunits determine G-protein interaction with transmembrane receptors. Nature 358:424-426

Kleuss C, Scherübl H, Hescheler J, Schultz G, Wittig B (1993) Selectivity of signal transduction determined by γ subunits of heterotrimeric G proteins. Science 259:832-834

Lee SB, Rhee, SG (1995) Significance of PIP_2 hydrolysis and regulation of phospholipase C isozymes. Curr Opinion Cell Biol 7:183-189

Lods JS, Rossignol B, Dreux C, Morisset J (1995) Phosphoinositide synthesis in desensitized rat pancreatic acinar cells. Am J Physiol 95:G1043-G1050

Marrero MB, Schieffer B, Ma H, Bernstein KE, Ling BN (1996) ANG II-induced tyrosine phosphorylation stimulates phospholipase C-γ1 and Cl^- channels in mesangial cells. Am J Physiol 270:C1834-C1842

Nishizuka Y (1995) Protein kinase C and lipid signaling for sustained cellular responses. FASEB J 9:484-496

Rhee SG, Choi, KD (1992) Regulation of inositol phospholipid-specific phospholipase C isozymes. J Biol Chem 267:12393-12396

Schmidt M, Fasselt B, Rümenapp U, Bienek C, Wieland T, van Koppen CJ, Jakobs KH (1995) Rapid and persistent desensitization of m3 muscarinic acetylcholine receptor-stimulated phospholipase D. Concomitant sensitization of phospholipase C. J Biol Chem 270:19949-19956

Schmidt M, Bienek C, Rümenapp U, Zhang C, Lümmen G, Jakobs KH, Just I, Aktories K, Moos M, von Eichel-Streiber C (1996a) A role for Rho in receptor- and G protein-stimulated phospholipase C. Reduction in phosphatidylinositol 4,5-bisphosphate by *Clostridium difficile* toxin B. Naunyn-Schmiedeberg's Arch Pharmacol 353:1-8

Schmidt M, Nehls C, Rümenapp U, Jakobs KH (1996b) m3 Muscarinic receptor-induced and G_i protein-mediated heterologous potentiation of phospholipase C stimulation: Role of phosphoinositide synthesis. Mol Pharmacol: in press

Stephens L, Jackson TR, Hawkins PT (1993) Activation of phosphatidylinositol 4,5-bisphosphate supply by agonists and non-hydrolysable GTP analogues. Biochem J 296:481-488

Thomas GMH, Cunningham E, Fensome A, Ball A, Totty NF, Truong O, Hsuan JJ, Cockcroft S (1993) An essential role for phosphatidylinositol transfer protein in phospholipase C-mediated inositol lipid signaling. Cell 74:919-928

Tolias KF, Cantley LC, Carpenter CL (1995) Rho family GTPases bind to phosphoinositide kinases. J Biol Chem 270:17665-17659

Ueda N, Iñiguez-Lluhi JA, Lee E, Smrcka AV, Robishaw, JD, Gilman AG (1994) G proteins βγ subunits. Simplified purification and properties of novel isoforms. J Biol Chem 269:4388-4395

Watson AJ, Katz A, Simon MI (1994) A fifth member of the mammalian G protein β-subunit family. Expression in brain and activation of the β2 isotype of phospholipase C. J Biol Chem 269:22150-22156

Wu L, Niemeyer B, Colley N, Socolich M, Zuker CS (1995) Regulation of PLC-mediated signalling *in vivo* by CDP-diacylglycerol synthase. Nature 373:216-222

Yang L, Baffy G, Rhee SG, Manning D, Hansen CA, Williamson, JR (1991) Pertussis toxin-sensitive G_i protein involvement in epidermal growth factor-induced activation of phospholipase-γ in rat hepatocytes J Biol Chem 266:22451-22458

Zhang C, Schmidt M, von Eichel-Streiber C, Jakobs KH (1996) Inhibition by toxin B of inositol phosphate formation induced by G protein-coupled and tyrosine kinase receptors in N1E-115 neuroblastoma cells: Involvement of Rho proteins. Mol. Pharmacol: in press

Phospholipase D

Balboa MA, Insel PA (1995) Nuclear phospholipase D in Madin-Darby canine kidney cells. Guanosine 5'-*O*-(thiotriphosphate)-stimulated activation is mediated by RhoA and downstream of protein kinase C. J Biol Chem 270:29843-29847

Billah MB (1993) Phospholipase D and cell signaling. Curr Opinion Immunol 5:114-123

Boarder MR (1994) A role for phospholipase D in control of mitogenesis. TiPS 15:57-62

Bowman EP, Uhlinger DJ, Lambeth JD (1993) Neutrophil phospholipase D is activated by a membrane-associated Rho family small molecular weight GTP-binding protein. J Biol Chem 268:21509-21512

Brown HA, Gutowski S, Moomaw CR, Slaughter C, Sternweis PC (1993) ADP-ribosylation factor, a small GTP-dependent regulatory protein stimulates phospholipase D activity. Cell 75:1137-1144

Brown HA, Gutowski S, Kahn RA, Sternweis PC (1995) Partial purification and characterization of ARF-sensitive phospholipase D from porcine brain. J Biol Chem 270:14935-14943

Chant J, Stowers L (1995) GTPase cascades choreographing cellular behavior: movement, morphogenesis, and more. Cell 81:1-4

Chuang TH, Bohl BP, Bokoch GM (1993) Biologically active lipids are regulators of Rac·GDI complexation. J Biol Chem 268:26206-26211

Cockcroft S (1992) G-protein-regulated phospholipases C, D and A_2-mediated signalling in neutrophils. Biochim Biophys Acta 1113:135-160

Cockcroft S, Thomas GMH, Fensome A, Geny B, Cunningham E, Gout I, Hiles I, Totty NF, Truong O, Hsuan JJ (1994) Phospholipase D: A downstream effector of ARF in granulocytes. Science 263:523-526

Conricode K, Brewer KA, Exton JH (1992) Activation of phospholipase D by protein kinase C. Evidence for a phosphorylation-independent mechanism. J Biol Chem 267:7199-7202

Cook SJ, Wakelam MJ (1992) Epidermal growth factor increases sn-1,2-diacylglycerol levels and activates phospholipase D-catalyzed phosphatidylcholine breakdown in Swiss 3T3 cells in the absence of inositol-lipid hydrolysis. Biochem J 285:247-253

Cross MJ, Roberts S, Ridley AJ, Hodgkin MN, Stewart A, Claesson-Welsh L, Wakelam MJO (1996) Stimulation of actin fibre formation mediated by activation of phospholipase D. Curr Biol 6:588-597

Dubyak GR, Schomisch SJ, Kusner DJ, Xie M (1993) Phospholipase D activity in phagocytic leucocytes is synergistically regulated by G-protein and tyrosine kinase-based mechanism. Biochem J 292:121-128

Exton JH (1994) Phosphatidylcholine breakdown and signal transduction. Biochim. Biophys. Acta 212:26-42

Ghosh S, Strum JC, Sciorra VA, Daniel L, Bell RM (1996) Raf-1 kinase possesses distinct binding domains for phosphatidylserine and phosphatidic acid. Phosphatidic acid regulates the translocation of Raf-1 in 12-*O*-tetradecanoylphorbol-13-acetate-stimulated Madin-Darby canine kidney cells. J Biol Chem 271:8472-8480

Ha KS, Exton JH (1993a) Activation of actin polymerization by phosphatidic acid derived from phosphatidylcholine in IIC9 fibroblasts. J Cell Biol 123:1789-1796

Ha KS, Exton JH (1993b) Differential translocation of protein kinase C isozymes by thrombin and platelet-derived growth factor. A possible function for phosphatidylcholine-derived diacylglycerol. J Biol Chem 268:10534-10539

Hammond, SM, Altshuller YM, Sung TC, Rudge SA, Rose K, Engebrecht J, Morris AJ, Frohman MA (1995) Human ADP-ribosylation factor-activated phosphatidylcholine-specific phospholipase D defines a new and highly conserved gene family. J. Biol. Chem. 270: 29640-29643

Jenkins GH, Fisette PL, Anderson RA (1994) Type I phosphatidylinositol 4-phosphate 5-kinase isoforms are specifically stimulated by phosphatidic acid. J Biol Chem 269:11547-11554

Jiang H, Luo JQ, Urano T, Frankel P, Lu Z, Foster DA, Feig LA (1995) Involvement of Ral GTPase in v-Src-induced phospholipase D activation. Nature 378:409-412

Ktistakis NT, Brown HA, Sternweis PC, Roth MG (1995) Phospholipase D is present on Golgi-enriched membranes and its activation by ADP ribosylation factor is sensitive to brefeldin A. Proc Natl Acad Sci USA 92:4952-4956

Kuribara H, Tago K, Yokozeki T, Sasaki T, Morii N, Narumiya S, Katada T, Kanaho Y (1995) Synergistic activation of rat brain phospholipase D by ADP-ribosylation factor and rhoA p21, and its inhibition by *Clostridium botulinum* C3 exoenzyme. J Biol Chem 270:25667-25671

Kwak JY, Lopez I, Uhlinger DJ, Ryu SH, Lambeth JD (1995) RhoA and a cytosolic 50-kDa factor reconstitute GTPγS-dependent phospholipase D activity in human neutrophil subcellular fractions. J Biol Chem 270:27093-27098

Lambeth JD, Kwak JY, Bowman EP, Perry D, Uhlinger DJ, Lopez I (1995) ADP-ribosylation factor functions synergistically with a 50-kDa factor in cell-free activation of human neutrophil phospholipase D. J Biol Chem 270:2431-2434

Lee YH, Kim HS, Pai JK, Ryu SH, Suh PG (1994) Activation of phospholipase D induced by platelet-derived growth factor is dependent upon the level of phospholipase C-γ1. J Biol Chem 269:26842-26847

Liscovitch M, Cantley LC (1995) Signal transduction and membrane traffic: the PITP/phosphoinositide connection. Cell 81:659-662

Liscovitch M, Chalifa V, Pertile P, Chen CS, Cantley LC (1994) Novel function of phosphatidylinositol 4,5-bisphophate as a cofactor for brain membrane phospholipase D. J. Biol. Chem. 269:21403-21406
Lopez I, Burns DJ, Lambeth JD (1995) Regulation of phospholipase D by protein kinase C in human neutrophils. Conventional isoforms of protein kinase C phosphorylate a phospholipase D-related component in the plasma membrane. J Biol Chem 270:19465-19472
Malcolm KC, Ross AH, Qui RG, Symons, M, Exton JH (1994) Activation of rat liver phospholipase D by the small molecular GTP-binding protein RhoA. J Biol Chem 259:25951-25954
Massenburg D, Han JS, Liyanage M, Patton WA, Rhee SG, Moss J, Vaughan M (1994) Activation of rat brain phospholipase D by ADP-ribosylation factors 1, 5, and 6: Separation of ADP-ribosylation factor-dependent and oleate-dependent enzymes. Proc Natl Acad Sci USA 91:11718-11722
Moritz A, De Graan PNE, Gispen WH, Wirtz KWA (1992) Phosphatidic acid is a specific activator of phosphatidylinositol-4-phosphate kinase. J Biol Chem 267:7207-7210
Morris AJ, Engebrecht JA, Frohman MA (1996) Structure and regulation of phospholipase D. Trends Pharmacol Sci 17:182-185
Nobes CD, Hall A (1995) Rho, Rac, and Cdc42 GTPases regulate the assembly of multimolecular focal complexes associated with actin stress fibers, lamellipodia, and filopodia. Cell 81:53-62
Offermanns S, Wieland T, Homann D, Sandmann J, Bombien E, Spicher E, Schultz G, Jakobs KH (1994) Transfected muscarinic acetylcholine receptors selectively couple to G_i-type G proteins and $G_{q/11}$. Mol Pharmacol 45:890-898
Pertile P, Liscovitch M, Chalifa C, Cantley LC (1995) Phosphatidylinositol 4,5-bisphosphate synthesis is required for activation of phospholipase D in U937 cells. J Biol Chem 270:5130-5135
Randazzo PA, Kahn RA (1994) GTP hydrolysis by ADP-ribosylation factor is dependent on both an ADP-ribosylation factor GTPase-activating protein and acid phospholipids. J Biol Chem 269:10758-10763
Rümenapp U, Geiszt M, Wahn F, Schmidt M, Jakobs KH (1995) Evidence for ADP-ribosylation-factor-mediated activation of phospholipase D by m3 muscarinic acetylcholine receptor. Eur J Biochem 234:240-244
Schmidt M, Hüwe SM, Fasselt B, Homann D, Rümenapp U, Sandmann J, Jakobs KH (1994) Mechanisms of phospholipase D stimulation by m3 muscarinic acetylcholine receptors. Evidence for involvement of tyrosine phosphorylation. Eur J Biochem 225:667-675
Schmidt M, Rümenapp U, Bienek C, Keller J, von Eichel-Streiber C, Jakobs KH (1996a) Inhibition of receptor signaling to phospholipase D by *Clostridium difficile* toxin B. Role of Rho proteins. J Biol Chem 271:2422-2426
Schmidt M, Rümenapp U, Nehls C, Ott S, Keller J, von Eichel-Streiber C, Jakobs KH (1996b) Restoration of *Clostridium difficile* toxin B-inhibited phospholipase D by phosphatidylinositol 4,5-bisphosphate. Eur J Biochem: in press
Siddiqi AR, Smith JL, Ross AH, Qui RG, Symons M, Exton JH (1995) Regulation of phospholipase D in HL-60 cells. Evidence for a cytosolic phospholipase D. J Biol Chem 270:8466-8473
Singer WD, Brown A, Bokoch GM, Sternweis PC (1995) Resolved phospholipase D activity is modulated by cytosolic factors other than ARF. J Biol Chem 270:14944-14950
Singer WA, Brown HA, Jiang X, Sternweis PC (1996) Regulation of phospholipase D by protein kinase C is synergistic with ADP-ribosylation factor and independent of protein kinase activity. J Biol Chem 271:4504-4510
Uings IJ, Thompson NT, Randall RW, Spacey GD, Bonser RW, Hudson AT, Garland LG (1992) Tyrosine phosphorylation is involved in receptor coupling to phospholipase D but not phospholipase C in the human neutrophil. Biochem J 281:597-600
Waksman M, Eli, Y, Liscovitch M, Gerst JE (1996) Identification and characterization of a gene encoding phospholipase D activity in yeast. J Biol Chem 271:2361-2364
Ward DT, Ohanian J, Heagerty AM, Ohanian V (1995) Phospholipase D-induced phosphatidate production in intact small arteries during noradrenaline stimulation:

involvement of both G-protein and tyrosine-phosphorylation-linked pathways. Biochem J 307:451-456
Yeo EJ, Kazlauskas A, Exton JH (1994) Activation of phospholipase C-γ is necessary for stimulation of phospholipase D by platelet-derived growth factor. J Biol Chem 269:27823-27826
Yeo EJ, Exton JH (1995) Stimulation of phospholipase D by epidermal growth factor requires protein kinase C activation in Swiss 3T3 cells. J Biol Chem 270:3980-3988
Yokozeki T, Kuribara H, Katada T, Touhara K, Kanaho Y (1996) Partially purified RhoA-stimulated phospholipase D activity specifically binds to phosphatidylinositol 4,5-bisphosphate. J Neurochem 66:1234-1239

Ceramide: A Central Regulator of The Cellular Response to Injury and Stress

Ghassan S. Dbaibo
Department of Pediatrics
American University of Beirut
Beirut, Lebanon

Yusuf A. Hannun[1]
Department of Medicine
Duke University Medical Center
Durham, NC 27710, USA

Introduction

During evolution, a variety of responses to stress and injury were developed and preserved. These stress responses allow the cell or the organism to survive a wide range of stresses such as genotoxic damage, extreme physical conditions, infection, nutritional starvation or others. The nature of the stressful condition determines the type of response generated by the organism. For example, nutritional starvation results in metabolic adaptation with catabolism of new substrates and slowing down of the metabolic rate, genotoxic damage results in the induction of several genes which orchestrate the repair of damaged DNA, while severely damaged or infected cells are eliminated by a process of programmed cell death, known as apoptosis. The latter helps rid the organism of defective cells that may compromise its survival. The specific biochemical events which control the various stress responses are poorly understood (1-3).

Recently, ceramide, the structural backbone of sphingolipids, has emerged as a candidate modulator of the stress response. Ceramide has been shown to have predominantly growth-suppressive activities (4,5) including the induction of cell cycle arrest (6,7) and apoptosis

[1] Send correspondance to: Yusuf A. Hannun, Duke University Medical Center, P.O. Box 3355, Durham, NC, 27710, USA

NATO ASI Series, Vol. H 101
Molecular Mechanisms of Signalling
and Membrane Transport
Edited by Karel W. A. Wirtz

(8,9). Additionally, ceramide has been shown to activate the stress activated protein kinase (SAPK) known as c-Jun N-terminal protein kinase (JNK) which is thought to play a central role in a variety of stress responses (10). Moreover, the tumor suppressor p53, one of the central regulators of the response to DNA damage, has recently been shown to induce ceramide accumulation upon its induction (Dbaibo et al, submitted). In this chapter we will review the evidence which supports a role for ceramide in the stress response

Inducers of Ceramide Accumulation

Many inducers of ceramide accumulation have been described. These generally fall into three major categories: 1) cytokines and components of the immune system involved in inflammation, host defense, and clonal deletion [e.g. tumor necrosis factor (TNF) alpha, interleukin-1 beta, gamma-interferon, CD28, Fas, and complement] (11-19), 2) cytotoxic agents such as chemotherapeutic agents and irradiation (20-24), or 3) stressful conditions such as serum deprivation of cultured cells and heat shock (6,25,26). These inducers cause either cytotoxicity (mostly apoptosis) in target cells or cell cycle arrest. As discussed below, there is considerable evidence that ceramide may mediate both biologic effects.

The kinetics of ceramide generation following these inducers depend on the inducer, its dose, and the target cell type and range from minutes to hours (6,19,27). However, in most systems studied significant ceramide generation (2-fold and more) occurs only after several hours of stimulation.

The tumor suppressor p53 as an inducer of ceramide

p53 has been proposed as the "guardian of the genome", activated in response to DNA-damaging agents and required for cell cycle arrest and/or apoptosis in response to these agents (28). Genotoxic insults such as chemotherapeutic agents, gamma irradiation, or ultraviolet irradiation, have been shown to induce p53 expression. p53 appears to function by either inducing cell cycle arrest, allowing the repair of damaged DNA, or by inducing apoptosis. The similarities between the functions of p53 and ceramide, led to the examination of their relationship. It was found that, in a p53-dependent pathway, e.g. induction of apoptosis of

Molt-4 cells by low concentrations of actinomycin D, induction of p53 was followed by a dramatic rise in endogenous ceramide occurring over several hours (Dbaibo et al, submitted) . In turn, ceramide accumulation was followed by apoptosis and cell cycle arrest. Exogenous cell-permeable ceramide analogs simulated the effects of actinomycin D on apoptosis but not on p53 induction. This indicated that p53 functions upstream of ceramide on the apoptotic pathway stimulated by low concentrations of actinomycin D. These results indicate that ceramide may mediate some of the effects of p53 in the response to genotoxic damage. Moreover, the presence of p53-independent pathways of apoptosis and cell cycle arrest which are mediated by ceramide, e.g. TNF alpha, indicates that ceramide may be more universally involved in the response to stress compared to p53.

The Mechanism of Action of Ceramide

Cell-permeable ceramide analogs were shown to reproduce the biologic effects of the specific inducer in a number of different systems. More importantly, analogs of the naturally occurring D-*erythro*-ceramide, were the most potent of the ceramide stereoisomers in producing growth suppression. Additionally, D-*erythro*-dihydroceramide, the metabolic precursor of ceramide which differs structurally only by the absence of the *trans*-4,5 double bond, was completely inert in producing any effects attributable to ceramide (8,29,30). These specific structural and stereochemical requirements suggested that a specific biochemical target is directly activated by ceramide to produce its biologic effects. Several candidate effector targets have been proposed including ceramide-activated protein kinase (31) and protein kinase ζ (32) but their role in mediating the biologic effects of ceramide remains to be proven. Most importantly, ceramide has been shown to specifically activate a phosphatase termed ceramide-activated protein phosphatase (CAPP) (33). This phosphatase is a member of the 2A family of serine/threonine protein phosphatases (34) and has been shown to be an effector molecule for ceramide-induced biology (35). The detection of this activity in the yeast *saccharomyces cerevisiae* (36) underscores the central importance of this pathway which has necessitated its evolutionary conservation.

Activation of JNK by ceramide

The stress-activated protein kinases (SAPKs) are presumably involved in the cellular response to intra- and extracellular stress such as heat shock, protein synthesis inhibition, ultraviolet light, and TNFα. A prototype of this family is JNK which preferentially phosphorylates the c-Jun and ATF2 transcription factors (37). This phosphorylation results in an increase in their transcriptional activating potential. This activation allows c-Jun and ATF2 to stimulate the expression of a variety of genes which presumably allow the organism to rapidly adapt to stressful situations or undergo apoptosis. An essential role for JNK and related kinases in apoptosis was recently proposed when inhibition of the upstream kinases responsible for phosphorylating and activating JNK, by expressing dominant negative mutants, protected cells from apoptosis induced by NGF withdrawal (38).

Since some of the inducers of JNK were known to activate the ceramide pathway, the role of ceramide was investigated. Exogenous bacterial sphingomyelinase, which results in the hydrolysis of membrane sphingomyelin and generation of ceramide, was found to reproduce the effects of TNFα on JNK activation in HepG2 cells (37). C_2-ceramide and sphingomyelinase, but not dihydroceramide, activated JNK in HL-60 cells within 5 minutes (10). Moreover, expression of dominant negative protein kinases upstream of JNK which effectively blocks the sequential activation which occurs in this pathway, rendered transfected cells resistant to apoptosis induced by TNF alpha or C_2-ceramide (39). This protection occurred without interference with ceramide generation. Therefore, these studies indicated that activation of the JNK pathway was potentially required for the ceramide-mediated stress response.

Ceramide-Mediated Biology:
1. Role in cell cycle arrest

Control of the cell cycle is a fundamental process in biology with paramount importance. The intricate mechanisms by which the cell cycle is regulated are slowly being discovered. The retinoblastoma protein (Rb) is now recognized as a central regulator of the cell cycle (40-42). Rb is a phosphoprotein whose function is determined by its phosphorylation status (42-45). The hypophosphorylated form is the active form and predominates in G_0/G_1. As Rb becomes gradually phosphorylated towards the end of G_1, it loses its growth-suppressive effects. The mechanism of action of Rb appears to be through the ability of the hypophosphorylated form to bind and inactivate a number of transcription factors which are crucial for the transcription of growth promoting genes. These include the E2F family of transcription factors as well as Elf-1, Myo-D, PU.1, ATF-2, and c-Abl (46,47). The hypophosphorylated form of Rb is targeted by a number of viral proteins such as E1A of adenovirus, large T antigen of simian virus 40 (SV40), and E7 of papillomavirus (46-51). These viral proteins bind and inactivate Rb allowing the release of transcription factors important for viral replication.

A key mechanism by which cell cycle progression is regulated is the control of the phosphorylation state of Rb. Cyclin-dependent kinases (cdk), particularly those regulated by the D and E cyclins, phosphorlyate Rb at more than a dozen serine and threonine residues. Rb can be maintained in a hypophosphorylated state when the activity of these cyclin-cdk complexes is inhibited. Several proteins have been described which specifically inhibit cdk function including p21, p16, p15, and p27. These inhibitors are induced by a number of growth-suppressive signals such as transforming growth factor beta and p53 (42). Alternatively, Rb can be dephosphorylated by the action of the recently described Rb phosphatase which functions independently of p53, p21, and the cyclin-cdk complexes (52). Therefore, cell cycle control is partly achieved by a balance of signals which regulate the phosphorylation status of Rb.

A role for ceramide in the regulation of Rb phosphorylation was suggested following studies utilizing the serum deprivation model. When Molt-4 cells were serum deprived, a specific G_0/G_1 cell cycle arrest occurred (6). These effects were first seen at 24 hours following serum deprivation but became maximal (> 96 % in G_0/G_1) by 48 hours. This was accompanied by a gradual rise in endogenous ceramide which reached 8- and 15-fold over control levels at 48 and 96 hours, respectively. The possibility that ceramide may be mediating the effects of serum deprivation on cell cycle arrest was tested. Treatment of Molt-4 cells with cell-permeable ceramides, but not dihydroceramide, resulted in a specific arrest of cells in the G_0/G_1 phase of the cell cycle. These effects were evident following 16 hours of treatment. Similarly, experiments done with glucosylceramide synthase inhibitor, PDMP (threo-1-phenyl-2-decanoylamino-3-morpholino-1-propanol), which blocks the synthesis of more complex sphingolipids and results in the accumulation of endogenous ceramide (53) supported the studies utilizing ceramide analogs. Treatment of NIH 3T3 cells overexpressing the insulin-like growth factor-1 receptor with PDMP resulted in a time dependent accumulation of these cells in G_0/G_1 and G_2 which was attributed to the noted increase in endogenous ceramide. This was a reversible effect since the removal of PDMP allowed the cells to progress through the cell cycle after ceramide levels returned to normal.

Naturally, the effects of serum deprivation and ceramide on cell cycle arrest led to the examination of the role of Rb in mediating these effects. It was found that following serum deprivation of Molt-4 cells, Rb became dephosphorylated starting at 24 hours and was maximal by 96 hours. Next, the effects of ceramide on Rb phosphorylation were examined. Treatment of Molt-4 cells with cell-permeable ceramides resulted in a dose- and time-dependent dephosphorylation of Rb with significant effects seen after 4-6 hours using 10 μM of ceramide (7). The role of Rb in mediating ceramide-induced growth suppression was examined using cell lines which lack functional Rb due to mutation or expression of viral proteins which inactivate Rb. The effects of ceramide on these cells were compared with its effects on cells expressing functional Rb. Rb-deficient cells were found to be relatively resistant to the growth suppressive effects of ceramide. Significantly, expression of the adenoviral E1A protein in Molt-4 cells, which effectively abrogates Rb function, rendered these cells resistant to the specific effects of ceramide on the G_0/G_1 cell cycle when compared to vector cells. Therefore, these studies suggest that ceramide may be an important regulator of Rb activity. In turn, Rb appears to partially mediate the growth suppressive effects of ceramide by inducing cell cycle arrest. Future studies aim at defining the molecular targets involved in this pathway; specifically, the effects of ceramide on the cyclin-cdk activities and the Rb phosphatase.

2. Role of ceramide in apoptosis:

Apoptosis is a highly conserved mechanism of cell suicide involved in development, oncogenesis, and the immune response (54,55). This process allows the organism to dispose of unwanted cells in order to achieve proper development and maintain homeostasis. Under stressful conditions such as infection, unrepairable genetic damage, or nutritional or growth factor deprivation, the sacrifice of a few cells allows the organism to survive and reproduce (56). When a cell receives a suicide signal, several enzymes are activated, particularly proteases and endonucleases, resulting in the organized packaging of the digested cell into small vesicles known as “apoptotic bodies” that are phagocytosed (57). This process spares the organism from the inflammatory response seen with necrotic cell death which is initiated by the release of intracellular molecules (55).

The molecular regulators of apoptosis are rapidly being identified. Two families of proteins appear to play an essential role in apoptosis. The first, is the family of cysteine proteases related to the Interleukin-1β converting enzyme (ICE) (57). This family of proteases includes the product of the *ced-3* death gene identified in the nematode *Caenorhabditis elegans*. These proteases cleave specific targets, including other members of the ICE family, following an aspartic acid residue which results in either activation or inactivation of the target protein

(58,59). Activation of a cascade of these proteases is believed to be a necessary event in the execution of apoptosis. The second family of apoptosis-regulating proteins is the Bcl-2 family which are homologous to an anti-apoptotic gene product from *C. elegans*, Ced-9. Members of this family either promote or inhibit apoptosis and can either homo- or heterodimerize through homologous binding domains. The ratio of death-promoting (Bax, Bad, Bak, Bag, and Bcl-x_S) versus death-inhibiting (e.g. Bcl-2 and Bcl-x_L) members in a cell determines the ultimate fate of the cell (60). Despite extensive study, the mechanism of action of these proteins remains unknown.

A number of the inducers of ceramide accumulation, particularly TNF alpha, were known to have "cytotoxic effects". These effects were later demonstrated to be due to apoptotic death. Similarly, ceramide was shown to be cytotoxic in several early studies (14,30,61). Therefore, it became important to determine whether ceramide can mediate the specific apoptotic effects of some of its inducers. Initial studies were done in the leukemia cell line, U937, which undergoes apoptosis in response to TNF alpha. Treatment of these cells with C_2-ceramide, resulted in DNA fragmentation on gel electrophoresis typical of apoptosis (8). These effects occurred within 1-3 hours and were specific to C_2-ceramide since treatment with dihydroceramide or other amphiphilic lipids was ineffective. Importantly, activation of the protein kinase C (PKC) pathway by phorbol ester resulted in attenuation of ceramide-induced apoptosis raising the possibility that the balance between these two opposing pathways may determine the fate of the cell (8). Later, ceramide was shown to induce apoptosis in many cell types mimicking the effects of a number of its inducers including serum deprivation, radiation, vincristine, and IgM antibody or Fas cross-linking (6,9,17,18,21,23,24).

The mechanism by which ceramide drives the apoptotic pathway was studied next. It became important to examine its relationship with the ICE and Bcl-2 families of proteins. It was hypothesized that ceramide may activate one or more proteases belonging to the ICE family and hence initiate the apoptotic cascade. In order to test this hypothesis, cleavage of a known substrate for these proteases was assessed following ceramide treatment. The substrate chosen was the enzyme poly(ADP-ribose) polymerase (PARP) which is cleaved by some, but not all, members of this family of proteases (62,63). PARP, which is involved in DNA repair, becomes cleaved during apoptosis to a signature 85 kDa fragment by the preferential action of the ICE-like protease CPP32 (prICE/Yama/apopain) (64,65) (66). C_2- or C_6-ceramide treatment of Molt-4 cells resulted in cleavage of PARP to its specific apoptotic fragment after 3-4 hours (67). These effects were specific to ceramide and were not seen after treatment with dihydroceramide or dioctanoylglycerol. Therefore these studies defined a novel downstream target of ceramide closely coupled to its apoptotic activities.

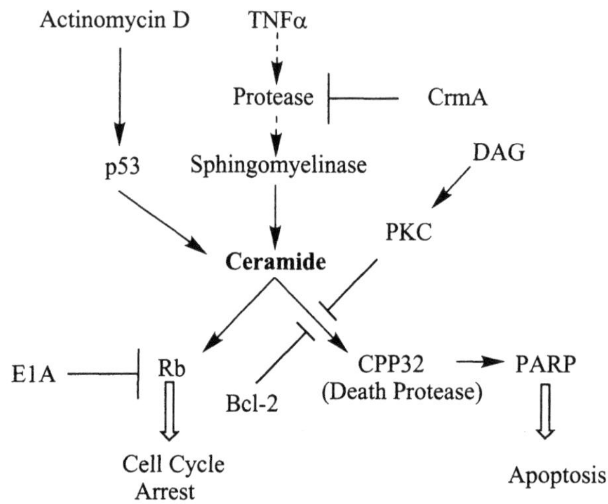

Figure: Schematic presentation of pathways and molecules interacting with the ceramide pathway in regulating apoptosis and cell cycle arrest.

Recently, the cowpox virus protein CrmA (cytokine response modifier) was found to be a potent inhibitor of ICE (68). In addition to shutting down Interleukin-1 beta production, CrmA was found to be a potent inhibitor of apoptosis (69). Its anti-apoptotic activity was presumed to be secondary to its ability to inhibit the ICE family of proteases. However, it was found that CrmA inhibited some members of this family more effectively than others. Specifically, it was a 1000-fold more potent in inhibiting ICE compared to CPP32. Therefore, it became important to determine whether expression of CrmA protected from ceramide-induced apoptosis. The TNF alpha sensitive MCF-7 breast carcinoma cell line was used to develop CrmA-expressing cells. Expression of CrmA in these cells rendered them remarkably resistant to apoptosis induced by TNF alpha when compared to vector cells. Surprisingly, CrmA-expressing cells remained as sensitive to ceramide-induced apoptosis as the vector cells. More importantly, PARP was cleaved in response to ceramide equally whether CrmA was expressed or not (Dbaibo et. al., submitted). These results suggested that CrmA does not function by inhibiting the PARP-cleaving ICE-like protease, rather, an alternative mechanism was involved. Therefore, the accumulation of ceramide following TNF alpha was studied in these cells. In vector-transfected cells, TNF alpha resulted in a several fold accumulation of endogenous ceramide starting at 7-9 hours and reaching 4-5 fold by 24 hours. In contrast, CrmA expression completely inhibited ceramide accumulation and subsequent apoptosis. These studies suggested that CrmA was targeting a protease upstream of ceramide accumulation and that this was the major mechanism by which it was inhibiting apoptosis.

The interactions of ceramide and Bcl-2 were also studied in several laboratories. Using cells overexpressing Bcl-2, studies showed that these cells were resistant to apoptosis induced by ceramide and many of its inducers (21,70,71). For example, after treatment of Molt-4 cells with vincristine, a vinca alkaloid chemotherapeutic agent, ceramide accumulated and the cells subsequently underwent apoptosis (21). Overexpression of Bcl-2 in these cells protected them from the apoptotic effects of vincristine *and* cell permeable ceramide. Interestingly, ceramide accumulation was not inhibited indicating that Bcl-2 was functioning downstream of ceramide. Bcl-2 was found to inhibit ceramide-induced PARP cleavage indicating that it was targeting an event either at, or proximal to, CPP32 activation but distal to ceramide accumulation. Similar results were obtained in MCF-7 cells overexpressing Bcl-2 and treated with TNF alpha.

These results suggest that at least two ICE-like proteases are involved in the ceramide pathway: one upstream of ceramide generation inhibitable by CrmA and the other is ceramide-activated and, probably indirectly, inhibitable by Bcl-2 but not CrmA (see figure).

Conclusions

A central role for ceramide as a regulator of the stress response is emerging. A number of stressful stimuli appears to converge on ceramide which then functions as a "gauge" of cell stress. Ceramide accumulation is not a terminal event as indicated by studies where apoptosis was prevented by Bcl-2 expression, PKC activation, or inhibition of JNK activation. Instead, ceramide may function as a sensor which drives the cell towards adaptation, repair, or suicide depending on the degree of expression of proteins, such as Bcl-2, intracellular level of diglycerides, or the level of activation of the mitogen-activated protein kinase (MAPK) pathway which appears to counterbalance the JNK pathway (38). A role for ceramide in driving the repair of damaged cells is supported by its transcriptional activation of alpha B-crystallin, one of the heat shock proteins which are thought to function by repairing misfolded proteins (25) and by its induction of cell cycle arrest. Notably, neither Bcl-2 expression nor phorbol ester treatment interferes with ceramide-induced Rb dephosphorylation and cell cycle arrest. Under these conditions, ceramide elevation drives the cell towards a G0/G1 arrest allowing the recovery from the imposed stress.

Therefore, understanding the pathways which regulate ceramide generation and identifying the target molecules which are activated by ceramide will not only improve our understanding of the cellular response to stress and the regulation of cell life and death but may also provide novel targets for intervention in a variety of human diseases.

References

1. Hilt, W., and Wolf, D. H. (1992). Stress-induced proteolysis in yeast. Mol. Microbiol. 6, 2437-2442.
2. Fornace, A. J. Jr, Jackman, J., Hollander, M. C., Hoffman-Liebermann, B., and Liebermann, D. A. (1992). Genotoxic-stress-response genes and growth-arrest genes. gadd, MyD, and other genes induced by treatments eliciting growth arrest. Ann. N. Y. Acad. Sci. 663, 139-153.
3. Canman, C. E., Chen, C. Y., Lee, M. H., and Kastan, M. B. (1994). DNA damage responses: p53 induction, cell cycle perturbations, and apoptosis. Cold Spring Harb. Symp. Quant. Biol. 59, 277-286.
4. Hannun, Y. A., and Linardic, C. M. (1993). Sphingolipid breakdown products: anti-proliferative and tumor-suppressor lipids. Biochim. Biophys. Acta Bio-Membr. 1154, 223-236.
5. Pushkareva, M., Obeid, L. M., and Hannun, Y. A. (1995). Ceramide: an endogenous regulator of apoptosis and growth suppression. Immunol. Today. 16, 294-297.
6. Jayadev, S., Liu, B., Bielawska, A. E., Lee, J. Y., Nazaire, F., Pushkareva, M. Y. u., Obeid, L. M., and Hannun, Y. A. (1995). Role for ceramide in cell cycle arrest. J. Biol. Chem. 270, 2047-2052.
7. Dbaibo, G. S., Pushkareva, M. Y., Jayadev, S., Schwarz, J. K., Horowitz, J. M., Obeid, L. M., and Hannun, Y. A. (1995). Retinoblastoma gene product as a downstream target for a ceramide-dependent pathway of growth arrest. Proc. Natl. Acad. Sci. U S A. 92, 1347-1351.
8. Obeid, L. M., Linardic, C. M., Karolak, L. A., and Hannun, Y. A. (1993). Programmed cell death induced by ceramide. Science. 259, 1769-1771.
9. Jarvis, W. D., Kolesnick, R. N., Fornari, F. A., Traylor, R. S., Gewirtz, D. A., and Grant, S. (1994). Induction of apoptotic DNA damage and cell death by activation of the sphingomyelin pathway. Proc. Natl. Acad. Sci. USA. 91, 73-77.
10. Westwick, J. K., Bielawska, A. E., Dbaibo, G., Hannun, Y. A., and Brenner, D. A. (1995). Ceramide activates the stress-activated protein kinases. J. Biol. Chem. 270, 22689-22692.
11. Kim, M. -Y, Linardic, C., Obeid, L., and Hannun, Y. (1991). Identification of sphingomyelin turnover as an effector mechanism for the action of tumor necrosis factor alpha and gamma-interferon. Specific role in cell differentiation. J. Biol. Chem. 266, 484-489.
12. Niculescu, F., Rus, H., Shin, S., Lang, T., and Shin, M. L. (1993). Generation of diacylglycerol and ceramide during homologous complement activation. J. Immunol. 150, 214-224.
13. Okazaki, T., Bell, R. M., and Hannun, Y. A. (1989). Sphingomyelin turnover induced by vitamin D_3 in HL-60 cells. Role in cell differentiation. J. Biol. Chem. 264, 19076-19080.
14. Okazaki, T., Bielawska, A., Bell, R. M., and Hannun, Y. A. (1990). Role of ceramide as a lipid mediator of 1 alpha,25-dihydroxyvitamin D_3-induced HL-60 cell differentiation. J. Biol. Chem. 265, 15823-15831.
15. Ballou, L. R., Chao, C. P., Holness, M. A., Barker, S. C., and Raghow, R. (1992). Interleukin-1-mediated PGE2 production and sphingomyelin metabolism. Evidence for the regulation of cyclooxygenase gene expression by sphingosine and ceramide. J. Biol. Chem. 267, 20044-20050.
16. Boucher, L. M., Wiegmann, K., Futterer, A., Pfeffer, K., Machleidt, T., Schütze, S., Mak, T. W., and Krönke, M. (1995). CD28 signals through acidic sphingomyelinase. J. Exp. Med. 181, 2059-2068.
17. Cifone, M. G., De Maria, R., Roncaioli, P., Rippo, M. R., Azuma, M., Lanier, L. L., Santoni, A., and Testi, R. (1994). Apoptotic signaling through CD95 (Fas/Apo-1) activates an acidic sphingomyelinase. J. Exp. Med. 180, 1547-1552.

18. Tepper, C. G., Jayadev, S., Liu, B., Bielawska, A., Wolff, R., Yonehara, S., Hannun, Y. A., and Seldin, M. F. (1995). Role of ceramide as an endogenous mediator of Fas-induced cytotoxicity. Proc. Natl. Acad. Sci. USA. 92, 8443-8447.
19. Mathias, S., Younes, A., Kan, C. -C, Orlow, I., Joseph, C., and Kolesnick, R. N. (1993). Activation of the sphingomyelin signaling pathway in intact EL4 cells and in a cell-free system by IL-1 beta. Science. 259, 519-522.
20. Strum, J. C., Small, G. W., Pauig, S. B., and Daniel, L. W. (1994). 1-b-D-arabinofuranosylcytosine stimulates ceramide and diglyceride formation in HL-60 cells. J. Biol. Chem. 269, 15493-15497.
21. Zhang, J., Alter, N., Reed, J. C., Borner, C., Obeid, L. M., and Hannun, Y. A. (1996). Bcl-2 interrupts the ceramide-mediated pathway of cell death. Proc. Natl. Acad. Sci., USA. 93, 5325-5328.
22. Bose, R., Verheij, M., Haimovitz-Friedman, A., Scotto, K., Fuks, Z., and Kolesnick, R. N. (1996). Ceramide synthase mediates daunorubicin-induced apoptosis: an alternative mechanism for generating death signals. Cell. 82, 405-414.
23. Haimovitz-Friedman, A., Kan, C. C., Ehleiter, D., Persaud, R. S., McLoughlin, M., Fuks, Z., and Kolesnick, R. N. (1994). Ionizing radiation acts on cellular membranes to generate ceramide and initiate apoptosis. J. Exp. Med. 180, 525-535.
24. Quintans, J., Kilkus, J., McShan, C. L., Gottschalk, A. R., and Dawson, G. (1994). Ceramide mediates the apoptotic response of WEHI 231 cells to anti-immunoglobulin, corticosteroids and irradiation. Biochem. Biophys. Res. Commun. 202, 710-714.
25. Chang, Y., Abe, A., and Shayman, J. A. (1995). Ceramide formation during heat shock: A potential mediator of alpha B-crystallin transcription. Proc. Natl. Acad. Sci., USA. 92, 12275-12279.
26. Pronk, G. J., Ramer, K., Amiri, P., and Williams, L. T. (1996). Requirement of an ICE-like protease for induction of apoptosis and ceramide generation by REAPER. Science. 271, 808-810.
27. Schütze, S., Potthoff, K., Machleidt, T., Berkovic, D., Wiegmann, K., and Krönke, M. (1992). TNF activates NF-kabba B by phosphatidylcholine-specific phospholipase C-induced "acidic" sphingomyelin breakdown. Cell. 71, 765-776.
28. Lane, D. P. (1992). p53, guardian of the genome. Nature. 358, 15-16.
29. Bielawska, A., Crane, H. M., Liotta, D., Obeid, L. M., and Hannun, Y. A. (1993). Selectivity of ceramide-mediated biology: lack of activity of erythro-dihydroceramide. J. Biol. Chem. 268, 26226-26232.
30. Dbaibo, G. S., Obeid, L. M., and Hannun, Y. A. (1993). TNFa signal transduction through ceramide: dissociation of growth inhibitory effects of TNFa from activation of NF-kB. J. Biol. Chem. 268, 17762-17766.
31. Liu, J., Mathias, S., Yang, Z., and Kolesnick, R. N. (1994). Renaturation and tumor necrosis factor-alpha stimulation of a 97-kDa ceramide-activated protein kinase. J. Biol. Chem. 269, 3047-3052.
32. Lozano, J., Berra, E., Municio, M. M., Diaz-Meco, M. T., Dominguez, I., Sanz, L., and Moscat, J. (1994). Protein kinase C z isoform is critical for kB-dependent promoter activation by sphingomyelinase. J. Biol. Chem. 269, 19200-19202.
33. Dobrowsky, R. T., and Hannun, Y. A. (1992). Ceramide stimulates a cytosolic protein phosphatase. J. Biol. Chem. 267, 5048- 5051.
34. Dobrowsky, R. T., Kamibayashi, C., Mumby, M. C., and Hannun, Y. A. (1993). Caramide activates heterotrimeric protein phosphatase 2A. J. Biol. Chem. 268, 15523-15530.
35. Wolff, R. A., Dobrowsky, R. T., Bielawska, A., Obeid, L. M., and Hannun, Y. A. (1994). Role of ceramide-activated protein phosphatase in ceramide-mediated signal transduction. J. Biol. Chem. 269, 19605-19609.
36. Fishbein, J. D., Dobrowsky, R. T., Bielawska, A., Garrett, S., and Hannun, Y. A. (1993). Ceramide-mediated biology and CAPP are conserved in Saccharomyces cerevisiae. J. Biol. Chem. 268, 9255-9261.

37. Kyriakis, J. M., Banerjee, P., Nikolakaki, E., Dai, T., Rubie, E. A., Ahmad, M. F., Avruch, J., and Woodgett, J. R. (1994). The stress-activated protein kinase subfamily of c-Jun kinases. Nature. 369, 156-160.
38. Xia, Z., Dickens, M., Raingeaud, J., Davis, R. J., and Greenberg, M. E. (1995). Opposing effects of ERK and JNK-p38 MAP kinases on apoptosis. Science. 270, 1326-1331.
39. Verheij, M., Bose, R., Lin, X. H., Yao, B., Jarvis, W. D., Grant, S., Birrer, M. J., Szabo, E., Zon, L. I., Kyriakis, J. M., Haimovitz-Friedman, A., Fuks, Z., and Kolesnick, R. N. (1996). Requirement for ceramide-initiated SAPK/JNK signalling in stress-induced apoptosis. Nature. 380, 75-79.
40. Weinberg, R. A. (1990). The retinoblastoma gene and cell growth control. Trends Biochem. Sci. 15, 199-202.
41. Harlow, E. (1992). For our eyes only. Nature. 359, 270-271.
42. Weinberg, R. A. (1995). The retinoblastoma protein and cell cycle control. Cell. 81, 323-330.
43. Ludlow, J. W., Shon, J., Pipas, J. M., Livingston, D. M., and DeCaprio, J. A. (1990). The retinoblastoma susceptibility gene product undergoes cell cycle-dependent dephosphorylation and binding to and release from SV40 large T. Cell. 60, 387-396.
44. Goodrich, D. W., Wang, N. P., Qian, Y. -W, Lee, E. Y. -H.P., and Lee, W. -H (1991). The retinoblastoma gene product regulates progression through the G1 phase of the cell cycle. Cell. 67, 293-302.
45. DeCaprio, J. A., Furukawa, Y., Ajchenbaum, F., Griffin, J. D., and Livingston, D. M. (1992). The retinoblastoma-susceptibility gene product becomes phosphorylated in multiple stages during cell cycle entry and progression. Proc. Natl. Acad. Sci. USA. 89, 1795-1798.
46. Nevins, J. R. (1992). E2F: a link between the Rb tumor suppressor protein and viral oncoproteins. Science. 258, 424-429.
47. Wang, J. Y., Knudsen, E. S., and Welch, P. J. (1994). The retinoblastoma tumor suppressor protein. Adv. Cancer Res. 64, 25-85.
48. Bagchi, S., Raychaudhuri, P., and Nevins, J. R. (1990). Adenovirus E1A proteins can dissociate heteromeric complexes involving the E2F transcription factor: a novel mechanism for E1A trans-activation. Cell. 62, 659-669.
49. Chellappan, S., Kraus, V. B., Kroger, B., Munger, K., Howley, P. M., Phelps, W. C., and Nevins, J. R. (1992). Adenovirus E1A, simian virus 40 tumor antigen, and human papillomavirus E7 protein share the capacity to disrupt the interaction between transcription factor E2F and the retinoblastoma gene product. Proc. Natl. Acad. Sci. USA. 89, 4549-4553.
50. Ludlow, J. W. (1993). Interactions between SV40 large-tumor antigen and the growth suppressor proteins pRB and p53. FASEB J. 7, 866-871.
51. Vousden, K. (1993). Interactions of human papillomavirus transforming proteins with the products of tumor suppressor genes. FASEB J. 7, 872-879.
52. Dou, Q. P., An, B., and Will, P. L. (1995). Induction of a retinoblastoma phosphatase activity by anticancer drugs accompanies p53-independent G1 arrest and apoptosis. Proc. Natl. Acad. Sci. U S A. 92, 9019-9023.
53. Rani, C. S., Abe, A., Chang, Y., Rosenzweig, N., Saltiel, A. R., Radin, N. S., and Shayman, J. A. (1995). Cell cycle arrest induced by an inhibitor of glucosylceramide synthase. Correlation with cyclin-dependent kinases. J. Biol. Chem. 270, 2859-2867.
54. Wyllie, A. H. (1980). Glucocorticoid-induced thymocyte apoptosis is associated with endogenous endonuclease activation. Nature. 284, 555-556.
55. Gerschenson, L. E., and Rotello, R. J. (1992). Apoptosis: a different type of cell death. FASEB J. 6, 2450-2455.
56. Vaux, D. L., and Strasser, A. (1996). The molecular biology of apoptosis. Proc. Natl. Acad. Sci., USA. 93, 2239-2244.
57. Martin, S. J., and Green, D. R. (1995). Protease activation during apoptosis: death by a thousand cuts?. Cell. 82, 349-352.

58. Chinnaiyan, A. M., Tepper, C. G., Seldin, M. F., O'Rourke, K., Kischkel, F. C., Hellbardt, S., Krammer, P. H., Peter, M. E., and Dixit, V. M. (1996). FADD/MORT1 is a common mediator of CD 95 (Fas/APO-1) and tumor necrosis factor receptor-induced apoptosis. J. Biol. Chem. 271, 4961-4965.
59. Enari, M., Talanian, R. V., Wong, W. W., and Nagata, S. (1996). Sequential activation of ICE-like and CPP32-like proteases during Fas-mediated apoptosis. Nature. 380, 723-726.
60. Oltvai, Z. N., and Korsmeyer, S. J. (1994). Checkpoints of dueling dimers foil death wishes. Cell. 79, 189-192.
61. Bielawska, A., Linardic, C. M., and Hannun, Y. A. (1992). Ceramide-mediated biology: determination of structural and stereospecific requirements through the use of N-acyl-phenylaminoalcohol analogs. J. Biol. Chem. 267, 18493-18497.
62. Kaufmann, S. H., Desnoyers, S., Ottaviano, Y., Davidson, N. E., and Poirier, G. G. (1993). Specific proteolytic cleavage of poly(ADP-ribose) polymerase: an early marker of chemotherapy-induced apoptosis. Cancer Res. 53, 3976-3985.
63. Lazebnik, Y. A., Kaufmann, S. H., Desnoyers, S., Poirier, G. G., and Earnshaw, W. C. (1994). Cleavage of poly(ADP-ribose) polymerase by a proteinase with properties like ICE. Nature. 371, 346-347.
64. Fernandes-Alnemri, T., Litwack, G., and Alnemri, E. S. (1994). CPP32, a novel human apoptotic protein with homology to Caenorhabditis elegans cell death protein Ced-3 and mammalian interleukin-1 beta-converting enzyme. J. Biol. Chem. 269, 30761-30764.
65. Tewari, M., Quan, L. T., O'Rourke, K., Desnoyers, S., Zeng, Z., Beidler, D. R., Poirier, G. G., Salvesen, G. S., and Dixit, V. M. (1995). Yama/CPP32 beta, a mammalian homolog of CED-3, is a CrmA-inhibitable protease that cleaves the death substrate poly(ADP-ribose) polymerase. Cell. 81, 801-809.
66. Nicholson, D. W., Ali, A., Thornberry, N. A., Vaillancourt, J. P., Ding, C. K., Gallant, M., Gareau, Y., Griffin, P. R., Labelle, M., Lazebnik, Y. A., Munday, N. A., Raju, S. M., Smulson, M. E., Yamin, T., Yu, V. L., and Miller, D. K. (1995). Identification and inhibition of the ICE/CED-3 protease necessary for mammalian apoptosis. Nature. 376, 37-43.
67. Smyth, M. J., Perry, D. K., Zhang, J., Poirier, G. G., Hannun, Y. A., and Obeid, L. M. (1996). prICE: a downstream target for ceramide-induced apoptosis and for the inhibitory action of bcl-2. Biochem. J. in press.
68. Ray, C. A., Black, R. A., Kronheim, S. R., Greenstreet, T. A., Sleath, P. R., Salvesen, G. S., and Pickup, D. J. (1992). Viral inhibition of inflammation: cowpox virus encodes an inhibitor of the interleukin-1beta converting enzyme. Cell. 69, 597-604.
69. Tewari, M., and Dixit, V. M. (1995). Fas- and tumor necrosis factor-induced apoptosis is inhibited by the poxvirus crmA gene product. J. Biol. Chem. 270, 3255-3260.
70. Martin, S. J., Takayama, S., McGahon, A. J., Miyashita, T., Corbeil, J., Kolesnick, R. N., Reed, J. C., and Green, D. R. (1995). Inhibition of ceramide-induced apoptosis by Bcl-2. Cell Death Differ. 2, 253-257.
71. Karasavvas, N., Erukulla, R. K., Bittman, R., Lockshin, R., Hockenbery, D., and Zakeri, Z. (1996). BCL-2 suppresses ceramide-induced cell killing. Cell Death Differ. 3, 149-151.

Ceramide changes during FAS (CD95/APO-1) mediated programmed cell death are blocked with the ICE protease inhibitor zVAD.FMK[1]

Evidence against an upstream role for ceramide in apoptosis

Daniel J. Sillence, Mike D. Jacobson and David Allan

Department of Physiology and Laboratory for Molecular Cell Biology
University College London
Rockefeller Building
University St
London WC1E 6JJ

SUMMARY

We sought to investigate whether the changes in ceramide that occur duing FAS mediated programmed cell death (PCD) are upstream or downstream of the activation of ICE-like proteases. Changes in ceramide have been studied after challenge with the anti-FAS antibody, anti APO-1 in SKW 6.4 and U937 cells. Challenge with anti APO-1 leads to a 50% drop in cell survival as judged by nuclear morphology after 4 hours of incubation. In both U937 and SKW 6.4 cultures labelled for 48 hours with ^{14}C acetate, the amount of ceramide approximately doubled after 24h incubation with anti APO-1 but the time course of ceramide changes was slower than the time course of anti APO-1

Abbreviations:
zVAD.fmk, N-benzyloxycarbonyl-Val-Ala-Asp-(O-methyl)fluoromethylketone;
zFA-fmk, N-benzyloxycarbonyl-Phe-Ala-fluoromethylketone;
ICE, interleukin-1ß converting enzyme; PCD, programmed cell death.[1]

NATO ASI Series, Vol. H 101
Molecular Mechanisms of Signalling
and Membrane Transport
Edited by Karel W. A. Wirtz

induced cell death. Complete inhibition of the effects of anti APO-1 on cell death **and** on ceramide production was observed when the ICE protease inhibitor zVAD.fmk but not zFA.fmk (a structurally similar but inactive peptide) were added together with anti-APO-1. These results suggests that the activation of sphingomyelin hydrolysis occurs downstream of the receptor-linked activation of members of the ICE protease family. Moreover, these results suggest that ceramide is not an upstream messenger in apoptosis and may instead be produced as a consequence of cell death.

INTRODUCTION

Programmed cell death (PCD) or apoptosis, where cells activate an intrinsic death programme, has recently been recognised as a widespread and physiologically important phenomenon. One of the important primary mediators of apoptosis in mammals is the FAS receptor [Nagata and Golstein (1995)]. Dimerisation of FAS leads it to activate a series of aspartate-directed proteases which have homology to the ICE family [Fraser and Evans, (1996); Chinnaiyan et al, (1996)]. Other evidence has also suggested that FAS ligation activates a plasma membrane sphingomyelinase, leading to the generation of ceramide (Fig.1.[Cifone et al, (1994); Martin et al, (1995); Tepper et al, (1995)] which may play an integral part in the signalling processess that lead to PCD [Linardic and Hannun, (1994); Pronk et al, (1996); Brugg et al, (1996); Chang et al, (1995); Karrasavvas et al, (1996); Latinis and Koretsky, (1996); Nickels and Broach, (1996); Saba et al, (1996); Sawai et al, (1996); Wiesner and Dawson, (1996); Wright et al, (1996); Venable et al, (1995); Verheij et al, (1996); Zhang et al., (1995)]. Such speculations have been supported by the observation that exogenously added sphingomyelinase or short chain ceramides promote apoptosis in various cell types, although these findings do not by themselves confirm an essential role for ceramide in apoptosis. One thing which has been unclear is the temporal relationship between activation of ICE-like protease and activation of sphingomyelinase; clearly it is important to know which event comes first.

We sought to characterise FAS-mediated ceramide changes in absence and the presence of the ICE protease inhibitor, zVAD.fmk which is able to block PCD in SKW and U937 cells [Jacobson et al, (1996)] . Here we show that ceramide does increase following FAS ligation, but this effect is slow (hours) and is blocked by zVAD.fmk but not zFA.fmk a structurally similar but inactive dipeptide. The results are consistent with the downstream activation of a sphingomyelinase following the activation of ICE-like proteases in FAS-mediated apoptosis. Moreover, these observations raise the possibility that ceramide production may be a consequence of cell death rather than an upstream regulator of the apoptotic process.

RESULTS

Time course for ceramide changes during FAS mediated apoptosis- The basal level of ceramide in SKW cells was 1.5±0.1% of total cholesterol levels, or 0.4% of total phospholipid label. This value is similar to that reported previously in this cell type using the DAG kinase mass assay [Tepper et al, (1995)]. This value rose by 50% after 8h and by 100% after 24h following the addition of anti APO-1 whereas in the absence of anti APO-1 ceramide rose only by only 20% in 24h. Ceramide changes in treated and control samples paralleled changes in cell death which increased to 50% at 4h and close to 100% at 24h. Thus the kinetics of ceramide production seemed to be significantly slower than those of cell death and contrasted sharply with the rapidity of ceramide formation relative to cell death reported by other investigators [Cifone et al, (1994); Verheij et al, (1996); Obeid et al, (1993)]. We repeated this experiment with U937 cells which have previously been reported to respond rapidly to anti APO-1 with a large increase in the mass of ceramide measured by *Escherichia.coli* DAG kinase [Cifone et al, (1994)]. However, like Garmen et al, (1996) we were unable to see any fast changes in ceramide (Fig. 1) and saw rises in ceramide only at later times when cell death had increased markedly. Other investigators have only been able to observe slow changes during in ceramide during PCD and the reasons for the descrepancies are not clear (Pronk et al, (1996); Tepper et al, (1995)]. Indeed one group failed to find increases in ceramide at all [Betts et al, (1994)].

Blockade of the FAS-mediated increases in ceramide by the ICE protease inhibitor, zVAD.fmk. - Since the kinetics of the ceramide response to FAS ligation were slower than that of the induction of PCD it was thought that the induction of increases in ceramide may be the result of the activation of PCD and therefore downstream of the initial signalling events. In order to test this idea we reasoned that a downstream ceramide response would be blocked by the ice protease inhibitor zVAD.fmk. Figure 2 shows that indeed inhibition of PCD by the addition of zVAD.fmk also blocks FAS induced increases in ceramide. The possibility that this effect is due to some non-specific action of zVAD.fmk is made less likely by the observation that the inactive dipeptide zFA.fmk has no effect (figure 2).

CONCLUSION

One of the main expectations of a cell signalling pathway is that receptor activation should rapidly induce the production of the signal compared with the physiological effect. In this study we have shown that ceramide mass does not rapidly increase following FAS receptor ligation in SKW 6.4 cells but increases with kinetics behind those for cell death. The results obtained with the ICE protease inhibitor clearly indicate that sphingomyelinase activaton is downstream of the activation of an ICE-like protease and therefore not proximal to FAS receptor ligation. These studies put into doubt the current model for the activation of sphingomyelinase as a key initial signal transduction event in receptor activated programmed cell death.

1a)

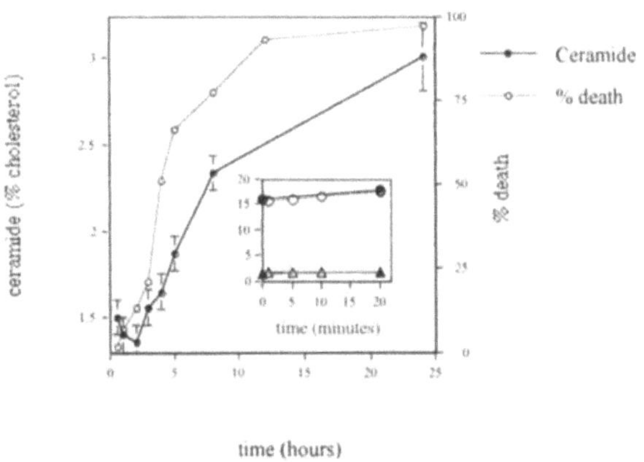

1b)

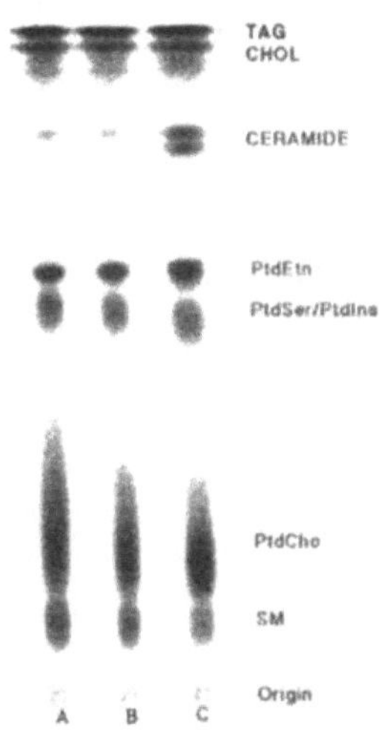

Figure 1- Ceramide increases and cell survival following anti APO-1 challenge in SKW 6.4 cells 1a) SKW cells labelled with ^{14}C acetate for 48 hours were incubated with and without addition of 50ng/ml anti APO-1 in the growth medium. At the indicated times aliquots were taken and cell death and ceramide levels were determined as described in the Materials and methods section. Insert - ceramide changes in U937 and SKW cells at shorter time points. Ceramide data are expressed as a % of cholesterol radioactivity.

1b) Autoradiogram of some of the results for SKW cells in figure 1a); All treatments were for 10 minutes as follows: **A**- Control; **B**- 50ng/ml of anti APO-1; **C**- 100mU/ml *S. Aureus* Shingomyelinase.

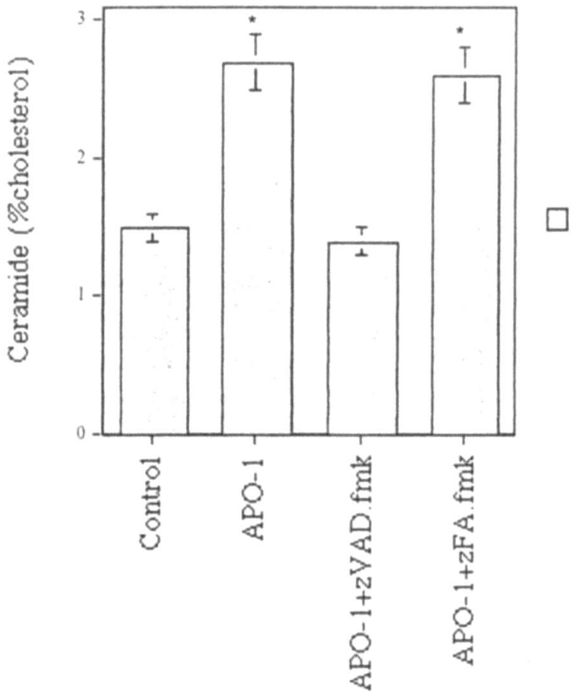

Figure 2. zVAD.fmk inhibition of FAS mediated ceramide changes during apoptosis ^{14}C labelled SKW cells were were challenged with 50 ng/ml anti APO-1 in the presence and absence of 20μM zVAD.fmk or zFA.fmk for 15 hours. All incubations were done in the presence of 0.2% DMSO.*p<0.01

REFERENCES

Betts, J. C., Agranoff, A. B., Nabel, G. J., Shayman, J. A. (1994) *J. Biol. Chem.* **269,** 8455-8

Brugg, B., Michel, P. P., Agid, Y. and Ruberg, M. (1996) *J Neurochem.* **66,** 733-9

Chang, Y., Abe, A. and Shayman, J. A. (1995) *Proc Natl Acad Sci U S A.* **92,** 12275-9

Chinnaiyan, A. M., Tepper, C. G., Seldin, M. F., O'Rourke, K., Kischkel, M. C., Hellbardt, H., Krammer, P. H., Peter, M. E. and Dixit, V. M. (1996) *J. Biol. Chem.* **271,** 4961-4965

Cifone, M. G., De, M. R., Roncaioli, P., Rippo, M. R., Azuma, M., Lanier, L. L., Santoni, A. and Testi, R. (1994) *J Exp Med.* **180,** 1547-52

Fraser, A. and Evan, G. (1996) *Cell.* **85,** 781-784

Garmen, S., Marzo, I., Anel, A., Pineiro, A. and Naval, J. (1996) *FEBS Lett.* **390,** 233-37

Jacobson, M. D., Weil, M. and Raff, M. C. (1996) *J. Cell Biol.* **133,** 1041-51

Karasavvas, N., Erukulla, R. K., Bittman, R., Lockshin, R. and Zakeri, Z. (1996) *Eur J Biochem.* **236,** 729-37

Latinis, K. M. and Koretzky, G. A. (1996) *Blood.* **87,** 871-5

Linardic, C. M. and Hannun, Y. A. (1994) *J Biol Chem.* **269,** 23530-7

Martin, S. J., Newmeyer, D. D., Mathias, S., Farschon, D. M., Wang, H. G., Reed, J. C., Kolesnick, R. N. and Green, D. R. (1995) *Embo J.* **14,** 5191-200

Nagata, S. and Golstein, P. (1995) *Science.* **267,** 1449-1456

Nickels, J. T. and Broach, J. R. (1996) *Genes Dev.* **10,** 382-94

Obeid, L. M., Linardic, C. M., Karolak, L. A. and Hannun, Y. A. (1993) *Science,* **259,** 1769-71

Pronk, G. J., Ramer, K., Amiri, P. and Williams, L. T. (1996) *Science.* **271,** 808-10

Saba, J. D., Obeid, L. M. and Hannun, Y. A. (1996) *Philos Trans R Soc Lond B Biol Sci.* **351,** 233-40

Sawai, H., Okazaki, T., Yamamoto, H., Okano, H., Takeda, Y., Tashima, M., Sawada, H., Okuma, M., Ishikura, H., Umehara, H. and et al. (1995) *J Biol Chem.* **270,** 27326-31

Tepper, C. G., Jayadev, S., Liu, B., Bielawska, A., Wolff, R., Yonehara, S., Hannun, Y. A. and Seldin, M. F. (1995) *Proc Natl Acad Sci U S A.* **92,** 8443-7

Wiesner, D. A. and Dawson, G. (1996) *J Neurochem.* **66,** 1418-25

Wright, S. C., Zheng, H. and Zhong, J. (1996) *Faseb J.* **10,** 325-32

Venable, M. E., Lee, J. Y., Smyth, M. J., Bielawska, A. and Obeid, L. M. (1995) *J Biol Chem.* **270,** 30701-8

Verheij, M., Bose, R., Lin, X. H., Yao, B., Jarvis, W. D., Grant, S., Birrer, M. J., Szabo, E., Zon, L. I., Kyriakis, J. M., Haimovitz, F. A., Fuks, Z. and Kolesnick, R. N. (1996) *Nature.* **380,** 75-9

Zhang, J., Alter, N., Reed, J. C., Borner, C., Obeid, L. M. and Hannun, Y. A. (1996) *Proc Natl Acad Sci U S A.* **93,** 5325-8

The structure, biosynthesis and function of GPI membrane anchors

Michael A. J. Ferguson
Department of Biochemistry,
University of Dundee,
Dundee DD1 4HN,
United Kingdom.

Introduction

Glycosyl-phosphatidylinositol (GPI) membrane anchors are present in organisms at most stages of eukaryotic evolution, including protozoa, yeast, slime moulds, invertebrates and vertebrates, and are found on a diverse range of proteins. They are primarily responsible for the anchoring of cell-surface proteins in the plasma membrane (or in some cases to the topologically equivalent lumenal surface of secretory vesicles) and may be considered as an alternative to the hydrophobic transmembrane polypeptide anchor of type-1 integral membrane proteins (Fig. 1).

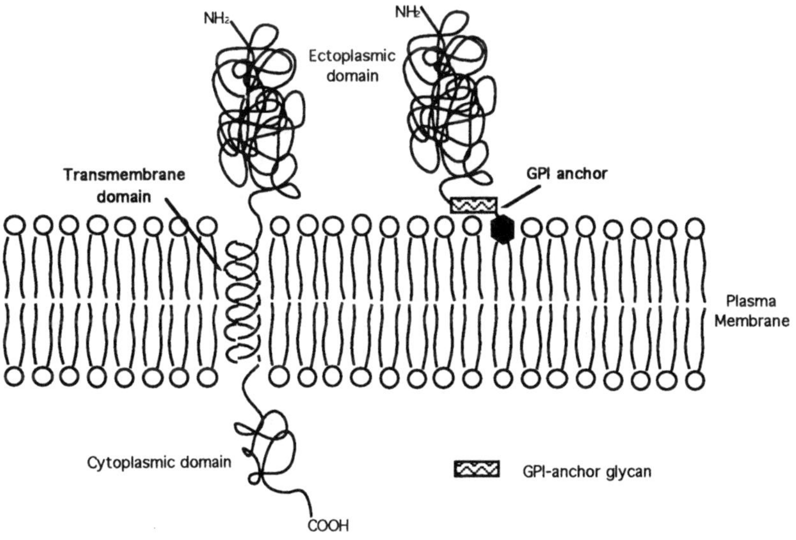

Fig. 1. Comparison between a type-1 transmembrane protein and a GPI-anchored protein

NATO ASI Series, Vol. H 101
Molecular Mechanisms of Signalling
and Membrane Transport
Edited by Karel W. A. Wirtz

Many other functions have been proposed (though some remain controversial) for GPI anchors, including roles in intracellular sorting, transmembrane signalling, and novel endocytic processes. Most of these proposed functions are dependent on the hypothesis that a GPI anchor might allow proteins to associate in specialised membrane microdomains. These putative functions, as well as the structure, biosynthesis and distribution of GPI anchors, have been extensively reviewed (Englund, 1993; McConville and Ferguson, 1993; Brown, 1993; Anderson, 1993; Lisanti, 1994; Ferguson, 1994; Stevens, 1995).

GPI structure

Although over 100 examples of GPI-anchored proteins have been described relatively few GPI anchor structures have been characterised in detail. To date partial or complete structures have been described for a variety of protozoal proteins, including *Trypanosoma brucei* variant surface glycoprotein (VSG) (Ferguson et al., 1988) and procyclic acidic repetitive protein (PARP) (Field et al., 1991; Ferguson et al., 1993), *Leishmania major* promastigote surface protease (Schneider et al., 1990), *Trypanosoma cruzi* 1G7 antigen (Güther et al., 1992; Heise et al., 1995), *T. cruzi* Tc85 glycoprotein (Couto et al. 1993; Abuin et al., 1996) and *T. cruzi* mucin-like glycoproteins (Previato et al., 1995; Serrano et al, 1995), *Plasmodium falciparum* antigens (Gerold et al., 1996), *Toxoplasma gondii* gp23 (Tomavo et al., 1993) and *Paramecium primaurelia* surface antigen (Azzouz et al., 1996). They have also been characterised in *Saccharomyces cerevisiae* glycoproteins (Fankhauser et al., 1993), *Dictyostelium discoideum* prespore-specific antigen (Haynes et al., 1993) and *Torpedo* electric organ acetylcholinesterase (Mehlert et al., 1993). To date eight GPI structures on mammalian proteins have been characterised: rat brain Thy-1 antigen (Homans et al., 1988), human erythrocyte acetylcholinesterase (Roberts et al., 1988a,b; Deeg et al., 1992), hamster brain scrapie prion protein (Stahl et al., 1992), bovine liver 5'-nucleotidase (Taguchi et al., 1994), human placental alkaline phosphatase (Redman et al., 1994), human urine CD59 (Nakano et al., 1994; Meri et al., 1996), pig and human membrane dipeptidase (Brewis et al., 1995) and human CD52 antigen (Treumann et al., 1995). These chemical structures are summarised in Fig. 2.

From the structural studies it is apparent that the GPI anchor consists of a highly conserved core structure of: ethanolamine-PO_4-6Manα1-2Manα1-6Manα1-4GlcN$H_2\alpha$1-6*myo*-inositol-1-PO_4-lipid. Often attached to this conserved core are variable side-chains which may be protein, tissue and/or species specific (McConville and Ferguson, 1993). Examples include an α-galactose branch on *T. brucei* VSG molecules and additional α-mannose residue(s) on a number of structures (including *T.cruzi*, yeast, slime-mould and mammalian glycoproteins). A

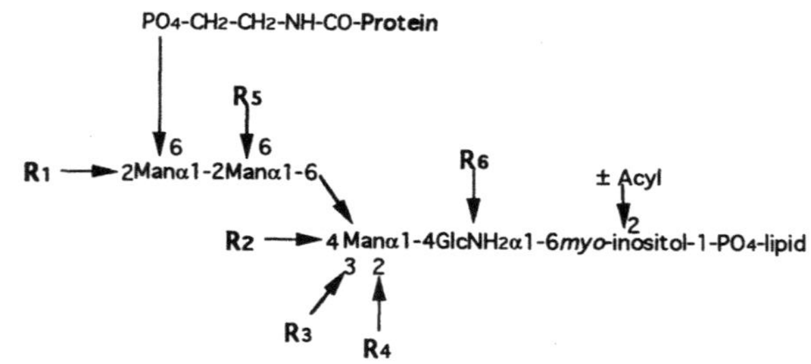

Protein	R1	R2	R3	R4	R5	R6	Lipid	Inositol-acyl
T.brucei VSG			αGal_{2-4}				diacylglycerol	
T.brucei PARP		[sialylated polylactosamine sidechain]					*lyso*-acylglycerol	+
T. cruzi 1G7 antigen	αMan						alkylacylglycerol and ceramide	
T.cruzi Tc85							alkylacylglycerol	palmitoyl
T.cruzi mucins*	αMan	[± Hex]				AEP	alkylacylglycerol and ceramide	
Leishmania PSP							alkylacylglycerol	
Toxoplasma gp23	βGalNAc						diacylglycerol	
Plasmodium	αMan						diacylglycerol	± myristoyl
Paramecium		[Manα - PO4 -]					ceramide	
S.cerevisiae anchors	αMan_{1-2}						ceramide and diacylglycerol	
D.discoideum PsA	±αMan	[1-2 EtNPO4, positions unknown]					ceramide	
Torpedo AChE	αGlc	±βGalNAc		EtNPO4	±EtNPO4?		diacylglycerol	
Rat Thy-1	±αMan	βGalNAc		EtNPO4			n.d.	
Hamster Prp	±αMan	[±NANA-Hex-HexNAc / EtNPO4]					n.d.	
Mouse NCAM	±αMan	±βGalNAc		n.d.	n.d.		n.d.	
Bovine 5'-NT	±αMan	±HexNAc?		EtNPO4?	±EtNPO4?		n.d.	
Porcine MDP		±Galβ1-3GalNAc		EtNPO4	±EtNPO4?		diacylglycerol	
Human MDP	±αMan	±Galβ1-3GalNAc		n.d.	n.d.		n.d.	
Human AChE				EtNPO4?	±EtNPO4		alkylacylglycerol	palmitoyl
Human PLAP		[		EtNPO4	]		alkylacylglycerol	
Human CD52	±αMan			EtNPO4?	±EtNPO4?		diacylglycerol	±palmitoyl
Human CD59	±αMan	±βGalNAc	[	EtNPO4	]		n.d.	

Fig. 2. GPI anchor structures
All GPI anchors attached to protein contain the conserved structure shown above with various substituents (R1-R5) and lipids, as indicated. Some structures contain an additional fatty acyl chain attached to the 2-position of the *myo*-inositol ring. All metazoan organisms contain at least one, and sometimes two, extra ethanolamine phosphate (EtNPO4) substituents in addition to the one used as a bridge to the protein C-terminal amino acid. When a substituent is known to be attached to a certain sugar residue but the linkage position is unknown, this is indicated by a question mark. Square brackets are used to show substituents of which the site of attachment has not been determined. The ± symbol indicates that the associated residue is found on only a proportion of the structures. AEP is 2-aminoethylphosphononate.

single sialic acid (N-acetylneuraminic acid; NANA) residue is present on a proportion (30%) of the scrapie prion protein anchors whilst an average of five sialic acid residues are found on the *T. brucei* PARP anchor.

In contrast to lower eukaryote GPI anchor structures, all metazoan GPI structures studied to date contain at least one additional ethanolamine phosphate residue, although the exact position(s) have only been determined in rat brain Thy-1, human erythrocyte acetylcholinesterase and human CD52. The function, if any, of side-chain modifications in general remains obscure, although the α-galactose branch in VSG has been suggested to be involved in the dense packing of the protective surface coat of the trypanosome (Homans et al., 1989).

From the lipid moieties that have been determined so far, it appears that this part of the anchor structure can be quite variable. They include ceramides in slime-moulds and some yeast, *T.cruzi* and *Paramecium* anchors, *sn*-1-alkyl-2-acylglycerols in human and bovine erythrocyte acetylcholinesterase, *L. major* PSP, some T. *cruzi* glycoproteins and human placental alkaline phosphatase, *sn*-1,2-diacylglycerols in *T. brucei* VSG, *Torpedo* acetylcholinesterase (Bütikofer et al., 1990), human CD52 and pig membrane dipeptidase and *sn*-1-acyl-2-*lyso*-glycerol in *T. brucei* PARP. Some GPI anchors contain an additional fatty acid (generally palmitate) in hydroxyester linkage to the 2-position of the *myo*-inositol ring. This has the property of rendering them resistant to the action of PI-PLC and GPI-PLC enzymes. Examples include human erythrocyte acetylcholinesterase, *T.brucei* PARP and one form of human CD52.

Biosynthesis

The mRNAs of GPI-anchored proteins encode an N-terminal signal sequence, to direct the protein to the endoplasmic reticulum (ER), and a C-terminal GPI-attachment signal sequence. This sequence is cleaved with the concomitant addition of a preassembled GPI precursor. The anchor attachment site (ω) may be one of six amino acid residues, all of which have small side-chains (Ala, Asn, Asp, Cys, Gly or Ser) (Moran et al., 1991; Gerber et al., 1992). In addition, the residue at the ω + 2 position is restricted to Ala, Gly or Ser (Gerber et al., 1992). The transfer of GPI precursor to protein occurs in the ER and GPI-anchored proteins are subsequently transported through the Golgi stacks to (generally) the plasma membrane.

Many of the details of the biosynthesis of GPI precursors have come from studying African trypanosomes. The tsetse-fly transmitted African trypanosomes, which cause human sleeping sickness and a variety of livestock diseases, are able to survive in the mammalian bloodstream by virtue of their dense cell-surface coat. This coat consists of 10 million copies of a 55 kDa GPI anchored glycoprotein called the variant surface glycoprotein (VSG) (Cross, 1990). The relative abundance of the VSG protein in *Trypanosoma brucei* has made this organism extremely useful for the study of GPI anchor biosynthesis. The structure of the VSG GPI anchor is known (Ferguson et al., 1988) and the principal features of the GPI biosynthetic pathway in trypanosomes were elucidated using a cell-free system based on washed trypanosome membranes (Masterson et al., 1989, 1990; Menon et al., 1990b). The first step in the pathway involves the transfer of GlcNAc from UDP-GlcNAc to endogenous phosphatidylinositol (PI), via a sulphydryl-dependent GlcNAc-transferase (Milne et al., 1992), to form GlcNAc-PI which is rapidly de-N-acetylated (Doering et al., 1989; Milne et al., 1994) to form glucosaminyl-PI (GlcN-PI). Three α-mannose residues are sequentially transferred onto GlcN-PI from dolichol-phosphate-mannose (Dol-P-Man) (Menon et al., 1990a) to form the intermediate Manα1-2Manα1-6Manα1-4GlcN-PI. At least this much of the pathway is believed to occur on the cytoplasmic face of the endoplasmic reticulum (Vidugiriene and Menon, 1993, 1994). Ethanolamine phosphate (EtNP) is then transferred from phosphatidylethanolamine (Menon and Stevens, 1992; Menon et al., 1993) to the terminal mannose residue to form EtNP-6Manα1-2Manα1-6Manα1-4GlcN-PI (known as glycolipid A'). This species then undergoes a series of fatty acid remodelling reactions (Masterson et al., 1990), whereby the fatty acids of the PI moiety are removed and replaced with myristate, to yield the mature GPI precursor glycolipid A. Concomitant with the formation of glycolipid A is the formation of glycolipid C (the inositol-acylated version of glycolipid A). Both glycolipid A and glycolipid C have been shown to be competent for transfer to VSG polypeptide when added exogenously to a trypanosome cell-free system (Mayor et al., 1991), although there is no evidence for the transfer of glycolipid C *in vivo*. Although only glycolipids A and C can be observed by labelling living trypanosomes with radiolabelled monosaccharides, a range of intermediates can be seen when the cell-free system is labelled with radiolabelled sugar-nucleotides. Some of these intermediates (from Manα1-4GlcN-PI onwards) are also inositol-acylated (Masterson et al., 1989; Menon et al., 1990b; Güther et al., 1994; Güther and Ferguson, 1995). Recent data suggest that, *in vivo,* the inositol-acylated and de-acylated forms of the intermediates in *T.brucei* bloodstream forms are in dynamic equilibrium, through the action of a phenylmethanesulphonic acid-sensitive inositol-acyltransferase and a diisopropylfluorophosphate-sensitive inositol-deacylase (Güther et al., 1994; Güther and Ferguson, 1995).

The GPI biosynthetic pathways in mammalian cells (Sugiyama et al., 1991; Lemansky et al., 1991, Hirose et al.,1991, 1992a,b; Kamitani et al., 1992; Puoti et al., 1991; Puoti and Conzelmann, 1992,1993; Mohney et al., 1994) and yeast (Costello and Orlean, 1992; Sipos et al., 1994), as well as in other protozoan organisms such as *Toxoplasma* (Tomavo et al., 1992a,b) and *Plasmodium falciparum* (Gerold et al., 1994), appear to be broadly similar to that described above for the bloodstream form of *T.brucei.* Some notable differences in the mammalian GPI pathway include the almost quantitative inositol-acylation of all GPI

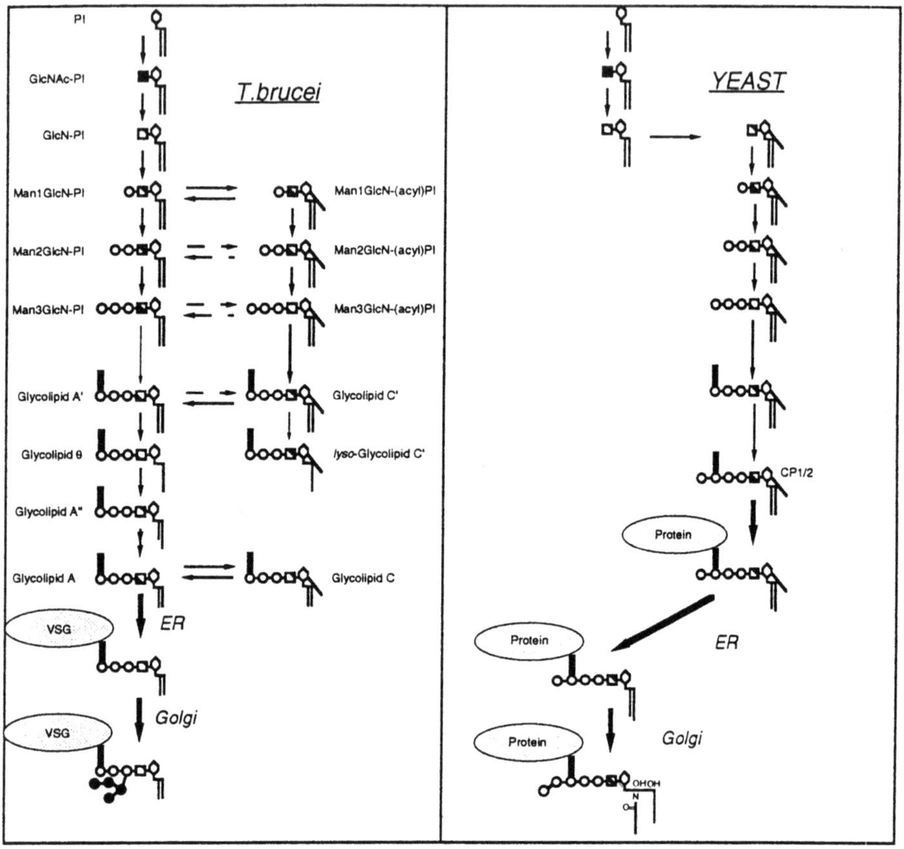

Fig. 3. Comparison of the GPI pathways of bloodstrem form T.brucei and yeast

The model for *T.brucei* GPI biosynthesis is adapted from Güther and Ferguson (1995) and contains data from the groups of Englund, Cross, Menon and Ferguson. The model for yeast GPI biosynthesis is based on the data of Conzelmann.

Notes: 1. The fatty acid remodelling reactions (A' -> A) are unique to *T.brucei*. The acylation and deacylation enzymes maintain a dynamic equilibrium between glycolipid A (the true precursor) and glycolipid C. Once transferred to protein, the fatty acids are maintained and αGal residues are added in the Golgi.

2. The yeast GPI intermediates are all inositol-acylated diacyl-PI species up to and including CP1 and CP2 (that probably differ in C26 fatty acid hydroxylation). The deacylation reaction probably occurs in the ER once the GPI precursor is attached to protein. Most protieins exchange the diacyl-PI for ceramide-PI in the Golgi. An additional Manα1-2 or Manα1-3 residue is attached after transfer of the GPI to protein. The location of this modification is unknown.

intermediates from GlcN-PI onwards and the addition of extra ethanolamine phosphate groups. Similarly, the yeast GPI intermediates are also inositol-acylated from GlcN-PI onwards. The fatty acid remodelling reactions, as described above, appear to be unique to bloodstream form African trypanosomes. However, in mammalian cells there is evidence for the exchange of diacyglycerol for alkylacylglycerol (Singh et al., 1994) and in yeast the exchange of diacyglycerol for ceramide (after the addition of the GPI anchor to protein) has been described in detail (Conzelmann et al., 1992; Sipos et al., 1994). Interestingly, inhibition of ceramide synthesis in yeast retards the transport of GPI-anchored proteins out of the ER to the Golgi (Horvath et al., 1994). However, there is evidence to suggest that the ceramides of ceramide-GPI anchors are acquired in the *cis* Golgi rather than in the ER (Conzelmann et al., 1992) and that the inhibition of transport is due to the need for sphingolipids to assist GPI-protein transport (Horvath et al., 1994). The GPI pathways of *T.brucei* and yeast are compared in Fig. 3.

GPI function

Many properties and functions have been suggested for GPI anchors. These include: high lateral mobility, sequestration into caveolae and/or other membrane microdomains, transmembrane signalling, intracellular sorting and inter-cellular exchange. Some of the literature pertaining to these functions will be reviewed below.

Lateral mobility: The literature values for the lateral diffusion coefficients of GPI-anchored proteins show that these proteins can vary from being very mobile to essentially immobile in the lipid bilayer. A good example of this is the GPI-anchored sperm PH-20 glycoprotein that displays freely diffusing, intermediate and restricted lateral mobility at different stages of sperm development (Phelps et al., 1988). Studies using protein chimeras, where the same protein is engineered to have a GPI anchor or a transmembrane domain, indicate that interactions of the protein component with neighbouring molecules in biological membranes is likely to be the main factor dictating their average mobility (Zhang et al., 1991). Thus, while GPI anchored proteins have the potential to have high lateral mobilities, the local membrane environment often dictates otherwise.

Sequestration into membrane microdomains, transmembrane signalling and intracellular sorting: The punctate cell surface staining of many GPI-anchored proteins and their apparent sequestration around membrane invaginations called caveolae (Rothberg et al., 1990) has been called into question by using monovalent antibody fragments and more rigorous cell fixation conditions (Mayor et al., 1994). Nevertheless, the location of GPI-anchored proteins in specialised membrane microdomains (rich in sphingolipids, glycosphingolipids and cholesterol) is still a popular notion since GPI-anchored proteins are found in insoluble complexes following the extraction of cells with cold neutral detergents such as Triton X-100, reviewed recently in (Parton and Simons, 1995). This has lead to the idea that (i) these putative microdomains or "lipid rafts" may be involved in the known intracellular sorting of GPI anchored proteins, and certain glycosphingolipids, to the apical membranes of polarised epithelial cells and (ii) that these putative microdomains may, in some instances, form functional units capable of signal transduction events involving associated non-receptor protein tyrosine kinases, heterotrimeric G proteins and small G proteins, reviewed in (Brown et al., 1993; Lisanti et al., 1994). While these ideas are attractive, it should be noted that there is little evidence that these domains exist prior to detergent extraction (Kurzchalia et al., 1995). Indeed, it has been shown that GPI-anchored proteins are not clustered with each other (Mayor et al., 1994; Xue et al., 1994) or with themselves (Mayor and Maxfield, 1995) or with glycosphingolipids (Fra et al., 1994) prior to the extraction of biological membranes with detergents. Thus, it is possible that the detergent resistant complexes represent the artifactual coalescence of the most hydrophobic components within a given lipid bilayer upon detergent extraction of the less hydrophobic/lower-melting temperature phospholipid species. In this regard, it may be relevant that most GPI anchor PI moieties are significantly more hydrophobic than the average cellular phospholipid. The least hydrophobic mammalian GPI PI moieties reported so far, for human leukocyte CD52-1 and pig kidney membrane dipeptidase, both contain two saturated $C_{18:0}$ stearate fatty acids (Treumann et al., 1995; Brewis et al., 1995) and the other mammalian GPI anchors that have been characterised, i.e., human erythrocyte acetylcholinesterase (Roberts et al., 1988a) and human CD52-2 (Treumann et al., 1995), contain $C_{16:0}$ palmitate fatty acid attached to the inositol ring in addition to one fatty acid and an ether-linked alkyl chain or two fatty acids, respectively. Furthermore, non-receptor protein tyrosine kinases, such as the $p56^{lck}$ and $p60^{src}$, require the presence of two saturated fatty acids, an N-terminal $C_{14:0}$ myristate and a thioester-linked $C_{16:0}$ palmitate on Cys3, in order to become associated with the detergent-resistant domains (Shenoy et al., 1994). Such observations are perhaps consistent with the idea of coalescence upon removal of the phospholipids by low temperature detergent extraction

Ultimately the arguments for whether or not microdomain "lipid-rafts" exist prior to detergent extraction become rather circular. For example, it has been shown that depleting cells of sphingolipids and/or cholesterol reduces the formation of GPI-protein containing detergent-resistant complexes (Hanada et al., 1995; Futerman et al., 1995). However, one could argue that reducing these hydrophobic components of the plasma membrane may simply prevent coalescence of the other components upon detergent extraction. Indeed, at least one GPI-function, i.e., apical targeting of proteins in MDCK cells, was not affected by cholesterol depletion even though the formation of detergent insoluble complexes was significantly decreased, suggesting that the ability to form these complexes and intracellular targeting are not inextricably linked (Hannan and Edidin, 1996). Similarly, the *in vitro* studies showing that sphingolipids, glycosphingolipids, cholesterol and GPI-anchored protein will form detergent-resistant domains when liposomes containing these components plus phospholipid are stripped of phospholipid by detergent (Schroeder et al., 1994) does not predict whether microdomains preexisted in the liposomes or whether they were induced by the detergent extraction.

The ability of glycosphingolipids such as GM1 to cluster within bilayers (Thompson and Tillack, 1985) is widely quoted as the basis for the existence of "lipid rafts" or "glycosphingolipid-enriched complexes" in biological membranes. However, the elegant studies by Thompson, Tillack and colleagues have only shown clustering of about 15 molecules of GM1 in artificial bilayers (Rock et al., 1990) and that glycosphingolipids will accumulate in the gel-phase of mixed two-phase artificial membranes in a ceramide-structure dependent manner (Palestini et al., 1995). However, since (a) the size of the reported clusters are so small, (b) another report suggest that GM1 does not cluster at all (Antes et al., 1992) and (c) the existence of gel-phases in biological membranes at physiological temperatures is very uncertain, it is perhaps dangerous to extrapolate too much from these findings.

The arguments outlined above do not help to explain the many examples of transmembrane signalling via the ligation with antibody and second antibody of GPI-anchored proteins on various cells, particularly T-cells. These signalling events are clearly dependent on the presence of a GPI anchor (Robinson, 1991). However, it should be noted that, as yet, there is no known physiological receptor/ligand pair that signals directly via a GPI-anchored protein and therefore the physiological significance of these observations is unclear. In any case, the ability of a GPI-protein to interact laterally with a transmembrane (signal transducing) protein need not be dependent on the concept of "lipid rafts". It is probably significant that those GPI-anchored proteins that are known to be involved in physiologically relevant transmembrane signalling events, e.g., the cilary neurotrophic factor (CNTF) receptor-α and the glial cell line derived factor (GDNF) receptor-α, require association with one or two transmembrane β–co-receptors to transmit their signals (Massagué, 1996). Indeed, in the case of the CNTF receptor-α, the GPI anchor can be removed without loss of overall receptor function (Stahl and Yancopoulos, 1993). Similarly, other signal transducing GPI-anchored receptors like CD14 (the LPS/LPS binding protein receptor) function equally well with a GPI anchor or with a spliced transmembrane domain, suggesting a requirement for a transmembrane partner that interacts laterally with CD14 (Lee et al., 1993).

While the comments made above do not in any way rule out the existence of specialised membrane microdomains containing GPI-anchored proteins and accessory transmembrane elements, they do at least serve as a reminder that our understanding of molecular arrangements in biological membranes are far from clear.

Intercellular exchange: There are several examples of the the exchange of GPI-anchored protein from one cell surface to another, see (Kooyman et al., 1995; Anderson et al., 1996;

Ilangumaran et al., 1996; Medof et al., 1996 and references therein). While the precise mechanism of this exchange is unknown, it is clear that purified GPI-anchored proteins will spontaneously insert into lipid bilayers. The key difference between GPI-anchored proteins and transmembrane proteins in this regard must be the lack of any cytoplasmic domain. This property of GPI anchored proteins is being exploited experimentally to apply exogenous proteins to cell surfaces (Medof et al., 1996) but the physiological significance, if any, of GPI-protein exchange is still uncertain.

Acknowledgements

MAJF is supported by the Wellcome Trust and is a Howard Hughes Medical Institute International Research Scholar.

References

Abuin, G., Couto, A.S., Lederkremer, de, R.M., Casal, O.L., Galli, C., Colli, W. and Alves, M.J. M. (1996) Exp. Parasitol. 82, 290-297.
Anderson, R.G.W. (1993) Proc. Natl. Acad. Sci. USA 90, 1090-10913.
Anderson, S.M., Gang, Y., Giattina, M. and Miller, J.L. (1996) Proc. Natl. Acad. Sci. USA 93, 5894-5898.
Antes, P., Schwarzmann, G. and Sandhoff, K. (1992) Chemistry & Physics of Lipids 62, 269-280.
Azzouz, N., Striepen, B., Gerold, P. Capdeville, Y. and Schwarz, R.T. (1995) EMBO 14, 4422-4433.
Brewis, I.A., Ferguson, M.A.J., Mehlert, A., Turner, A.J. and Hooper, N.M. (1995) 270, 22946-22956.
Brown, D. A. (1993) Curr. Opin. Immunol. 5, 338-343.
Bütikofer, P., Kuypers, F.A., Shackleton, C., Brodbeck, U and Stieger, S. (1990) J. Biol. Chem. 265, 18983-18987.
Conzelmann, A., Puoti, A., Lester, R. L. and Desponds, C. (1992) EMBO J. 11, 457-66.
Costello, L. C. and Orlean, P. (1992) J. Biol. Chem. 267, 8599-8603.
Cross, G. A. M. (1990) Annu. Rev. Immunol. 8, 83-100.
Cuoto, A. S., Lederkremer, R. M., Colli, W. and Alves, M. J. M. (1993) Eur. J. Biochem. 217, 597-602.
Deeg, M. A., Humphrey, D. R., Yang, S. H., Ferguson, T. R., Reinhold, V. N. and Rosenberry, T. L. (1992) J. Biol. Chem. 267, 18573-18580
Doering, T. L., Masterson, W. J., Englund, P. T. and Hart, G. W. (1989) J. Biol. Chem. 264, 11168-11173.
Englund, P. T. (1993) Annu. Rev. Biochem. 62, 121-138
Fankhauser, C., Homans, S. W., Thomas-Oates, J. E., McConville, M. J., Desponds, C.,

Conzelman, A. and Ferguson, M. A. J. (1993) J. Biol. Chem. 268, 26365-26374.
Ferguson, M.A.J. (1994) Pasasitol. Today 10, 48-52.
Ferguson, M. A. J., Homans, S. W., Dwek, R. and Rademacher, T. W. (1988) Science 239, 753-759.
Ferguson, M. A. J., Murray, P., Rutherford, H. and McConville, M. J. (1993) Biochem. J. 291, 51-55.
Field, M. C., Menon, A. K. and Cross, G. A. M. (1991) EMBO J. 10, 2731-2739.
Fra, A.M., Williamson, E., Simons, K. and Parton, R.G. (1994) J. Biol. Chem. 269, 30745-30748.
Futerman, A.H. (1995) Trends in Cell Biol. 5, 377-380.
Gerber, L.D., Kodukula, K., and Udenfriend, S. (1992) J. Biol Chem. 267, 12168-12173.
Gerold, P., Dieckmann-Schuppert, A. and Schwartz, R.T. (1994) J. Biol. Chem. 269, 2597-2606.
Gerold, P., Schofield, L., Blackman, M.J., Holder, A.A., Schwarz, R.T. (1996) Mol. Biochem. Parasitol. 75, 131-143.
Güther, M. L. S. and Ferguson, M. A. J. (1995) EMBO J. 14, 3080-3093.
Güther, M. L. S., Cardoso de Almeida, M. L., Yoshida, N. and Ferguson, M. A. J. (1992) J. Biol Chem. 267, 6820-6828
Güther, M. L. S., Masterson, W. J. and Ferguson, M. A. J. (1994) J. Biol. Chem. 269, 18694-18701.
Hanada, K., Nishijima, M., Akamatsu, Y. and Pagano, R.E. (1995) J. Biol. Chem. 270, 6254-6260.
Hannan, L.A. and Edidin, M. (1996) J. Cell Biol. 133, 1265-1276.
Haynes, P. A., Gooley, A. A., Ferguson, M. A. J., Redmond, J. W. and Williams, K. L. (1993) Eur. J. Biochem. 216, 729-737.
Heise, N., Cardoso de Almeida, M.L. and Ferguson, M.A.J. (1995) Mol. Biochem. Parasitol. 70, 71-84.
Hirose, S., Prince, G. M., Sevlever, D., Ravi, L., Rosenberry, T. L., Ueda, E. and Medof, M. E. (1992a) J. Biol. Chem. 267, 16968-16974.
Hirose, S., Ravi, L., Hazra, S. V. and Medof, M. E. (1991) Proc. Natl. Acad. Sci. (USA) 88, 3762-3766.
Hirose, S., Ravi, L., Prince, G.M., Rosenfeld, M.G., Silber, R., Andresen, S.W., Hazra, S.V. and Medof, M.E. (1992b) Proc. Natl. Acad. Sci. USA 89, 6025-6029.
Homans, S.W., Ferguson, M.A.J., Dwek, R.A., Rademacher, T.W., Anand, R. and Williams, A.F. (1988) Nature 333, 269-272.
Homans, S. W., Edge, C. J., Ferguson, M. A. J., Dwek, R. A. and Rademacher, T. W. (1989) Biochemistry 28, 2881-2887.
Horvath, A., Sütterlin, C., Manning-Krieg, U., Movva, N.R. and Reizman, H. (1994) EMBO J. 13, 3687-3695.
Ilangumaran, S., Robinson, P.J. and Hoessli, D.C. (1996) Trends in Cell Biology 6, 163-167.
Kamitani, T., Menon, A. K., Hallaq, Y., Warren, C. D. and Yeh, E. T. H. (1992) J. Biol. Chem. 267, 24611-24619.
Kooyman, D.L., Byrne, G.W., McClellan, S., Nielsen, D., Tone, M., Waldmann, H., Coffman, T.M., McCurry, K.R., Platt, J.L., Logan, J.S. (1995) Science 269, 89-92.
Kurzchalia, T., Hartmann, E. and Dupree, P. (1995) Trends Cell Sci. 5, 187-189.
Lee, J.-D., Kravchenko, V., Kirkland, T.N., Han, J., Mackman, N., Moriarty, A., Leturcq, D., Tobias, P.S. and Ulevitch, R.J. (1993) Proc. Natl. Acad. Sci. USA 90, 9930-9934.
Lemansky, P., Gupta, D. K., Meyale, S., Tucker, G. and Tartakoff, A. M. (1991) Mol. Cell Biol. 11, 3879-3885.
Lisanti, M.P., Schere, P.E., Tang, Z. and Sargiacomo, M. (1994) Trends in Cell Biology 4, 231-235.
Massagué, J. (1996) Nature 382, 29-30.
Masterson, W. J., Doering, T. L., Hart, G. W. and Englund, P. T. (1989) Cell 56, 793-800.
Masterson, W.J., Raper, J., Doering, T.L., Hart, G.W. and Englund, P.T. (1990) Cell 62, 73-80.

Mayor, S. and Maxfield, F.R. (1995) Mol. Biol. Cell 6, 929-944.
Mayor, S., Menon, A. K. and Cross, G. A. M. (1991) J. Cell Biol. 114, 61-71.
Mayor, S., Rothberg, K.G. and Maxfield, F.R. (1994) Science 264, 1948-1951.
McConville, M. J. and Ferguson, M. A. J. (1993) Biochem. J. 294, 305-324.
Medof, M.E., Nagarajan, S., Tykocinski, M.L. (1996) FASEB J. 10, 574-586.
Mehlert, A., Varon, L., Silman, I., Homans, S. W. and Ferguson, M. A . J. (1993) Biochem. J. 296, 473-479
Menon, A. K., Eppinger, M., Mayor, S. and Schwarz, R. T. (1993) EMBO J. 12, 1907-1914.
Menon, A. K., Mayor, S. and Schwarz, R. T. (1990a) EMBO J. 9, 4249-4258.
Menon, A. K., Schwarz, R. T., Mayor, S. and Cross, G. A. M. (1990b) J. Biol. Chem. 265, 9033-9042.
Menon, A. K. and Stevens, V. L. (1992) J. Biol. Chem. 267,15277-15280.
Meri, S., Lehto, T., Sutton, C.W., Tyynela, J. and Baumann, M. (1996) Biochem. J. 316, 923-935.
Milne, K. G., Ferguson, M. A. J. and Masterson, W. J. (1992) Eur. J. Biochem. 208, 309-314.
Milne, K. G., Field, R.A., Masterson, W. J., Cottaz, S., Brimacombe, J.S. and Ferguson, M. A. J. (1994) J. Biol. Chem. 269, 16403-16408.
Mohney, R. P., Knez, J. J. Ravi, L., Sevlever, D., Rosenberry, T. L. Hirose, S. and Medof, M. E. (1994) J. Biol. Chem. 269, 6536-6542.
Moran, P., Raab, H., Kohr, W. J. and Caras, I. W. (1991) J. Biol. Chem. 266, 1250-1257
Nakano, Y., Noda, K., Endo, T., Kobata, A. and Tomita, M. (1994) Arch. Biochem. Biophys. 311, 117-126.
Palestini, P., Allietta, M., Sonnino, S., Tettamanti, G., Thompson, T.E. and Tillack, T.W. (1995) Biochim. Biophys. Acta. 1235, 221-230.
Parton, R.G. and Simons, K. (1995) Science 269, 1398-1399.
Phelps, B. M., Primakoff, P., Koppel, D. E., Low, M. G. and Mylews, D. G. (1988) Science 240, 1780-1782.
Previato, J.O., Jones, C., Xavier, M.T., Wait, R., Travassos, L.R., Parodi, A.J. and Mendonça-Previato, L. (1995) J. Biol. Chem. 270, 7241-7250.
Puoti, A. and Conzelmann, A. (1992) J. Biol. Chem. 267, 22673-22680.
Puoti, A. and Conzelmann, A. (1993) J. Biol. Chem. 268, 7215-7224.
Puoti, A., Desponds, C., Fankhauser, C. and Conzelmann, A. (1991) J. Biol. Chem. 266, 21051-21059.
Redman, C. A., Thomas-Oates, J. E., Ogata, S., Ikehara, Y. and Ferguson, M. A. J. (1994) Biochem. J. 302, 861-865
Roberts, W. L., Santikarn, S., Reinhold, V. N. and Rosenberry, T. L. (1988a) J. Biol. Chem 263, 18776-18784
Roberts, W.L., Myher, J.J. Kuksis, A., Low, M.G. and Rosenberry, T.L. (1988b) J. Biol. Chem. 263, 18766-18775.
Robinson, P. J. (1991) Immunol. Today 12, 35-41.
Rock, P., Allietta, M., Young, Jr., W.W., Thompson, T.E. and Tillack, T.W. (1990) Biochemistry 29, 8484-8490.
Rothberg, K. G., Ying, Y., Kolhouse, J. F., Kamen, B. A. and Anderson, R. G. W. (1990) J. Cell Biol. 110, 637-649.
Schneider, P., Ferguson, M. A. J., McConville, M. J., Mehlert, A., Homans, S. W. and Bordier, C. (1990) J. Biol. Chem. 265, 16955-16964.
Schroeder, R., London, E. and Brown, D. (1994) Proc. Natl. Acad. Sci. USA 91, 12130-12134.
Serrano, A.A., Schenkman, S., Yoshida, N., Mehlert, A., Richardson, J.M. and Ferguson, A.J. (1995) J. Biol. Chem. 270, 27244-27253.
Singh, N., Zoeller, R. A., Tykocinski, M. L., Lazarow, P. B. and Tartakoff, A. M. (1994) Mol. Cell. Biol. 14, 21-31.
Shenoy, S. A., Dietzen, D.J., Kwong, J., Link, D.C. and Lublin, D. M. (1994) J. Cell. Biol. 126, 353-363.

Sipos, G., Puoti, A. and Conzelmann, A. (1994) EMBO J. 13, 2789-2796.
Sugiyama, E., DeGasperi, R., Urakaze, M., Chang, H-M., Thomas, L. J., Hyman, R.,
Stahl, N., Baldwin, M. A., Hecker, R., Pan, K. -M., Burlingame, A. L. and Prusiner, S. B. (1992) Biochemistry 31, 5043-5053.
Stahl, N. and Yancopoulos, G.D. (1993) Cell 74, 587-590.
Stevens, V. L. (1995) Biochem. J. 310, 361-370.
Taguchi, R., Hamakawa, N., Harada-Nishida, M., Fukui, T., Nojima, K. and Ikezawa, H. (1994) Biochemistry 33, 1017-1022.
Thompson, T.E., Tillack, T.W. (1985) Ann. Rev. Biophys. Chem. 14, 361-386.
Tomavo, S., Dubremetz, J-F. and Schwarz, R. T. (1992a) J. Biol. Chem. 267, 11721-11728.
Tomavo, S., Dubremetz, J-F. and Schwarz, R. T. (1992b) J. Biol. Chem. 267, 21446-21458.
Tomavo, S., Dubremetz, J-F., Schwarz, R.T. (1993) Biol. Cell 78, 155-162.
Treumann, A., Lifely, M.R., Schneider, P. and Ferguson, M.A.J. (1995) J. Biol. Chem. 270, 6088-6099.
Vai, M., Popolo, L., Grandori, R., Lacana, E. and Alberghina, L. (1990) Biochim. Biophys. Acta. 1038, 277-285
Vidugiriene, J. and Menon, A. K. (1993) J. Cell Biol. 121, 987-996.
Vidugiriene, J. and Menon, A. K. (1994) J. Cell Biol. 127, 333-341.
Xue, W., Kindzelskii, A.L., Todd III, R.F.T. and Petty, H.R. (1994) J. of Immunol. 152, 4630-4640.
Zhang, F., Crise, B., Su, B., Hou, Y., Rose J.K., Bothwell, A., Jacobson, K. (1991) J. Cell. Biol. 115, 75-84 .

Enzyme Assisted Synthetic Approaches to the Inositol Phospholipid Pathway

P.Andersch, B.Jakob, R.Schiefer and Manfred P.Schneider*
FB 9 - Organische Chemie
Bergische Universität - GH - Wuppertal
D - 42097 Wuppertal, Germany

1. Introduction

Inositol phospholipids and their molecular constituents such as D-*myo* - inositol phosphates, 1,2 - *sn* -diglycerides and also arachidonic acid play an important role as second messengers in living cells with numerous functions as regulators and signal transducers (Billington DC 1993, Reitz AB 1991, Irvine RF 1990, Lodish H 1996). Many examples for their highly potent biological activities can be found throughout this volume. Unfortunately, however, frequently many of these molecules are only accessible with difficulties in minute amounts from scarce natural sources after laborious isolation and purification procedures. Clearly, the elucidation of their biological role would be greatly facilitated if these molecules and structural analogues thereof could be made available *via* facile synthetic routes. It is not surprising therefore that inositol phospholipids and their molecular constituents have become attractive targets for organic syntheses. In this sense we also embarked a few years ago on the exploration of new synthetic routes towards these molecules and structural analogues thereof with the aim of using enzymes for the introduction of chirality into the respective molecular backbones.

Retrosynthetic analysis of the title compounds using disconnections catalysed by specifically acting phospholipases (e.g. A_2 and C) as a guideline (Fig. 1), three classes of second messengers are formally obtained as fragments:

D-*myo*-inositol (poly)phosphates
1,2 -*sn* - diglycerides
arachidonic acid.

NATO ASI Series, Vol. H 101
Molecular Mechanisms of Signalling
and Membrane Transport
Edited by Karel W. A. Wirtz

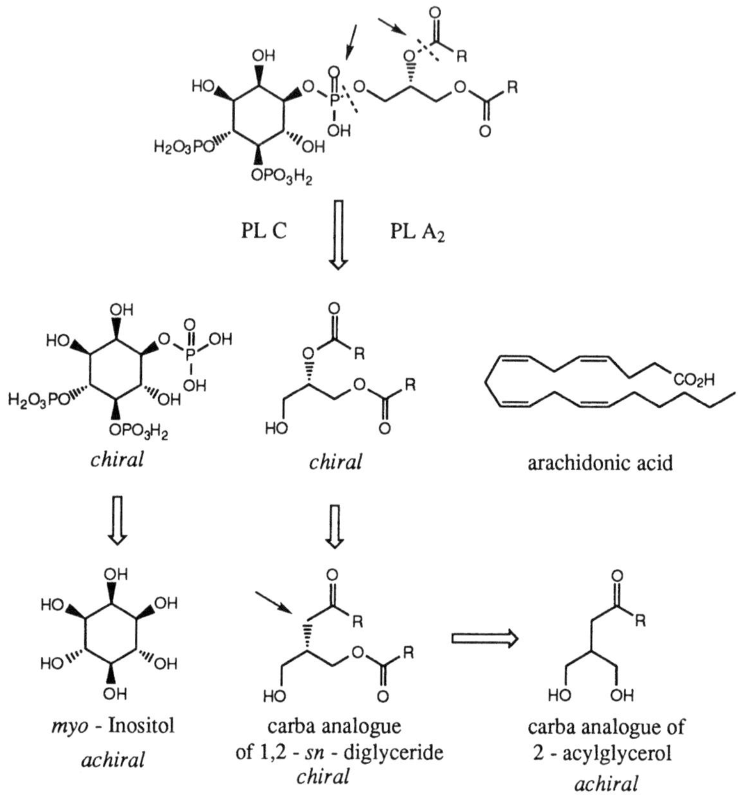

Fig. 1. D-*myo* -Inositol phospholipids - retrosynthetic analysis

Based on this retrosynthetic scheme we describe in the present article

(a) enzyme assisted syntheses of carba - analogous 1,2 - *sn* - diglycerides;

(b) enzyme assisted synthetic approaches towards D - *myo* - inositol phosphates;

(c) first experiments in connecting the above constituents *en route* to structural analogues of *myo* - inositol phospholipids.

1.1. Carba analogues of 1,2 - *sn* - diglycerides

Diglycerides are - next to D- *myo* - inositol phosphates - important G - protein linked second messengers and responsible for the activation of protein kinase C and thus the phosphorylation of proteins (Nishizuka Y 1992, Bell RM 1986). Interestingly - probably due to their rather low activities - their action is usually mimicked either by monoacyl glycerols or phorbol esters. It is tempting to speculate whether the low biological activities of "native" diglycerides are due either to their (a) rapid metabolism or (b) racemisation under physiological conditions, possibly both. Since 1,2 - *sn* - diglycerides are derived from natural inositol phospholipids with clearly defined absolute (*S*) - configuration one could assume with some confidence that the biological activities of these molecules are closely related to their natural 1,2 - *sn* - configuration. In contrast to the corresponding 1,3 - *sn* - isomers, diglycerides of this constitution are notoriously instable with a tendency to rapid racemisation due to facile acyl group migrations (compare Fig 2 below) expecially under the protic conditions of a physiological environment. If this assumption is in fact correct this could be also the reason for their fast inactivation *in vivo.* In order to prevent such acyl group migrations we have decided to synthesize a series of stereochemical analogues of such diglycerides in which the sp^3-oxygens of the acyl moieties are replaced by the corresponding carbon atoms leading to molecules of identical stereochemistry but at the same time with minimal deviations regarding bond angles and bond distances (compare Fig. 1,3).

1.2. D-*myo*-inositol phosphates

Following the retrosynthetic scheme outlined in Fig. 1 it should be appreciated that, while many inositol phosphates are chiral and only active in optically pure form, the parent compound *myo* - inositol is achiral, a *meso* - compound. Consequently, all synthetic approaches towards optically pure D - *myo* - inositol phosphates are centered around the problem of converting achiral *myo* - inositol or precursers thereof into enantiomerically pure building blocks for these target molecules. In the present article we describe such an approach in which the required introduction of chirality was achieved using enzymes.

2. Carba Analogues of Glycerides

As already pointed out in the introduction, optically pure 1,2 - *sn* - diglycerides are notoriously instable due to rapid acyl group migrations especially under protic conditions and at elevated temperatures causing immediate loss of optical purity (Fig. 2).

In view of this situation we felt that isosteric mimics would be attractive alternatives and we therefore decided to explore the synthesis of more stable C - analogues of these molecules using enzyme assisted routes. Replacement of the sp^3 - oxygen in an acyl group of a triglyceride by a sp^3 - carbon should lead to molecules which are very similiar to natural triglycerides with only minor deviations of bond angles and distances (Fig. 3).

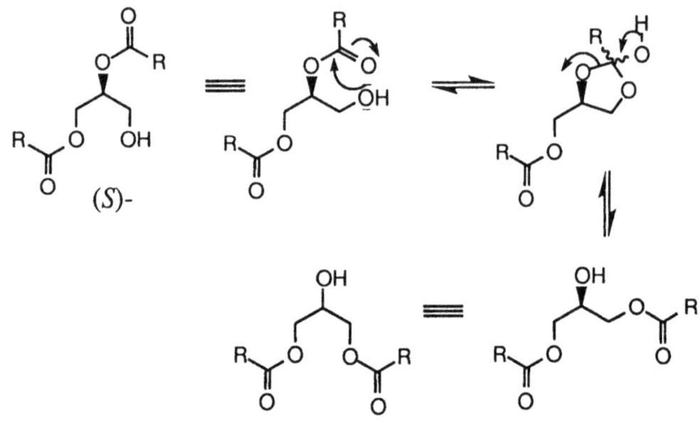

Fig. 2. Acyl group migrations in 1,2 - *sn* - diglycerides

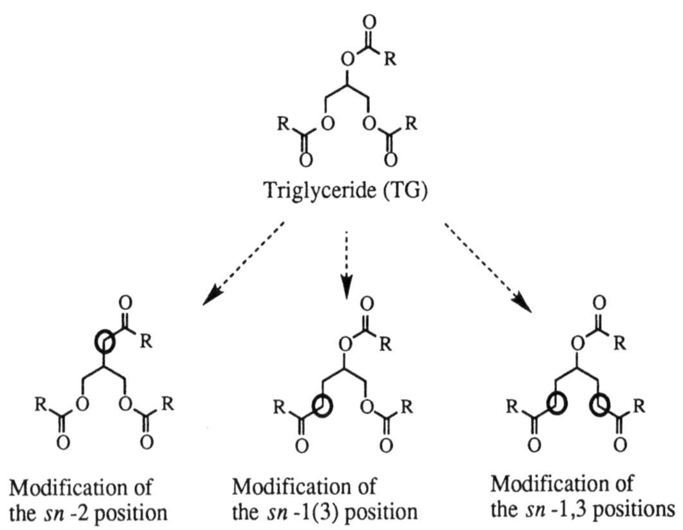

Fig. 3. Carba analogues of triglycerides as mimics of natural lipids

They should - with the exception of hydrolytic cleavage at the modified position - behave identical towards biological systems.

In order to test this hypothesis two synthetic routes for the synthesis of carba - analogues of triglycerides were developed. In the first route (Berger M 1994) (Fig. 4) an aldehyde with the desired chain length of C - atoms was converted *via* a Mannich reaction into the corresponding unsaturated aldehyde with the essential *exo* - methylene group in the desired position thus constituting the masked carbonyl group of the final product. Reduction, bromination, conversion to the corresponding malonic ester and its reduction provided the desired 1,3 - diol - the precurser of the final product. Esterification and ozonolysis resulted in the C - analogue of a triglyceride in which the 2 - position is blocked towards hydrolytic attacks - both chemically and enzymatically.

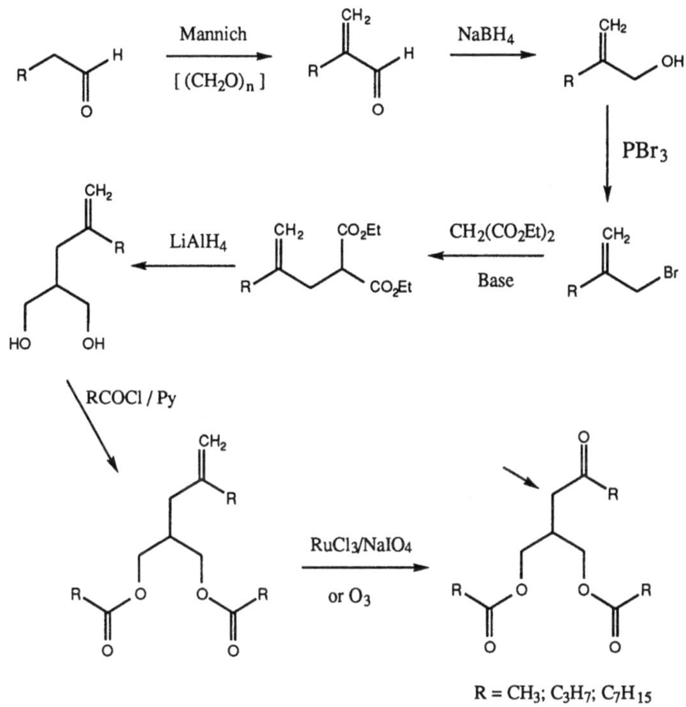

Fig.4. Carba Analogues of Triglycerides - Synthetic Route 1

Alternatively, and possibly more convenient due to the ready accessibility of practically every desired acid chloride the desired functionality can also be introduced *via* the corresponding diazoketones (Jakob B 1996) (Fig. 5).

Binding studies with lipases and both enzyme assisted hydrolyses and esterifications clearly demonstrated that native triglycerides and their C - analogues behave identical towards these biological systems (Berger et al 1994).

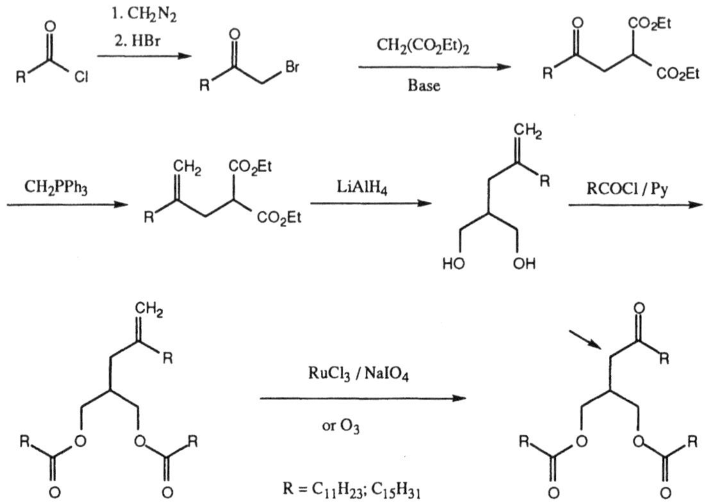

Fig. 5. Carba Analogues of Triglycerides - Synthetic Route 2

In order to provide C - analogues of 1,2 or 2,3 - *sn* - diglycerides in optically pure form the corresponding *exo* - methylene derivatives were hydrolyzed (Fig. 6) or esterified (Fig. 7) enantioselectively under the conditions of irreversible acyl transfer in presence of a lipase from *Pseudomonas species* *

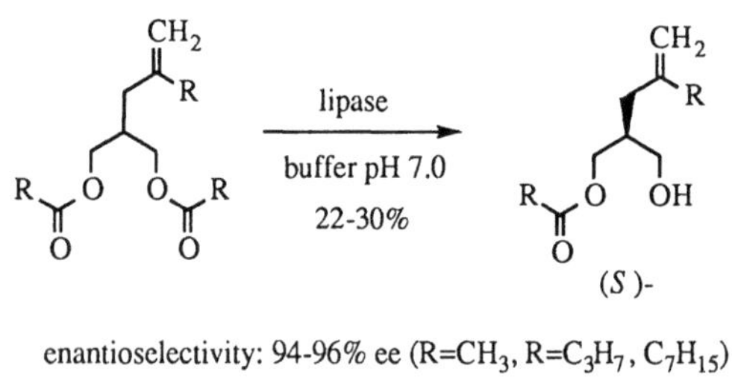

Fig. 6. Enzymatic hydrolysis of carba - analogous glycerides

Clearly, based on the high selectivity of this lipase towards the identical positions in both reaction modes the method provides access to both series of molecules with opposite absolute configurations (Jakob 1992).

*Lipase SAM-2, Amano Pharmaceuticals

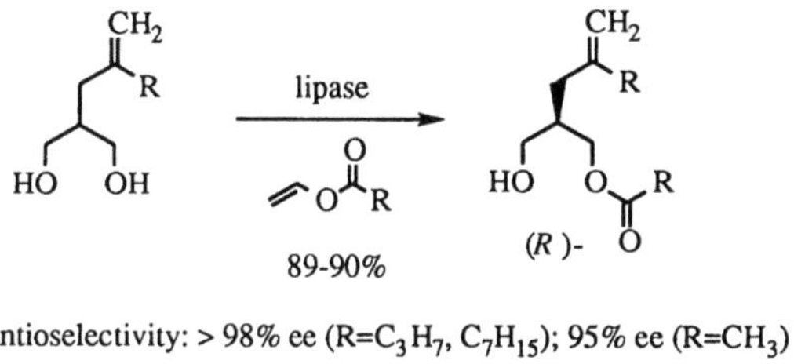

Fig. 7. Enzymatic esterification of carba - analogous glycerides

The desired monoesters were indeed obtained with high enantiomeric purities as determined by GC / HPLC with the higher chemical yields resulting from the esterification mode of reaction. The complementary absolute configurations resulting from these reaction modes were secured by chemical correlation with known molecules (Jakob B 1996).

In agreement with the retrosynthetic scheme outlined in Fig. 1, several carba - analogous 1,2 (2,3) - *sn* - diglycerides of high enantiomeric purity are now available for coupling reactions with the corresponding D-*myo*-inositol phosphates *en route* to the desired title compounds as outlined below.

Further isosteric carba - analogues of this kind, such as diglycerides with modifications in the *sn* - 1,2 - and *sn* - 1 - positions (Fig. 8) are presently synthesized in our laboratory using both

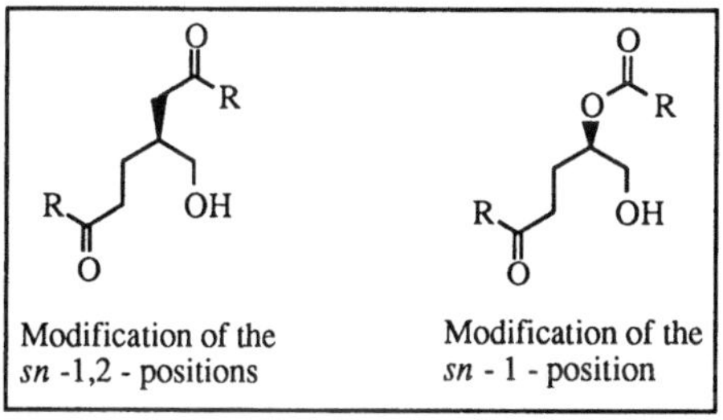

Fig. 8. Structural analogues of diglycerides

enzymatic methods and classical organic synthetic approaches (Jakob B 1996) . Upon completion of this work practically every isosteric mimic of a given diglyceride will be available for the anticipated coupling reactions or as substrates for testing their biological activities both *in vitro* and *in vivo*. We will then also be in a position to address the question whether these analogues of diglycerides can indeed be used more efficiently in replacement of phorbol esters in experiments aiming at the activation of protein kinase C.

3. D-*Myo* - Inositol phosphates

As clear from the above arguments (Introduction), enantiomerically pure *myo* - inositol phosphates have become highly attractive targets for synthetic organic chemists and numerous different approaches have been used for their synthesis in optically pure form. *Myo* - inositol itself, being derived from ubiquitous and abundantly available phytic acid is by far the most conveniently accessible and economical starting material. As already mentioned above *myo*- inositol is *achiral* while many *myo*- inositol phosphates are chiral and only biologically active in enantiomerically pure form. Consequently all known synthetic approaches are focussed on the problem of converting this *achiral* molecule into *enantiomerically pure* building blocks. The usually employed strategies are summarized in Fig. 9.

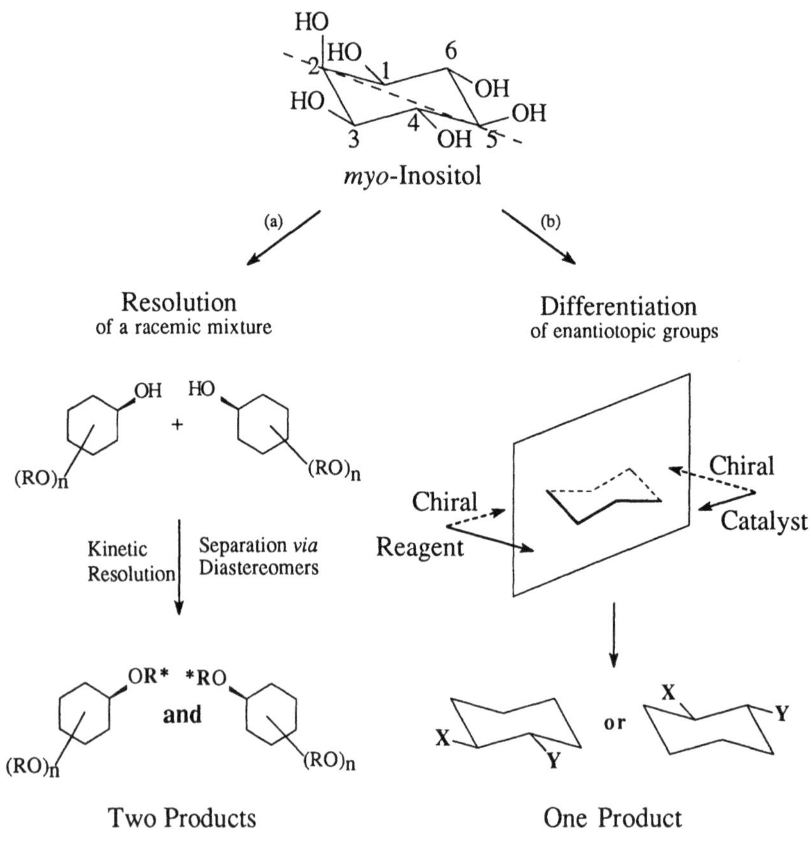

Fig. 9. Enantiomerically pure building blocks from *myo* - inositol - synthetic strategies

As clear from Fig. 9 *myo* - inositol has a plane of symmetry running through carbon atoms 2 and 5 thus bisecting the molecule into two halves which are mirror images of each other - the typical situation found in so - called *meso* - compounds. *Meso* - compounds of this kind can be transformed into the corresponding enantiomeric materials either *via* diastereomeric resolution of racemic derivatives using optically pure auxiliaries or kinetic resolutions using enzymes. Both approaches are illustrated below using two typical examples from the literature. As also shown in Fig. 9 the hydroxy groups in the 1(2) and 2(3) positions of *myo* - inositol are in an equatorial - axial relationship and can thus easily be converted into the corresponding ketals by reaction with ketones, e.g. cyclohexanone. Starting with *myo* - inositol - a *meso* - compound - this will automatically lead to a racemic or diastereomeric mixture of the corresponding ketals (compare Fig. 9 a and Fig. 10, 11).

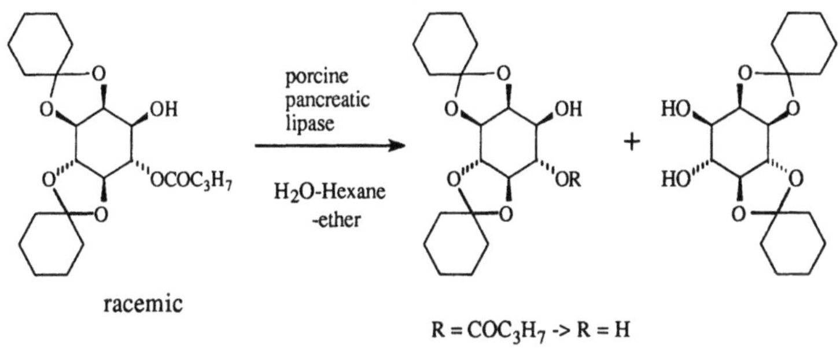

Fig. 10. Lipase catalyzed kinetic resolution of a racemic *myo* - inositol derivative

It was shown that the lipase catalyzed hydrolyses of esters derived from such racemic ketals was highly enantioselective with only one enantiomer being hydrolyzed, while the antipode remained largely untouched (Chen C-S 1994, 1996). The resulting diastereomers can be separated by chromatography and thus both enantiomeric series of *myo* - inositol derivatives are obtainable.

An alternative approach uses the direct formation of a diastereomeric ketals using the dimethylketal of (+) - campher as chiral auxiliary (Fig. 11) (Bruzik KS et al 1996).

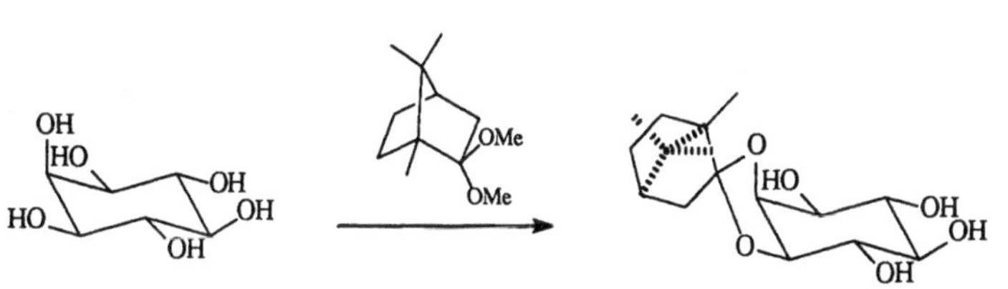

Fig. 11. Diastereomeric ketalisation of *myo* - inositol

This ketalisation, followed by chromatography and recrystallisation provides enantiomerically pure diastereomers which can be converted into the desired products using classical protection - deprotection steps. In all such cases - and inherent to the methods employed both enantiomers are always obtained and laborious separation and purification steps are always required.

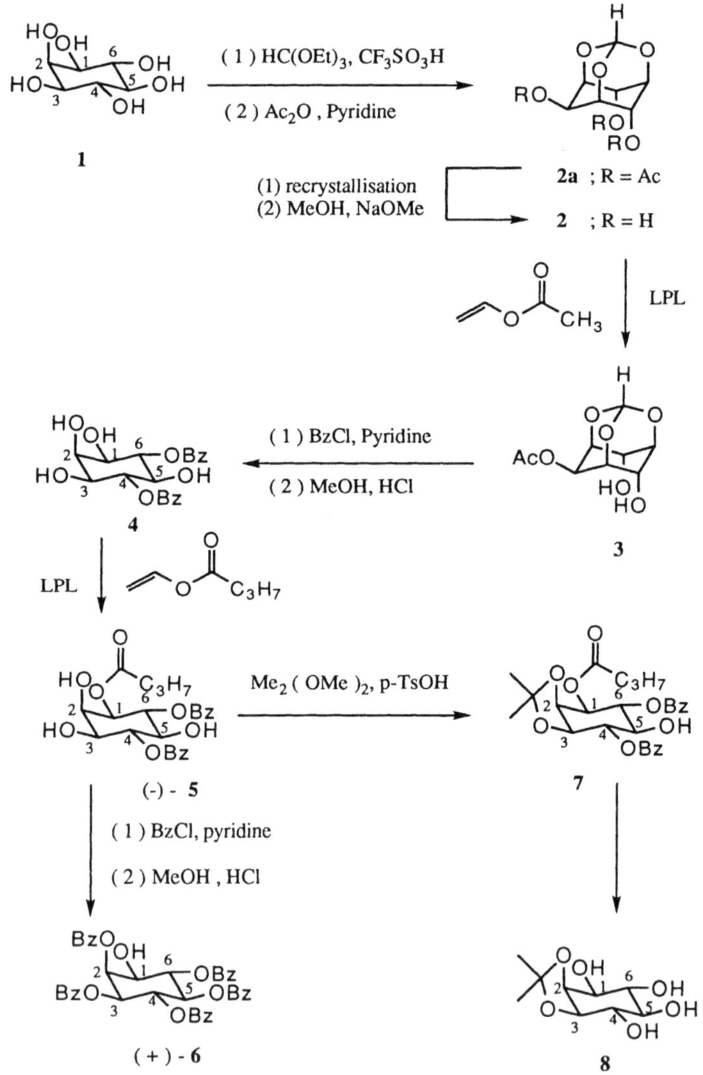

Fig. 12. Enzyme assisted syntheses of building blocks for D - *myo* - inositol phosphates

We therefore felt, that a much more elegant approach to the desired, enantiomerically pure buildung blocks could be the direct differentiation of the enantiotopic hydroxy groups in positions 1 and 3 or 4 and 6 of *myo* - inositol itself, or suitably functionalized derivatives thereof. This way the direct conversion of an *achiral* molecule into only *one enantiomer* with - theoretically - 100% optical and chemical yield could be achieved (compare Fig. 9 b). Moreover, if such an enantioselective transformation could indeed be achieved the resulting enantiomer would be synthetically useful regardless of its absolute configuration. Transformations of this kind are clearly the domaine of enzymes. In particular esterhydrolases and here lipases have been demonstrated to be highly effective in differentiating both enantiotopic hydroxy groups and esters derived thereof. The successful route using regio- and enantioselective reaction steps is outlined in Fig. 12 (Andersch P 1993, 1995).
In order to reduce the number of hydroxy groups in *myo* - inositol (**1**) this material was first converted into the corresponding orthoester (**2**) (Kishi Y 1985, Vasella A 1988). After protection of the equatorial hydroxy group using a regioselective, enzymatic esterification the resulting molecule **3** with two axial hydroxy groups in positions 4 and 6 was our first target for enzymatic esterifications and hydrolyses. Unfortunately, all our attempts to produce stable enantiomers of high optical purities were totally unsuccessful and we had to abandon this route. It was decided, however, to retain the general concept and to introduce benzoyl- and also benzyl protecting groups into these very positions. After removal of the orthoester moeity ("cap"), the resulting 4,6-protected *myo* -inositol (**4**) - again an achiral *meso*-compound with enantiotopic hydroxygroups in the 1 and 3-positions proved to be the molecule of choice for enantioselective esterifications. After screening numerous esterhydrolases (lipases) for their capability of selectively converting this substrate we found a lipoprotein lipase from the portfolio of Boehringer Mannheim* which was able to convert this substrate in one step into a single enantiomer. This reaction, in which only one out of *four* different hydroxy groups is selectively esterified, demonstrates once more the power of enzymatic methods as applied to organic synthesis (Andersch P 1993, 1995; Ghisalba O 1994).
The absolute configuration of D-1-O-butyryl-4,6-O-dibenzoyl-myo-inositol (-)-**5** was determined unambigiously by chemical correlation [e.g. with (+)-**6**] as was the optical purity of this building block.
After this key step - the introduction of chirality - the obtained building blocks (Ar = Bz, Bn) can be converted further into numerous, selectively protected *myo*-inositol derivatives which are now ready for the required phosphorylation. The obtainable phosphates are indicated below the corresponding formular in Fig. 13.

---

*Lipoproteinlipase LPL from *Pseudomonas species* ; Boehringer Mannheim cat. no. 0734284

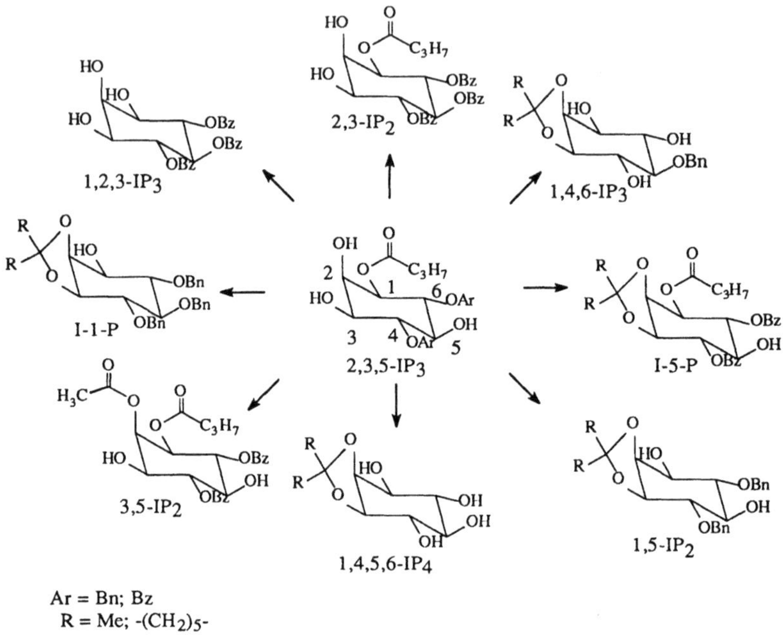

Fig. 13. Selectively protected *myo*-inositol derivatives

Using the preparation of I-1-P (Andersch P 1995) and 1,4,6-IP_3 (Andersch P 1995) as examples, typical synthetic sequences to this effect are shown in Fig 14 and 15.

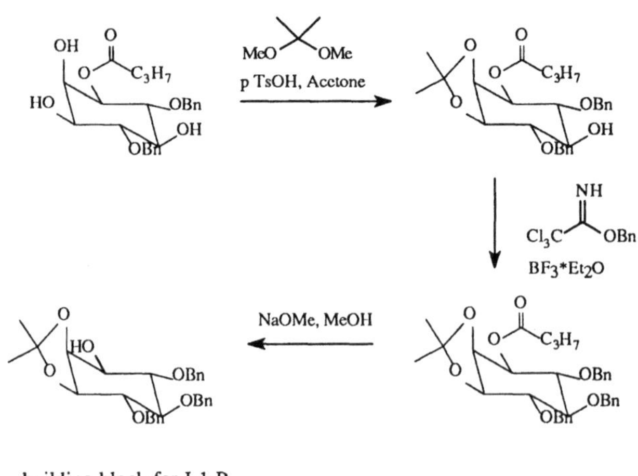

Fig.14. Synthesis of a building block for I-1-P

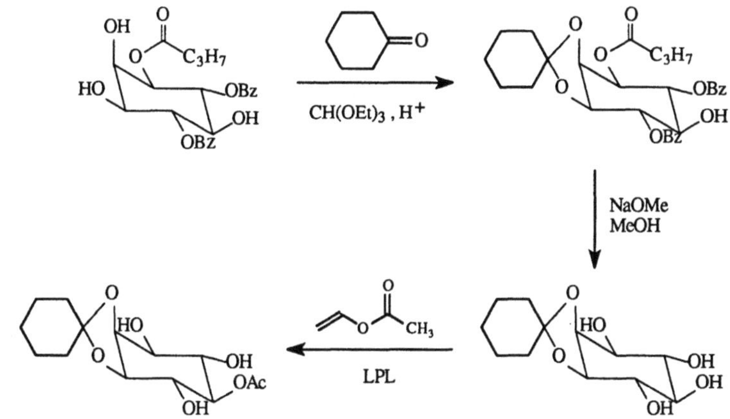

building block for 1,4,6-IP_3

Fig. 15. Synthesis of a building block for 1,4,6-IP_3

Among the methods available for the phosphorylation of such inositol derivatives we found the application of the trivalent phosphorous derivative N,N-dimethyldibenzylphosphoramidate $Me_2NP(OBn)_2$ best suited and also extremly convenient (Fig. 16). The required compound is conveniently prepared from hexamethyl phosphorous triamide and benzylalcohol (BnOH) (Andersch P 1995,1996).

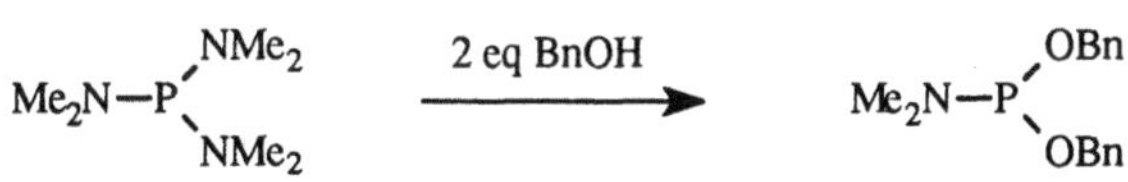

Fig. 16. Synthesis of the phosphorylation reagent $Me_2NP(OBn)_2$

Typically, these phosphorylations are carried out by reacting the inositol derivative with Me_2N-$P(OBn)_2$ in presence of tetrazole leading to the trivalent derivative. This, in turn is oxidized to the required pentavalent phosphorylate using metachloroperbenzoic acid (MCPBA) or *t*-BuOOH. Removal of the benzyl groups by catalytic hydrogenation, addition of NaOH, followed by ion exchange chromatography leads to the pure *myo*-inositol phosphates (Fig. 17) (Andersch P 1995, Schiefer R 1995).

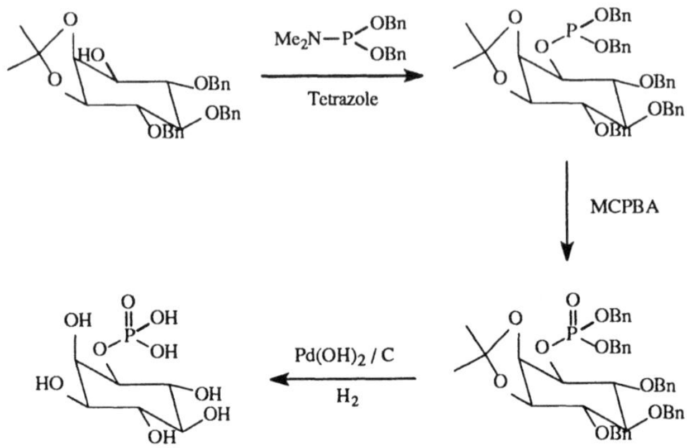

Fig. 17. Synthesis of D-*myo*-inositol-1-phosphate I-1-P

A complete synthetic sequence, exemplified here by the synthesis of D-*myo*-inositol-1,2,6-trisphosphate (1,2,6-IP$_3$; PP56; α-Trinositol), a novel experimental drug, is outlined in Fig. 18 (Andersch P 1996).

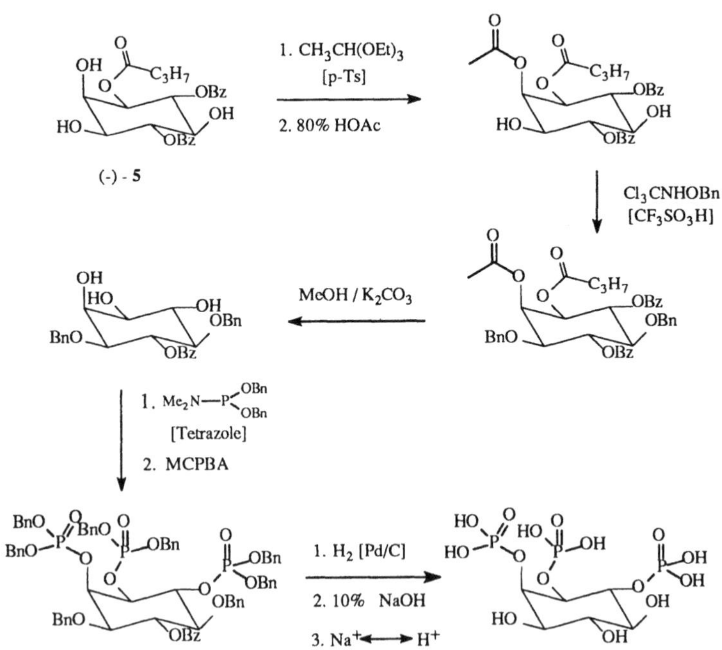

Fig. 18. Enzyme assisted synthesis of D-*myo* -inositol-1,2,6 trisphosphate 1,2,6-IP$_3$

Using the well known orthoester method (-)-**5** is converted selectively into the corresponding monoacetate in which only the axial hydroxy group in the 2-position becomes acylated. Benzylation of the equatorial hydroxy group at C_5 under acidic conditions leads to the fully protected inositol derivative. We were extremly pleased to find that the following removal of the ester functions was highly regioselective indeed, resulting in the rapid formation of the free hydroxy groups in the desired positions 1,2 and 6. While it is easily understandable that in the base catalysed methanolysis the acetate and butyrate functions are removed rapidly and faster than the more stable benzoate groups, it was somewhat surprising to find that in the progress of the reaction only one of the benzoate groups, exclusively the one in position 6 is removed selectively. The obtained triol can be phosphorylated as described above. Deprotection of the resulting trisphosphate ester with H_2 /Pd-C followed by saponification (NaOH, pH 11 -12) leads to 1,2,6-P_3 in nearly quantitative yield. All materials are obtained with very high isomeric purity as confirmed by ion exchange chromatography (Fig. 19) (Andersch P 1996).

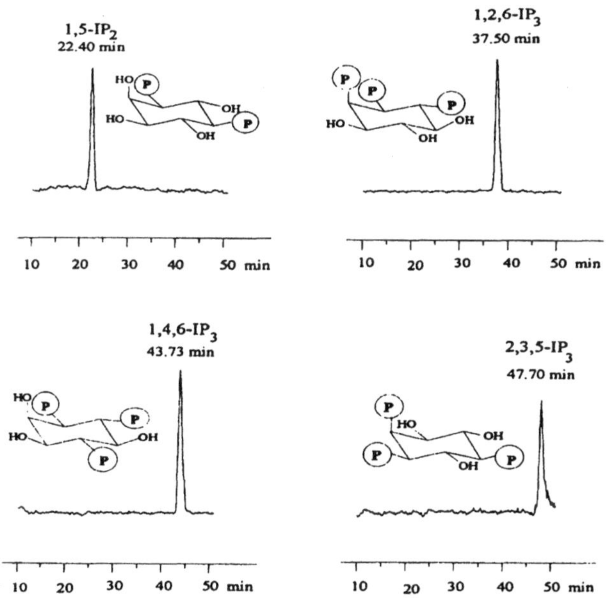

Fig. 19. Chromatographic analyses of several D-*myo*-inositol phosphates

HPLC-conditions: Mono Q-column (Pharmacia): Ø= 10 mm, l = 100 mm, 25 °C; buffer A: 50 mM Tris/HCl, pH 8.5; buffer B: 50 mM Tris/HCl 400 mM KCl, pH 8.5; buffer C: 2 mM ammoniumacetat, 30 µM YCl3, 200 µM PAR, pH 5.0; buffer A/B= 1.5 ml/min, buffer C: *postcolumn*=0.75 ml/min; detection: UV/VIS: 546 nm; Gradient time [min] (buffer B[%]): 0 (30), 2 (40), 16 (42), 20 (50), 38 (60), 48 (75), 50 (100), 51 (30)

In summary it can be concluded that, starting from *myo*-inositol itself and comploying selective enzymatic and chemical reactions enantiomerically pure building blocks can be obtained which are suitable for the synthesis of a variety of optically pure D-*myo*-inositol phosphates of defined absolute configuration.

4. *En route* to D-*myo*-inositol phospholipids

Following the retrosynthetic scheme outlined in Fig. 1 the above described building blocks, i.e. selectively protected D-*myo*-inositol derivatives and carba analogues of 1,2-*sn*-diglycerides have been made available *via* enzyme assisted synthetic routes and are now available to be linked *via* a phosphate bridge to produce the desired target molecules.
Although considerable improvements regarding the product yields are still required in this last step and the experimental proceduces are awaiting further optimisation we were pleased to find that the actual coupling reaction can indeed be carried out successfully. This is shown in Fig. 21 using the enantiomerically pure building block for D-*myo*-inositol-1-phosphate (Fig. 14,17) and one carba analogue of a diglyceride as coupling partners.
For this the above described phosphorylation method (see Fig. 16) was adapted by first preparing the corresponding trivalent phosphorylation reagent $(Me_2N)_2POBn$ (Fig. 20) by reacting $(Me_2N)_3P$ with only one equivalent of benzylalkohol leading to the desired reagent with two reactive functionalities.

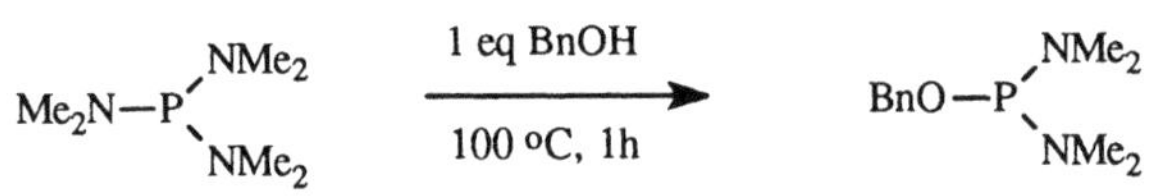

Fig. 20. Synthesis of the phosphorylation reagent $(Me_2N)_2POBn$

Sequential reactions with (a) the D-*myo* -inositol derivative and (b) the carba analogue of a diglyceride followed by oxidation with MCPBA leads to the corresponding fully protected D-*myo*-inositol phospholipid which can be deprotected by catalytical hydrogenation to produce the desired target molecule (Fig. 21) (Andersch P 1995, Schiefer R 1995).
These experiments are still at the beginning and far from being complete. Numerous problems have to be solved, expecially regarding the final deprotection step, the product yields and the scope of the reaction regarding the molecular constitutions of the coupling partners.

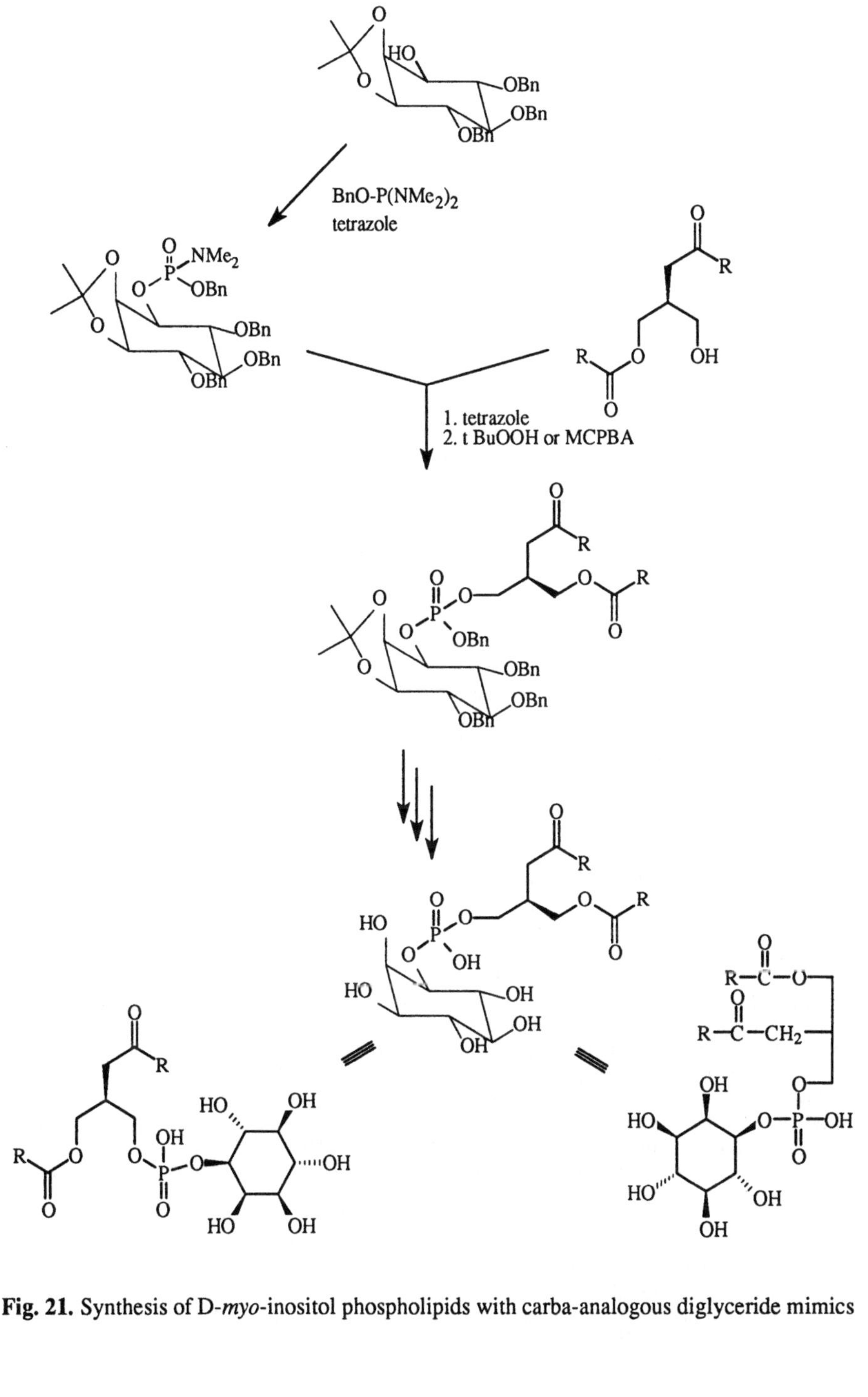

Fig. 21. Synthesis of D-*myo*-inositol phospholipids with carba-analogous diglyceride mimics

Summary

We are confident that some of the above described experiments are useful for the production of molecules involved in signal transduction processes and that the characterised and identified materials can be made available for an evalution of their biological activities and for comparison with naturally occuring compounds. In this sense we hope that synthetic organic chemistry can make useful contributions towards studies in cell biology.

Acknowledgements

We thank the *Fonds der Chemischen Industrie* for financial support of this work.

B.J. gratefully acknowledges a postgraduate stipend by the State of NRW, Germany

References

Andersch P, Schneider MP (1993) Enzyme Assisted Synthesis of Enantiomerically Pure *myo*-Inositol Derivatives - Chiral Building Blocks for Inositiol Polyphosphates. Tetrahedron Asymmetry 4: 2135-2138

Andersch P, Schneider MP (1996) Enzyme Assisted Synthesis of D-*myo*-Inositol-1,2,6-trisphosphate. Tetrahedron Asymmetry 7: 349-352

Andersch P (1995) Enzymunterstützte Synthesen von enantiomerenreinen *myo*-Inositolderivaten: Chirale Bausteine für die die Synthese von Inositolphosphaten und Phosphatidylinositolen. Dissertation UNI-GHS Wuppertal

Bell RB (1986) Protein Kinase C Activation by Diacylglycerol Second Messemgers. Cell 45: 631-632

Berger M et al (1994) Carba Analogues of Triglycerides - Isosteric Mimics for Natural Lipids. Novel Substrates for the Determination of Regio- and Enantioselectivities Displayed by Lipases. Bioorg & Med Chem 2(7): 573-588

Billington DC (1993) The Inositol Phosphates. VCH New York

Bruzik KS et al (1996) Synthesis of Inositol Phosphodiesters by Phospholipase C-Catalysed Transesterification. J Am Chem Soc 118: 7679-7688

Chen C-S et al (1994) Molecular Interactions of Endogenous D-*myo*-Inositol Phosphates with the Intracellular D-*myo*-Inositol 1,4,5-Triphosphate Recognition Site. Biochemistry 33: 11586-11597

Chen C-S, Wang Da-S (1996) Synthesis of the D-3 Series of Phosphatidylinositol Phosphates. J Org Chem 61: 5905-5910

Ghisalba O, Laumen K (1994) Preperative-scale Chemo-enzymatic Synthesis of Optically Pure D-*myo*-Inositol-1-phosphate. Biosci Biotech Biochem 58: (11) 2046-2049

Irvine RF (1990) Methods in Inosite Research Raven Press New York

Kishi Y, Lee HW (1985) Synthesis of Mono and Unsymmetrical Bis Ortho Esters of *scyllo*-Inositol. J Org Chem 50: 4402-4404

Jakob B (1992) Enantiomerenreine Monoester Carba-analoger Glyceride - Enzymatische Herstellung und analytische Charakterisierung. Diplomarbeit UNI-GHS Wuppertal

Jakob B (1996) - unpublished Ph.D thesis Bernd Jakob UNI-GHS Wuppertal

Lodish H.(1996) Molekulare Zellbiologie. 2. Auflage Walter de Gruyter s.951 ff; Molecular Cell Biology, Third Editon (1995) Scientific American Books

Nishizuka Y (1992) Intracellular signalling by hydrolysis of phospholipids and activation of proteinkinase C. Science 258: 607-614

Reitz AB (1991) Inositol Phosphates and Derivatives Synthesis, Biochemistry, and Therapeutic Potential ACS Symposium Series 463 Washington DC

Schiefer R (1995) Enzymunterstüzte Synthesen von *myo*-Inositolphosphaten. Diplomarbeit UNI-GHS Wuppertal

Vasella A et al (1988) A Synthesis of 1D- and 1L-*myo*-Inositol 1,3,4,5-Tetraphosphate. Helv Chim Act 1367-1378

COUPLING SIGNAL TRANSDUCTION TO TRANSCRIPTION: THE NUCLEAR RESPONSE TO cAMP

Emmanuel Zazopoulos, Dario De Cesare, Nicholas S. Foulkes, Cristina Mazzucchelli, Monica Lamas, Katherine Tamai, Enzo Lalli, Gianmaria Fimia, David Whitmore, Estelle Heitz and Paolo Sassone-Corsi

Institut de Génétique et de Biologie Moléculaire et Cellulaire
Centre National de la Recherche Scientifique
B. P. 163, 67404 Illkirch, Strasbourg
France

INTRODUCTION

The structural organization of most transcription factors is intrinsically modular, in most cases including a DNA binding domain and an activation domain. It has been shown that these domains can be interchanged between different factors and still retain their functional properties. This modularity suggests that, during evolution, increasing complexity of gene expression may have resulted not only by duplication and divergence of existing genes, but also by a domain shuffling process to generate factors with novel properties (Harrison, 1991).

An important step forward in the study of transcription factors has been the discovery that many constitute final targets of specific signal transduction pathways. The

NATO ASI Series, Vol. H 101
Molecular Mechanisms of Signalling
and Membrane Transport
Edited by Karel W. A. Wirtz

two major signal transduction systems are those including cAMP and diacylglycerol (DAG) as secondary messengers (Nishizuka, 1986). Each pathway is also characterized by specific protein kinases (Protein Kinase A and Protein Kinase C, respectively) and its ultimate target DNA control element (cAMP-responsive element (CRE) and TPA-responsive element (TRE), respectively). Although initially characterized as distinct systems, accumulating evidence points towards extensive cross-talk between these two pathways (Cambier et al., 1987; Yoshimasa et al., 1987; Masquilier and Sassone-Corsi, 1992).

Intracellular levels of cAMP are regulated primarily by adenylate cyclase. This enzyme is in turn modulated by various extracellular stimuli mediated by receptors and their interaction with G proteins (McKnight et al., 1988). cAMP binds cooperatively to two sites on the regulatory subunits of protein kinase-A (PKA), releasing the active catalytic subunits (Roesler et al., 1988; Lalli and Sassone-Corsi, 1994). These are translocated from cytoplasmic and Golgi complex anchoring sites and phosphorylate a number of cytoplasmic and nuclear proteins on serines in the context X-Arg-Arg-X-Ser-X (Lalli and Sassone-Corsi, 1994). In the nucleus, PKA-mediated phosphorylation ultimately influences the transcriptional regulation of various genes through distinct, cAMP-inducible promoter responsive sites (Ziff, 1990; Borrelli et al., 1992).

INDUCTION OF THE cAMP RESPONSIVE TRANSCRIPTION

The consensus cAMP-responsive element (CRE) is constituted by an 8 bp palindromic sequence (TGACGTCA) with a higher conservation in the 5' half of the palindrome than the 3' sequence. Several genes which are regulated by a variety of endocrine stimuli contain similar sequences in their promoter regions although at different positions (Sassone-Corsi, 1988; Borrelli et al., 1992).

The first CRE-binding factor to be characterised was CREB (CRE-binding protein; Hoeffler et al., 1988) but subsequently at least ten additional CRE-binding factor cDNAs have been cloned. They were obtained by screening a variety of cDNA expression libraries, with CRE and ATF sites (Hai et al., 1989; Foulkes et al., 1991). These proteins belong to the bZip transcription factor class.

The different factors are able to heterodimerize with each other but only in certain combinations. A "dimerization code" exists which seems to be a property of the leucine zipper structure of each factor. Some ATF/CREB factors are able to heterodimerize with Fos and Jun, and this may change the specific affinity of binding to a CRE with respect to a Fos-Jun binding site (Hai and Curran, 1991). This property resides in the similarity between the CRE (TGACGTCA) and TRE (TGACTCA) sequences (Sassone-Corsi et al.,

1990; Masquilier and Sassone-Corsi, 1992) and demonstrates the versatility of the transcriptional response to signal transduction.

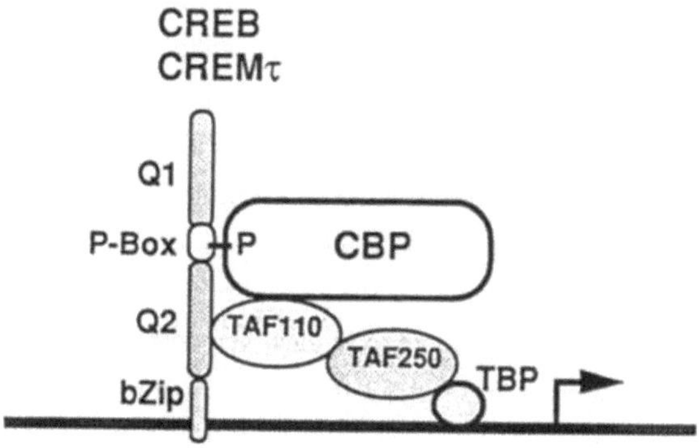

Figure 1. Scheme of the possible interaction between the transcriptional activators CREB and CREM with the co-activator CBP (Chrivia et al., 1993). Interaction is phosphorylation-dependent and allows the contact with other elements of the transcriptional machinery, such as TAF110, TAF250 and the TATA-binding protein TBP.

There are both activators and repressors of cAMP-responsive transcription. Some alternatively spliced CREM isoforms act as antagonists of cAMP-induced transcription. The cAMP-inducible ICER product deserves special mention since it is generated from an alternative promoter of the CREM gene and is responsible for its early response inducibility which is unique amongst CRE-binding factors (Molina et al., 1993; Stehle et al., 1993).

THE KEY ROLE OF PHOSPHORYLATION

The characterisation of the transcriptional activators CREB and CREM Gonzalez and Montminy, 1989; Foulkes et al., 1992) has helped the elucidation of the molecular mechanisms involved in transcriptional activation. These factors contain a transcriptional activation domain which is divided into two independent regions (Lalli and Sassone-Corsi, 1994). The first, known as the phosphorylation box (P-box) contains several consensus phosphorylation sites for various kinases, such as PKA, PKC, p34cdc2, glycogen synthase kinase-3 and casein kinases (CK) I and II (Gonzalez and Montminy, 1998; Lee et al., 1990; de Groot et al., 1993a; 1993b). The second region flanks the P-box and is constituted by domains rich in glutamine residues (Lalli and Sassone-Corsi, 1994).

Upon activation of the adenylyl cyclase pathway, a serine residue at position 133 of CREB and at position 117 of CREM is phosphorylated by PKA (Gonzalez and Montminy, 1989; de Groot et al., 1993a). The major effect of phosphorylation is to convert CREB and CREM into powerful transcriptional activators. Within the P-box, serine 133/117 is located in a region of about 50 amino acids containing an abundance of phosphorylatable serines and acidic residues which was shown to be essential for transactivation by CREB and CREM (Lee et al., 1990; de Groot et al., 1993a).

Interestingly, in PC12 cells, increases in the levels of intracellular Ca^{2+} caused by membrane depolarization have been shown to induce the phosphorylation of serine 133 in CREB and a concomitant activation of c-*fos* gene expression mediated by a CRE in the promoter (Sassone-Corsi et al., 1988; Sheng et al., 1990). Although Ca^{2+}-dependent CamK was shown to be able to phosphorylate serine 133 in vitro (Dash et al., 1991), the in vivo significance remains unclear, since PKA also seems to be necessary for c-*fos* induction mediated by Ca^{2+} influx in PC12 cells (Ginty et al., 1991).

An important finding that reveals the complexity of the transcriptional response elicited by these factors concerns the mitogen-induced p70 S6 kinase, which phosphorylates and activates CREM (de Groot et al., 1994). This finding implicates $p70^{s6k}$, a kinase generally considered cytoplasmic, in the mitogenic response also at the nuclear level. Interestingly, since CREM and other factors of the CREB/ATF family represent the final targets of the cAMP-pathway, these results show that they may also act as effectors of converging signalling systems and possibly as mediators of pathway cross-talk (de Groot et al., 1994).

INTERACTION WITH CO-ACTIVATOR CBP

The two domains flanking the P-box contain about three-times more glutamine residues than in the remainder of the protein in both CREB and CREM. Glutamine-rich domains have been characterized in other factors, such as AP-2 and Sp1 (Courey and Tjian, 1989; Williams et al., 1988) as transcriptional activation domains. The current notion is that they constitute surfaces of the protein which can interact with other components of the transcriptional machinery. Indeed, further steps towards an understanding of the mechanism of action of the P-box has come with the identification of a 265K, 2441 aminoacid protein, CBP (CREB-binding protein) that is able to interact specifically with the phosphorylated CREB P-box domain (Chrivia et al., 1993). The CBP sequence reveals two zinc finger domains, a glutamine-rich domain at its C-terminus and a single consensus PKA recognition site. Phosphorylation of Ser-133 promotes binding to CBP and consequently the interaction with TFIIB, a general transcription factor involved in RNA polymerase II activity (Kwok et al., 1994). Thus, CBP may act as a link between CREB and the transcription preinitiation complex. This interaction may need some RNA polymerase II cofactors, such as TAF110 (Figure 1). Finally, the adenoviral E1A oncoprotein-associated p300, which is thought to play a role in preventing the cell cycle G0/G1 transition, is structurally very closely related to CBP (Arany et al., 1995). Both CBP and p300 appear to have intrinsic activating properties which are inhibited by the E1A protein (Arany et al., 1995). Thus, it is clear that studies of the transcriptional activation domain of CRE-binding bZip factors continues to provide important insights into the function of transcription factors in general.

CREM BELONGS TO THE EARLY RESPONSE CLASS OF GENES

During studies of CREM expression within the neuroendocrine system, an unexpected new facet emerged: namely the transcription of the CREM gene is inducible by cAMP (Molina et al., 1993). Furthermore, the kinetics of this induction are those of an early response gene (Verma and Sassone-Corsi, 1987). This important finding further reinforces the notion that CREM products play a fulcral role in the nuclear response to cAMP since the expression of no other CRE-binding factor has been shown to be inducible to date.

The demonstration that the CREM gene was cAMP inducible first came from the finding that adrenergic signals direct CREM transcription in the pineal gland (Stehle et al., 1993). The inducibility phenomenon was then characterised in detail in the pituitary corticotroph cell line AtT20. In unstimulated cells the level of CREM transcript is undetectable. However, upon treatment with forskolin (or other cAMP analogs), within 30 minutes there is a rapid increase in CREM transcript levels which peak after 2 hours and then progressively decline to basal levels by 5 hours. These characteristic kinetics classify CREM as an "early response gene" and thus directly implicate the cAMP pathway in the cell's early response for the first time. CREM inducibility is specific for the cAMP pathway since the gene is not inducible by TPA or dexamethasone treatment. The inducible CREM transcript corresponds to a truncated product, termed ICER (Inducible cAMP Early Repressor) (Molina et al., 1993; Stehle et al., 1993).

The 5' end of the ICER clones correspond to an alternative transcription start site. The start of transcription, which identifies the P2 promoter, is within the 10kb intron which is C-terminal to the Q2 glutamine-rich domain exon. In contrast to the promoter generating all the previously characterised CREM isoforms (P1) and which is GC-rich and not inducible by cAMP (N. S. Foulkes, unpublished), the P2 promoter has a normal A-T and G-C content and is strongly inducible by cAMP. It contains two pairs of closely-spaced CRE elements organized in tandem, where the separation between each pair is only three nucleotides (Figure 2). These features make P2 unique amongst cAMP-regulated promoters and are suggestive of cooperative interactions among the factors binding to these sites.

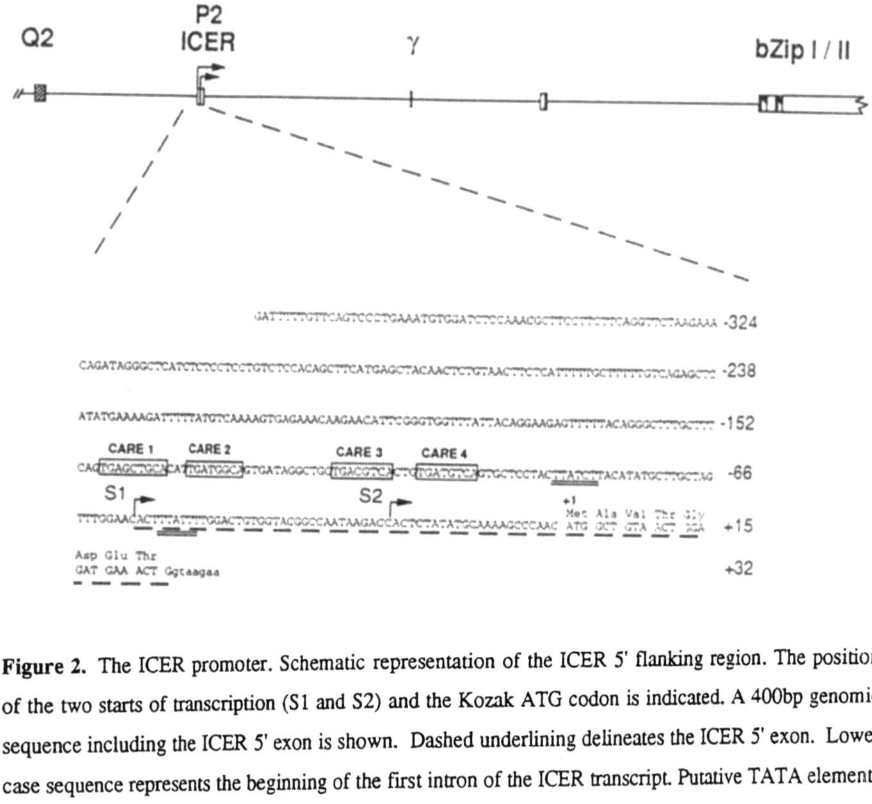

Figure 2. The ICER promoter. Schematic representation of the ICER 5' flanking region. The position of the two starts of transcription (S1 and S2) and the Kozak ATG codon is indicated. A 400bp genomic sequence including the ICER 5' exon is shown. Dashed underlining delineates the ICER 5' exon. Lower case sequence represents the beginning of the first intron of the ICER transcript. Putative TATA elements are indicated by double underlining while the four CRE-like elements (CAREs) are boxed and labeled. The position +1 corresponds to the A of the Kozak ATG initiation codon (Molina et al..1993).

The ICER open reading frame is constituted by the C-terminal segment of CREM. The predicted open reading frame encodes a small protein of 120 amino acids with a predicted molecular weight of 13.4kD. This protein, compared with the previously described CREM isoforms, essentially consists of only the DNA binding domain, which is constituted by the leucine zipper and basic region. The structure of ICER is suggestive of its function and makes it one of the smallest transcription factors ever described (Molina et al., 1993; Stehle et al., 1993).

The intact DNA binding domain directs specific ICER binding to a consensus CRE element. Importantly, ICER is able to heterodimerize with the other CREM proteins and with CREB. ICER functions as a powerful repressor of cAMP-induced transcription in transfection assays using an extensive range of reporter plasmids carrying individual CRE elements or cAMP-inducible promoter fragments (Molina et al., 1993). Interestingly, ICER-mediated repression is obtained at substoichiometric concentrations, similarly to the previously described CREM antagonists (Laoide et al., 1993). ICER escapes from PKA-dependent phosphorylation and thus constitutes a new category of CRE binding factor, for which the principle determinant of their activity is their intracellular concentration and not their degree of phosphorylation. Recent data implicates dynamic ICER expression as a more general feature of neuroendocrine systems (Lalli and Sassone-Corsi, 1995; Monaco et al., 1995).

ATTENUATION AND REFRACTORY PHASE

Dephosphorylation appears to represent a key mechanism in the negative regulation of CREB activation function. It has been proposed that a mechanism to explain the attenuation of CREB activity following induction by forskolin is dephosphorylation by specific phosphatases (Hagiwara et al., 1992; see Figure 3). After the initial burst of phosphorylation in response to cAMP, CREB is dephosphorylated in vivo by protein phosphatase-1 (PP-1). However, the situation is more complex since it has been shown that both PP-1 and PP-2A can dephosphorylate CREB in vitro (Nichols et al., 1992) resulting in an apparent decreased binding to low affinity CRE sites in vitro. Therefore, the precise role of PP-1 and PP-2A in the dephosphorylation of CREB remains to be determined.

Upon cotreatment with cycloheximide, the kinetics of CREM gene induction by forskolin are altered in that there is a significant delay in the post-induction decrease in the transcript; elevated levels persist for as long as 12 hours. This implicates a *de novo* synthesised factor which might downregulate CREM transcription (Molina et al., 1993). This observation combined with the presence of CRE elements in the P2 promoter, suggested that the transient nature of the inducibility could be due to ICER (Figure 3). Consistently, the CRE elements in the P2 promoter have been shown to bind to the ICER proteins. Detailed studies have demonstrated that the ICER promoter is indeed a target for ICER negative regulation (Molina et al., 1993). Thus, there exists a negative autoregulatory mechanism controlling ICER expression. The CREM feedback loop predicts the presence of a refractory inducibility period in the gene's transcription (Sassone-Corsi, 1994).

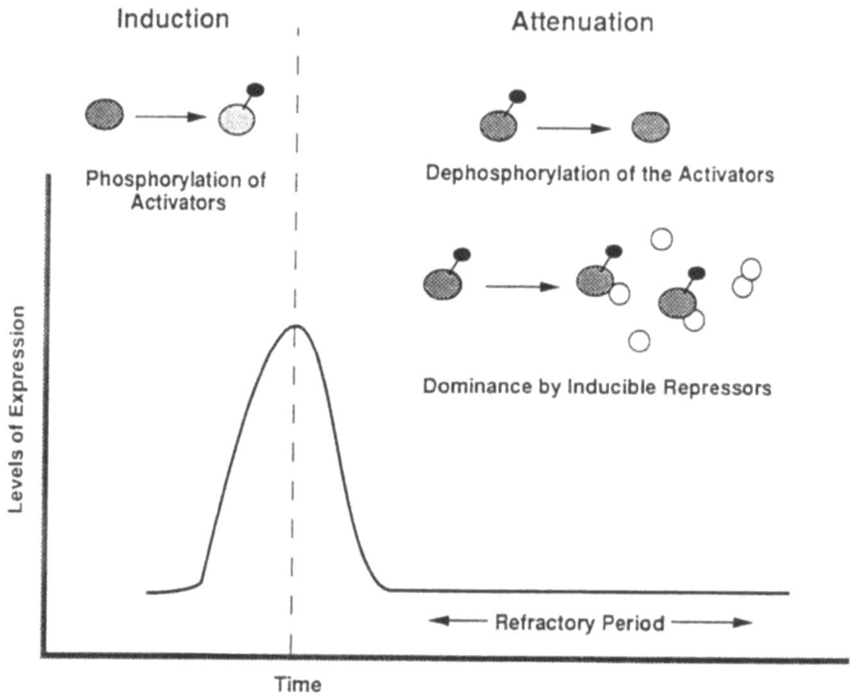

Figure 3. Kinetics of CREM inducibility. After the induction phase, due to the phosphorylation of the activators (i.e. CREB), expression is attenuated by at least two mechanisms: a) Dephosphorylation of the activators by some specific phosphatases. b) Negative autoregulation by the de novo synthesized ICER repressor on the P2 promoter (see Fig. 2) (Molina et al.,1993; Sassone-Corsi, 1994).

THE ROLE OF CREM IN SPERMATOGENESIS

CREM is a highly abundant transcript in adult testis while in prepubertal animals is expressed at very low levels. Thus, in testis CREM is the subject of a developmental switch in expression (Foulkes et al., 1992). Further charaterisation revealed that the abundant CREM transcript encodes exclusively the activator form, while in prepubertal testis only the repressor forms were detected at low levels. Thus, the developmental switch of CREM expression also constitiutes a reversal of function (Foulkes and Sassone-Corsi, 1992).

Spermatogenesis is a process occurring in a precise and coordinated manner within the seminiferous tubules (Jégou, 1993). During this entire developmental process the germ cells are maintained in intimate contact with the somatic Sertoli cells. As the spermatogonia mature, they move from the periphery towards the lumen of the tubule until the mature spermatozoa are conducted from the lumen to the collecting ducts.

A remarkable aspect of the CREM developmental switch in germ cells is constituted by its exquisite hormonal regulation. The spermatogenic differentiation program is under the tight control of the hypothalamic-pituitary axis (Jégou, 1993). The regulation of CREM function in testis seems to be intricately linked to FSH both at the level of the control of transcript processing and at the level of protein activity (Figure 4). For example, surgical removal of the pituitary gland leads to the loss of CREM expression in the rat adult testis (Foulkes et al., 1993). Furthermore hypophysectomisation in prepubertal animals, prevents the switch in CREM expression at the pachytene spermatocyte stage, thus implicating the pituitary directly in the maintenance of as well as the switch to high levels of CREM expression. Injection of FSH leads to a rapid and significant induction of the CREM transcript. The hormonal induction of CREM by FSH is not transcriptional, consistently with the housekeeping nature of the P1 promoter. Instead, by a mechanism of alternative polyadenylation, AUUUA destabilizer elements present in the 3' untranslated region of the gene are excluded, dramatically increasing the stability of the CREM message. CREM is the first example of a gene whose expression is modulated by a pituitary hormone during spermatogenesis (Foulkes et al., 1993). The implication of these findings is that hormones can regulate gene expression at the level of RNA processing and stability. Importantly the effect of FSH can not be direct since germ cells do not have FSH receptors. Recent data suggest that another hormonal message originating from the Sertoli cells upon FSH stimulation is mediating CREM activation in germ cells (L. Monaco, unpublished results).

CREM, A MASTER-SWITCH IN HAPLOID GERM CELLS

A first hint as to the role of CREM during spermatogenesis was indicated by its protein expression pattern. In the seminiferous epithelium, CREM transcripts accumulate in spermatocytes and spermatids, but CREM protein is detected only in haploid spermatids (Delmas et al., 1993). The absence of CREM protein in spermatocytes reflects a strict translational control and indicates multiple levels of regulation of gene expression in testis. It will be extremely important to analyze further the mechanism of this delay in translation and to define whether it is also hormonally dependent.

Phosphorylation by PKA activates CREM function allowing the relay of the hormonal signal from the cytoplasm to the nucleus (Lalli and Sassone-Corsi, 1994). The CREM activator is efficiently phosphorylated by cAMP-dependent PKA activity endogenous to the spermatids, indicating that the CREM protein is a nuclear target for the cAMP pathway in haploid spermatogenic cells (Delmas et al., 1993).

The expression of CREM activator protein in spermatids coincides with the transcriptional activation of several genes containing a CRE motif in their promoter region. These genes encode mainly structural proteins required for spermatozoon assembly (transition protein, protamine, etc.), suggesting a role for CREM in the activation of genes required for the late phase of spermatid differentiation (Figure 4). This observation implies that the transcription of some key structural genes is directly linked to hormonal control and consequently to the level of cAMP present in seminiferous epithelium.

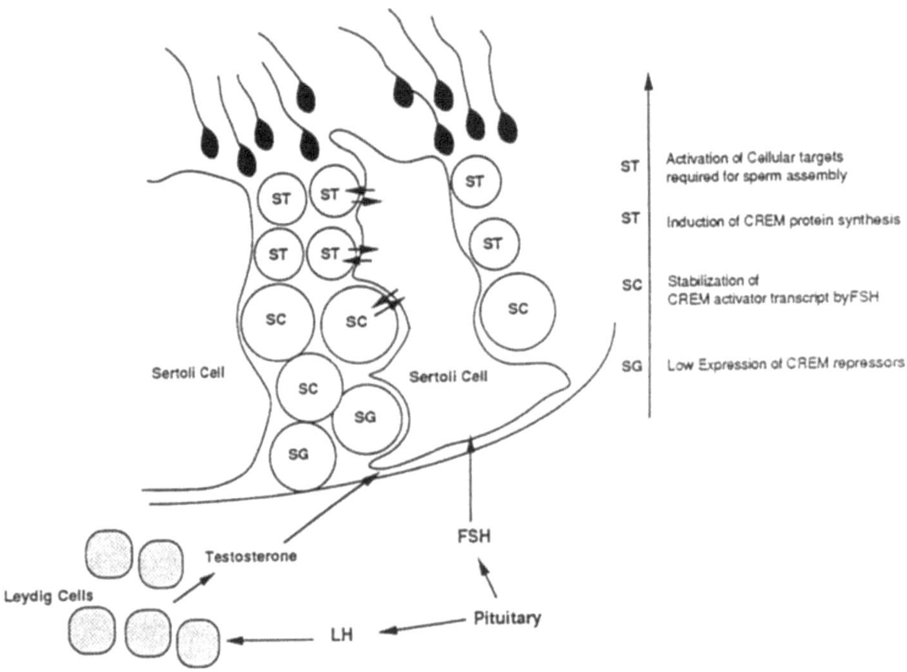

Figure 4. Schematic representation of a section of a seminiferous tubule where the CREM expression pattern is indicated. CREM expression is regulated at multiple levels during spermatogenesis. Premeiotic germ cells (spermatogonia, SG) express a low level of CREM repressor isoforms. During meiotic prophase, the pituitary Follicle Stimulating Hormone (FSH) is responsible for the stabilization of CREM activator transcripts in spermatocytes (SC); CREM protein, on the other hand, is detected only after meiosis in haploid spermatids (ST).

To date at least three genes, RT7 (Delmas et al., 1993), transition protein-1 (Kistler et al., 1994) and calspermin (Sun et al., 1995) have been shown to be targets of CREM-mediated transactivation in germ cells. A demonstration of the role of CREM in the expression of one of these genes, RT7, was shown using *in vitro* transcription experiments. A CREM-specific antibody blocks RT7 *in vitro* transcription with nuclear extracts from seminiferous tubules but not with extracts from liver (Delmas et al., 1993). In conclusion, CREM might participate in testis- and developmental-specific regulation of genes containing a CRE in their promoter region, by expressing the repressor isoforms before meiosis, and high levels of the activator after meiosis.

CONCLUSIONS AND PERSPECTIVES

To date much of the research in transcription factor biology has been devoted to understanding the structure and function relationship of these factors. Progress has been extremely rapid and now the basic principles of transcription factor function are close to being elucidated. As a result of this work, new questions have been raised and so it is clear we still have a long way to go before we understand completely how the promoters and enhancers of genes execute transcriptional control. However a much greater challenge lies ahead and that is to relate transcriptional control mechanisms to the physiology and biology of the organism. The use of homologous recombination to inactivate specific gene products offers a powerful tool to address such questions. Paradoxically however, in some cases it has complicated our understanding since it is clear that many important factors operate in the context of networks where there is considerable overlap of function. Thus, it is possible that the phenotype obtained by loss of a single factor could reflect more the compensatory adjustments made by other factors in its network rather than the function of the target factor itself. Importantly, recent results obtained with the ablation of the CREM gene in the mouse germline demonstrate its key role in the spermatogenesis differentiation process (Nantel et al., 1996). The levels of other CRE-binding proteins in these mice were not affected. CRE-binding proteins appear to play a central role in the physiology of the neuroendocrine system. Further studies of their molecular and functional characteristics will therefore represent another major step forward in our understanding of hormonal regulation and metabolism.

Acknowledgments

We wish to thank E. Borrelli and L. Monaco for help and support. This work was supported by grants from Centre National de la Recherche Scientifique, Institut National de la Santé et de la Recherche Médicale, Centre Hospitalier Universitaire Régional, Fondation de la Recherche Médicale, Université Louis Pasteur, Association pour la Recherche sur le Cancer and Rhône-Poulenc Rorer.

REFERENCES

Arany, Z., Newsome, D., Oldread, E., Livingston, D.M. and Eckner, R. 1995 A Family of Transcriptional Adaptor proteins targeted by the E1A Oncoprotein. *Nature* 374, 81-84.

Borrelli, E., Montmayeur, J. P., Foulkes, N. S. and Sassone-Corsi, P. 1992 Signal transduction and gene control: the cAMP pathway. *Critical Rev. Oncogenesis* 3, 321-338

Cambier, J. C., Newell, N. K., Justement, J. B., McGuire, J. C., Leach, K. L. and Chen, K.K. 1987 Ia binding ligands and cAMP stimulate nuclear translocation of PKC in b lymphocytes. *Nature* 327, 629-632.

Chrivia, J.C., Kwok, R.P.S., Lamb, N., Haniwawa, M., Montminy, M.R. and Goodman, R.H. 1993 Phosphorylated CREB binds specifically to the nuclear protein CBP. *Nature* 365, 855-859

Courey, A.J. and Tjian, R. 1989 Analysis of Sp1 in vivo reveals multiple transcriptional domains, including a novel glutamine activation motif. *Cell,* 55, 887-898

Dash, P.K., Karl, K.A., Colicos, M.A., Prywes, R. and Kandel, E.R. 1991 cAMP response element-binding protein is activated by Ca^{2+}/calmodulin- as well as cAMP-dependent protein kinase. *Proc. Natl. Acad. Sci. U.S.A.* 88, 5061-5065

de Groot, R.P., Ballou, L.M. and Sassone-Corsi, P. 1994 Positive Regulation of the cAMP-responsive Activator CREM by the p70 S6 Kinase: An Alternative Route to Mitogen-induced Gene expression. *Cell* 79, 81-91

de Groot, R.P., den Hertog, J., Vandenheede, J.R., Goris, J. and Sassone-Corsi, P. 1993a Multiple and cooperative phosphorylation events regulate the CREM activator function. *EMBO J.* 12, 3903-3911

de Groot, R.P., Derua, R., Goris, J. and Sassone-Corsi, P. 1993b Phosphorylation and negative regulation of the transcriptional activator CREM by p34cdc2. *Mol. Endocrinol.* 7, 1495-1501

Delmas, V., van der Hoorn, F., Mellström, B., Jégou, B. and Sassone-Corsi, P. 1993 Induction of CREM activator proteins in spermatids: downstream targets and implications for haploid germ cell differentiation. *Mol. Endocrinol.* 7, 1502-1514

Foulkes, N. S., Borrelli, E. and Sassone-Corsi, P. 1991 CREM gene: Use of alternative DNA binding domains generates multiple antagonists of cAMP-induced transcription. *Cell* 64, 739-749

Foulkes, N. S., Mellström, B., Benusiglio, E. and Sassone-Corsi, P. 1992 Developmental switch of CREM function during spermatogenesis: from antagonist to transcriptional activator. *Nature* 355, 80-84

Foulkes, N.S. and Sassone-Corsi, P. 1992 More is better: Activators and Repressors from the Same Gene. *Cell* 68, 411-414

Foulkes, N.S., Schlotter, F., Pévet, P. and Sassone-Corsi, P. 1993 Pituitary hormone FSH directs the CREM functional switch during spermatogenesis. *Nature* 362, 264-267

Ginty, D.D., Glowacka, D., Bader, D.S., Hidaka, H. and Wagner, J.A. 1991 Induction of immediate early genes by Ca^{2+} influx requires cAMP-dependent protein kinase in PC12 cells. *J. Biol. Chem.* 266, 17454-17458

Gonzalez, G.A. and Montminy, M.R. 1989 Cyclic AMP stimulates somatostatin gene transcription by phosphorylation of CREB at Ser 133. *Cell* 59, 675-680

Hagiwara, M., Alberts, A., Brindle, P., Meinkoth, J., Feramisco, J., Deng, T. and Montminy, M. 1992 Transcriptional attenuation following cAMP induction requires PP-1-mediated dephosphorylation of CREB. *Cell* 70, 105-113

Hai, T. Y. and Curran, T. 1991 Cross-family dimerization of transcription factors Fos:Jun and ATF/CREB alters DNA binding specificity. *Proc. Natl. Acad. Sci. U.S.A.* 88, 3720-3724

Hai, T. Y., Liu, F., Coukos, W. J. and Green, M. R. 1989 Transcription factor ATF cDNA clones: An extensive family of leucine zipper proteins able to selectively form DNA binding heterodimers. *Genes & Dev.* 3, 2083-2090

Harrison, S. C. 1991 A structural taxonomy of DNA-binding domains. *Nature* 353, 715-719

Hoeffler, J. P., Meyer, T. E., Yun, Y., Jameson, J. L. and Habener, J. F. 1988 Cyclic AMP-responsive DNA-binding Protein: Structure Based on a Cloned Placental cDNA. *Science* 242, 1430-1433

Jégou, B. 1993 The Sertoli-germ cell communication network in mammals. *Int. Rev. Cytol.* 147, 25-96

Kistler, M., Sassone-Corsi, P. and Kistler, S. W. 1994 Identification of a functional cAMP response element in the 5'-Flanking Region of the Gene for Transition Protein 1 (TP1), a basic Chromosomal Protein of Mammalian Spermatids. *Biol. Reprod.* 51, 1322-1329

Kwok, R.P., Lundblad, J.R., Chrivia, J.C., Richards, J.P., Bachinger, H.P., Brennan, R.G., Roberts, S.G., Green, M.R. and Goodman, R.H. 1994 Nuclear Protein CBP is a coactivator for the transcription factor CREB. *Nature* 370, 223-226

Lalli, E. and Sassone-Corsi, P. 1994 Signal Transduction and Gene Regulation: The nuclear Response to cAMP. *J. Biol. Chem.* 269, 17359-17362

Lalli, E. and Sassone-Corsi, P. 1995 Long-term desensitization of the TSH receptor involves TSH-directed induction of CREM in the thyroid gland. *Proc. Natl. Acad. Sci. USA* 92, 9633-9637.

Laoide, B. M., Foulkes, N.S., Schlotter, F. and Sassone-Corsi, P. 1993 The functional versatility of CREM is determined by its modular structure. *EMBO J.* 12, 1179-1191

Lee, C.Q., Yun, Y., Hoeffler, J.P. and Habener, J.F. 1990 Cyclic-AMP-responsive transcriptional activation involves interdependent phosphorylated subdomains. *EMBO J.* 9, 4455-4465

Masquilier, D. and Sassone-Corsi, P. 1992 Transcriptional cross-talk: nuclear factors CREM and CREB bind to AP-1 sites and inhibit activation by Jun. *J. Biol. Chem.* 267, 22460-22466.

McKnight, S. G., Clegg, C. H., Uhler, S. R., Chrivia, J. C., Cadd, G.G. and Correll, L. L. 1988 Analysis of the cAMP-dependent protein kinase system using molecular genetic approaches. *Rec. Progr. Horm. Res.* 44, 307-335

Molina, C.A., Foulkes, N.S., Lalli, E. and Sassone-Corsi, P. 1993 Inducibility and negative autoregulation of CREM: An alternative promoter directs the expression of ICER, an early response repressor. *Cell* 75, 875-886.

Monaco, L., Foulkes, N. S. and Sassone-Corsi, P. 1995 Pituitary follicle-stimulating hormone (FSH) induces CREM gene expression in Sertoli cells: Involvement in long-term desensitization of the FSH receptor. *Proc. Natl. Acad. Sci. USA* 92, 10673-10677

Nantel, F., Monaco, L., Foulkes, N.S., Masquilier, D., LeMeur, M., Hénriksen, K., Dierich, A., Parvinen, M. and Sassone-Corsi, P. 1996. Spermiogenesis deficiency and germ-cell apoptosis in CREM-mutant mice. *Nature* 380, 159-163.

Nichols, M., Weih, F., Schmid, W., DeVack, C., Kowenz-Leutz, E., Luckow, B. and Schutz, G. 1992 Phosphorylation of CREB affects its binding to high and low affinity sites: implications for cAMP induced gene transcription. *EMBO J.* 11, 3337-3346

Nishizuka, Y. 1986 Studies and perspectives of protein kinase C. *Science* 233, 305-312

Roesler, W. J., Vanderbark, G. R. and Hanson, R. W. 1988 Cyclic AMP and the induction of eukaryotic gene expression. *J. Biol. Chem.* 263, 9063-9066

Sassone-Corsi, P. 1988 Cyclic AMP induction of early adenovirus promoters involves sequences required for E1A-transactivation. *Proc. Natl. Acad. Sci. U.S.A.* 85, 7192-7196

Sassone-Corsi, P. 1994 Rhythmic Transcription and Autoregulatory Loops: Winding up the Biological Clock. *Cell* 78, 361-364

Sassone-Corsi, P., Ransone, L. J. and Verma, I. M. 1990 Cross-talk in signal transduction: TPA-inducible factor Jun/AP-1 activates cAMP responsive enhancer elements. *Oncogene*, 5, 427-431

Sassone-Corsi, P., Visvader, J., Ferland, L., Mellon, P. L. and Verma, I. M. 1988 Induction of proto-oncogene *fos* transcription through the adenylate cyclase pathway: characterization of a cAMP-responsive element. *Genes & Dev.* 2, 1529-1538

Sheng, M., McFadden, G. and Greenberg, M.E. 1990 Membrane depolarization and calcium induce c-fos transcription via phosphorylation of transcription factor CREB. *Neuron* 4, 571-582

Stehle, J.H., Foulkes, N.S., Molina, C.A., Simonneaux, V., Pévet P. and Sassone-Corsi, P. 1993 Adrenergic signals direct rhythmic expression of transcriptional repressor CREM in the pineal gland. *Nature* 365, 314-320

Sun, Z., Sassone-Corsi, P. and Means, A. 1995 Calspermin Gene Transcription is Regulated by two cyclic AMP response elements contained in an alternative Promoter in the Calmodulin Kinase IV gene. *Mol. Cell. Biol.* 15, 561-571

Verma, I.M. and Sassone-Corsi, P. 1987 Proto-oncogene fos: complex but versatile regulation. *Cell* 51, 513-514

Williams, T., Admon, A., Luscher, B. and Tjian, R. 1988 Cloning and Expression of AP-2, a cell-type-specific transcription factor that activates inducible enhancer elements. *Genes & Dev.* 2, 1557-1569

Yoshimasa, T., Sibley, D. R., Bouvier, M., Lefkowitz, R. J. and Caron, M. G. 1987 Cross-talk between cellular signalling pathways suggested by phorbol ester adenylate cyclase phosphorylation. *Nature* 327, 67-70

Ziff, E. B. 1990 Transcription factors: a new family gathers at the cAMP response site. *Trends Genet.* 6, 69-72

Vitamin E and the Metabolic Antioxidant Network

Lester Packer, Maurizio Podda, Manabu Kitazawa, Jens Thiele, Claude Saliou, Eric Witt, and Maret G. Traber
Department of Molecular and Cell Biology
251 Life Sciences Addition
University of California, Berkeley, CA 94720-3200, USA

Despite over 70 years of research since the discovery of vitamin E (Evans and Bishop 1922), its exact functions have not been fully elucidated; however, its antioxidant properties appear paramount to its function. Vitamin E does not work in isolation from other antioxidants, but rather is part of an interlinking set of antioxidant cycles, which has been termed the "antioxidant network". This antioxidant network is highlighted in the regulation of nuclear transcription. The linkage of vitamin E to other antioxidant substances, and their ability to regenerate vitamin E, means that other antioxidants (especially redox active thiols) may influence nuclear transcription not only directly through their own antioxidant effects, but indirectly by regenerating vitamin E as well. Hence, antioxidants are interlinked in a network that influences cell functions previously thought to be remote from antioxidant effects. The implications of such effects are just beginning to be understood.

In this paper we present evidence from our laboratory that supports the idea of an antioxidant network, not only in vitro, but in vivo, and that indicates that the network also operates in effects on gene transcription.

I. The antioxidant network

In this part we will summarize the in vitro evidence that laid the conceptual groundwork for the idea of an antioxidant network, and data, some over thirty years old and some just produced in our laboratory, which indicates that the network may also operate in vivo.

A. Theoretical basis

The necessity for the concept of an antioxidant network, which greatly bolsters vitamin E's effectiveness in membranes, is based on a discrepancy between the concentration of vitamin E in membranes and their possible rates of lipid peroxidation.

NATO ASI Series, Vol. H 101
Molecular Mechanisms of Signalling
and Membrane Transport
Edited by Karel W. A. Wirtz

Vitamin E is present in biological membranes at extremely low concentrations, usually less than 0.1 nmol per mg of membrane protein (about one molecule per 1000 to 2000 membrane phospholipid molecules). It is the major, if not the only, chain-breaking antioxidant in biological membranes (Burton and Ingold 1981). Lipid peroxyl radicals can be generated in membranes at the rate of 1 to 5 nmol per mg of membrane protein per minute, yet destructive oxidation of membrane lipids does not normally occur, nor is vitamin E rapidly depleted. Furthermore, vitamin E deficiency is remarkably difficult to induce in adult animals. These seeming paradoxes point to the existence of efficient mechanisms for permitting low concentrations of vitamin E to have such high efficiency in protecting membranes against damage and in supporting normal biological activity.

The vitamin E radical (a chromanoxyl radical) produced when vitamin E reacts with a peroxyl radical can either react with another radical (e.g., chromanoxyl, alkoxyl, or peroxyl) to produce unreactive products with no further free radical scavenging activity, or it can be reduced back to a functional vitamin E molecule, ready to undergo another round of peroxidation chain-breaking.

The reactions between water- and lipid-soluble substances by both nonenzymatic and enzymatic mechanisms, which regenerate vitamin E from its chromanoxyl radical back to vitamin E are the focus of this paper.

B. In vitro evidence for the antioxidant network

Abundant in vitro evidence indicates that vitamin C (ascorbate) can directly regenerate intact vitamin E from its radical form in solution, in membranes, and in lipoproteins. There is also a great deal of evidence that vitamin C can in turn be regenerated from the semidehydroascorbyl radical produced when it regenerates vitamin E, both by enzymatic and non-enzymatic means. Thiols, both endogenous and exogenous, figure prominently in this regeneration of ascorbate. These interactions are illustrated in Fig. 1.

Ascorbate As long ago as 1941, it was observed that vitamin C increased the antioxidant potency of vitamin E in lard and cottonseed oil (Golumbic and Mattill 1941), and in 1968 Tappel suggested that vitamin C could regenerate vitamin E from the vitamin E radical formed when vitamin E quenches a lipid peroxyl radical (Tappel 1968). In 1978, Packer et al. (Packer et al. 1979) confirmed this suggestion using electron spin resonance to detect ascorbyl and vitamin E radicals. Vitamin C also regenerated vitamin E radical formed by quenching peroxyl radicals generated in solution (Niki et al. 1984) and in liposomes (Niki et al. 1982; Doba et al. 1985). The liposome results are important in that they suggest that ascorbate, in the aqueous phase, can regenerate tocopherol from the tocopheroxyl radical, whose chromanoxyl head is near the surface of the membrane. Thus, in vitro, aqueous antioxidants and antioxidants in a lipid environment are linked.

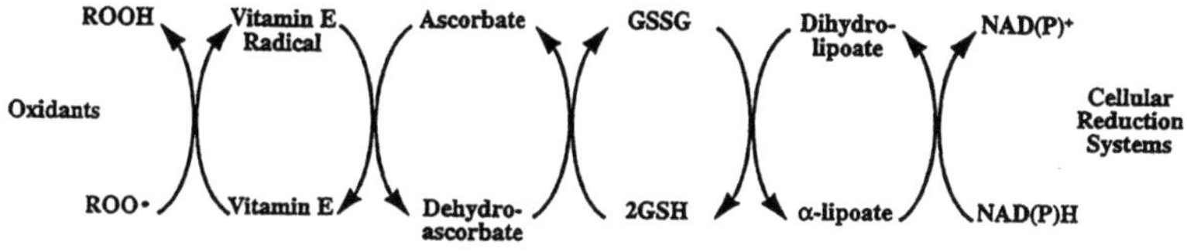

Fig. 1. The antioxidant network. The vitamin E radical, produced when vitamin E quenches lipid peroxidation (or directly produced by UV illumination in test systems), is reduced to vitamin E by vitamin C (ascorbate), producing the vitamin C (ascorbyl) radical. The vitamin C radical can be reduced to vitamin C by glutathione, and the resulting glutathione disulfide molecule can be reduced by cellular reducing systems or by dihydrolipoate, as shown. Dihydrolipoate can also directly reduce the vitamin C radical.

Thiols The semidehydroascorbyl radical decays almost entirely via disproportionation, to ascorbate and dehydroascorbate (the two-electron oxidation product of ascorbate). Dehydroascorbate can have several fates: it can irreversibly decompose to diketogluconic acid, or it can be converted to ascorbate in a thiol-dependent reaction. The latter occurs both enzymatically and non-enzymatically.

Dehydroascorbate is chemically converted back to ascorbate by glutathione in a coupled reaction (Hopkins and Morgan 1936). Dihydrolipoic acid, which is the reduced form of α–lipoate, can also directly reduce dehydroascorbate at an even more rapid rate than glutathione (Winkler et al. 1994). Dihydrolipoate may directly reduce the tocopheroxyl radical; we have found that the presence of dihydrolipoate reduces the tocopheroxyl radical ESR signal in liposomes exposed to UV light (unpublished data). Unlike glutathione, which is not absorbed in usable form when given as a dietary supplement, α–lipoate is absorbed, transported to tissues, and reduced to dihydrolipoate (Handelman et al. 1994; Podda et al. 1994; Podda et al. 1994). Hence it represents a thiol reductant of vitamin C that may be useful as a dietary supplement. We have concentrated much of our research on this thiol as a recycler of vitamin C and, indirectly, vitamin E. Evidence for vitamin E recycling by dihydrolipoate via ascorbate comes from microsomes (Scholich et al. 1989), liposomes (Kagan et al. 1992), erythrocyte membranes (Constantinescu et al. 1993), and human low density lipoproteins (Kagan et al. 1992) .

The antioxidant network is relatively easy to demonstrate in vitro, but because the reactions occur rapidly, and metabolism of products takes place, it is difficult to show in vivo. Therefore, we have undertaken studies to modify antioxidant status and then examine these interactions in model systems.

C. The antioxidant network in skin

Skin offers an easily accessible tissue in which to study the antioxidant network. All major antioxidants are present (Shindo et al. 1993), and oxidative stress can be provided by irradiation with ultraviolet (UV) light. In both epidermis and dermis UV-light depletes major antioxidants and causes concomitant formation of lipid hydroperoxides, an indicator of oxidative damage to lipids (Shindo et al. 1993; Shindo et al. 1994). Indeed, vitamin E and ascorbyl radicals are detectable by electron spin resonance (ESR) in UV-irradiated skin homogenates (Kagan and Packer 1994).

The precise mechanisms by which UV irradiation destroys skin antioxidants are unknown. Therefore, we have undertaken a careful assessment of the mechanism of radical formation using ESR detection and UV irradiation of skin homogenates.

Materials and methods Skin was obtained from female 8-week old hairless mice (Simonsen, strain SKH-1). After cervical dislocation, the skin samples were dissected immediately from the side and back. Adherent subcutis was gently removed by scraping the dermal side. Skin homogenates were prepared in a ratio of 1 g of epidermal/dermal sheet to 4 ml of 100 mM phosphate buffer, pH 7.4, using an UltraTurrax homogenizer (Ika-Works, Inc., Cincinnati, OH). UV light was generated by an Oriel (Stratford, CT) model 66021 solar light simulator with a 1000 W Xenon arc lamp (ozone free). The wavelength covered was 310 nm to 400 nm by using a dichroic mirror (Oriel model 66226) and a UVC cutoff filter. The power density of the light at the sample surface in the UVA region was 3.0 mW/cm2. The distance between the light source and the sample was 80 cm.

Ascorbate free radicals formed during UV irradiation were detected using an IBM ER 200 D-SRC electron spin resonance spectrometer. Instrument conditions were as follows: microwave power, 10 mW; modulation amplitude, 1.25 G; time constant, 0.2 sec; sweep scan rate, 1.0 G/sec; central field, 3475 G; scan range, 100 G; gain, 8.0 X 10^5.

The test compound at the indicated concentration was dissolved in 100 mM phosphate buffer solution (pH = 7.4) or methanol and was added to the skin homogenate. The mixture was transferred to P4518-50 capillary tubes (Baxter, McGaw Park, IL) (50 ul X 2). Each tube was placed in the ESR cavity and spectra recorded during UV irradiation at room temperature. The signal intensities were quantitated by measuring the peak height of the first signal of the doublet of the ascorbate free radical and the third signal of the quintet of the vitamin E free radical, respectively.

Results When mouse skin homogenates were irradiated with UV light, the ascorbyl radical signal was immediately observable (Fig. 2). The signal persisted for 50 to 60 min, reaching maximum signal intensity at 20 min from the beginning of UV irradiation. The ascorbyl radical signal was not detectable in the absence of UV irradiation, and other ESR signals were not detectable.

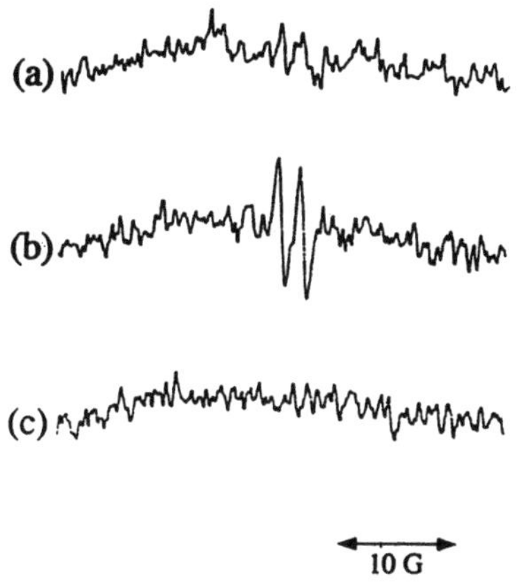

Fig. 2. ESR spectrum of the ascorbyl radical signal obtained from UV-irradiated mouse skin homogenates. (a) no UV irradiation; (b) immediately after the start of UV irradiation; (c) UV irradiation for 60 min

To study the interactions of vitamins E and C, the water-soluble homologues of tocopherol, Trolox (which has a carboxyl group in place of the hydrocarbon chain tail of α–tocopherol) and pentamethylchromane (PMC, which has a methyl group in place of the hydrocarbon chain tail of α–tocopherol), were used. These more hydrophilic compounds maximize interactions between ascorbate and the chromanol head group of vitamin E. Addition of 10 mM Trolox to the skin homogenate decreased the signal intensity of the ascorbyl radical and shortened the duration of the signal (compare Fig. 3 (a) with (b)). The Trolox free radical signal appeared after the ascorbyl radical signal disappeared (Fig. 3 (d)). Addition of 10 mM PMC similarly shortened the duration of the ascorbyl radical signal; the PMC free radical signal appeared after the disappearance of the ascorbyl radical signal (Fig. 4 (d)). In contrast to Trolox, the addition of PMC increased the ascorbyl radical signal intensity immediately upon UV irradiation. The ascorbyl signal arising from auto-oxidation was unambiguous in this case (Fig. 4 (a)), but its signal intensity and duration were weaker and longer than those observed with UV irradiation. In contrast, the signal of the PMC radical that appeared subsequently was less pronounced than that of Trolox (Figs. 3 (d) and 4 (d)). Doubling the PMC concentration from 10 mM to 20 mM increased the signal (Fig. 4 (e)). Typical time courses of signal intensities for ascorbyl and Trolox or PMC free radicals are shown in Figs. 5 (a) and (b), respectively.

The addition of lipoic acid or dihydrolipoic acid to skin homogenates with vitamin E homologues was also investigated (Table I). When 0.5 mM DHLA was added to the skin homogenate in the presence of 10 mM PMC, the duration of the ascorbyl radical signal was prolonged beyond 80 minutes, while the PMC radical was not detected for up to 2 hrs.

TABLE I
Influence of dihydrolipoic acid or α–lipoic acid with vitamin E homologues on UV-irradiated mouse skin homogenates[a]

Condition	ESR signal duration (min)	
	Ascorbyl radical	Vitamin E homologue radical
Vehicle	56.8 ± 7.5	—
PMC (10 mM)	13.0 ± 1.0	9.7 ± 0.6
+DHLA (0.5 mM)	> 80.0	ND[b]
+DHLA (0.2 mM)	23.6 ± 5.5	9.0 ± 1.0
+ α–Lipoic acid (0.5 mM)	14.7 ± 2.9	8.3 ± 1.5
+ α–Lipoic acid (0.2 mM)	11.3 ± 1.5	7.3 ± 3.5
Trolox (10 mM)	14.0 ± 1.0	14.0 ± 1.7
+ DHLA (0.5 mM)	42.0 ± 5.0	ND[b]
+DHLA (0.2 mM)	20.3 ± 2.5	6.7 ± 2.3
+ α–Lipoic acid (0.5 mM)	18.7± 1.2	7.3 ± 1.2
+ α–Lipoic acid (0.2 mM)	16.7 ± 3.1	9.3 ± 1.2

[a] n= 3 (Vehicle, n = 5); data is presented as mean ± SD

[b] ND; not detectable

DHLA similarly prolonged the duration of the ascorbyl radical signal in skin homogenates in the presence of 10 mM Trolox. Lipoic acid was without effect in either system.

Addition of riboflavin (2 uM) a known photoactive sensitizer (chromophore) (Shibamoto 1994), to the skin homogenate enhanced the ascorbyl radical signal intensity and shortened it duration (Fig. 6). Addition of both a vitamin E homologue and riboflavin significantly shortened the duration of the ascorbyl radical signal (Table II).

Discussion: Exposure of skin homogenates to UV light generated the ascorbyl radical signal, which was the only detectable radical. Thus, photosensitizing processes occurred. The ascorbyl radical was the only radical detected because of its much higher concentrations in mouse skin compared to vitamin E concentrations, and the rapid interactions of vitamin E radicals with vitamin C. The maximum signal intensity persisted for 20 min, attesting to the stability of ascorbyl radical.

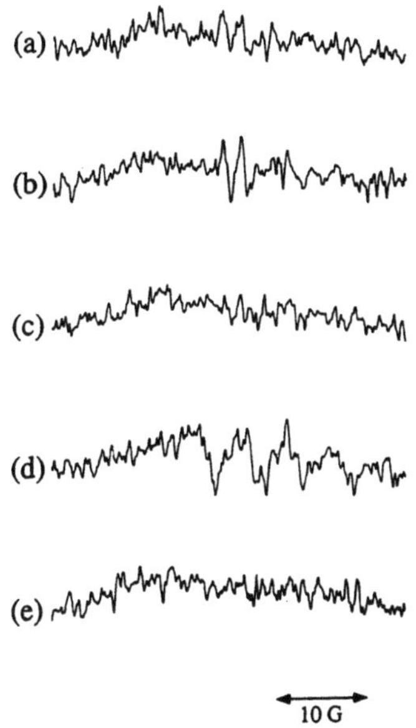

Fig. 3. ESR spectrum of the ascorbyl radical signal obtained from UV irradiated mouse skin homogenates in the presence of Trolox (10 mM). (a) no UV irradiation; (b) immediately after the start of UV irradiation; (c) UV irradiation for 13 min; (d) for 20 min; (e) for 30 min

TABLE II

Acceleration of UV-induced oxidation by the addition of riboflavin, a photosensitizer mouse skin homogenates

Condition	ESR ascorbyl radical signal duration (min)
Vehicle	54.7 ± 4.6
Riboflavin (2 uM)	4.8 ± 0.3[a]
Riboflavin (2 uM) + Trolox (10 mM)	3.2 ± 0.3[a,b]
Riboflavin (2 uM) + PMC (10 mM)	2.0 ± 0.0[a,b]

[a] Different from vehicle, $p < 0.05$

[b] Different from riboflavin alone, $p < 0.05$

Shortening of the ascorbyl radical signal duration by the addition of vitamin E homologues suggests that in this system vitamin E was acting as a pro-oxidant. The homologues accelerated consumption of ascorbate. This likely occurred because vitamin E homologues absorb UV light (absorbance maximum at 295 nm, with significant absorbance at the wavelengths employed in this study) and become free radicals (Mehlhorn et al. 1990; Kagan and Packer 1994). Hence, the rapid disappearance of the ascorbyl radical signal is consistent with the operation of the antioxidant network in this system, with ascorbate regenerating the vitamin E homologue from its radical. The amounts of vitamin E homologue free radicals remain sufficiently small to escape detection by ESR measurement as long as ascorbate is available to regenerate chromanol from the chromanoxyl radical. Upon ascorbate depletion, thechromanoxyl radical concentrations increase to detectable levels. These studies are a demonstration of the antioxidant network showing the close interactions between the antioxidant functional group of vitamin E (the chromanol ring) and vitamin C.

Both of the vitamin E homologues we used, PMC and Trolox, have similar redox potentials (+480 to +500 mV) and both are hydrophilic and can interact with ascorbate in the aqueous phase. Similarly, vitamin E dispersed in liposomes interacts with ascorbate (Niki et al. 1982). Although Trolox and PMC shortened the duration of ascorbyl radical appearance (Fig. 5), PMC is more effective in this regard. This difference may be explained by differences in their charges; Trolox is an anion at physiological pH whereas PMC is neutral. Since ascorbate is anionic at physiological pH one would expect less interaction with Trolox than with PMC due to electrostatic repulsion. Nonetheless, this electrostatic repulsion was not sufficient to negate interactions of the chromanoxyl radical of the vitamin E homologue with ascorbate. These observations also suggest that ascorbate radicals interact with tocopheroxyl radical chromanol head group in membranes, where it is accessible on the surface, and which would involve presumably a more thermodynamically favorable situation for the interactions of the hydrophobic tocopheroxyl radical with the hydrophilic ascorbate molecule.

To assess the interactions of the antioxidant network the addition of thiols to the skin homogenates was also investigated. Addition of the thiol antioxidant dihydrolipoic acid (DHLA) prolonged the ascorbyl radical lifetime and the time until appearance of the vitamin E homologue radical ESR signal. This is consistent with thiol recycling of ascorbate, which is then able to regenerate the vitamin E homologue and the entire process can continue for a longer period. Specifically, when DHLA is consumed, then ascorbate is consumed by interaction with the vitamin E homologue radical. This extension of the "lag phase" before disappearance of the ascorbyl radical is dependent on the concentration of DHLA used (Table I). Lipoic acid, the oxidized form of DHLA which cannot regenerate ascorbate, had no effect on the length of the lag time.

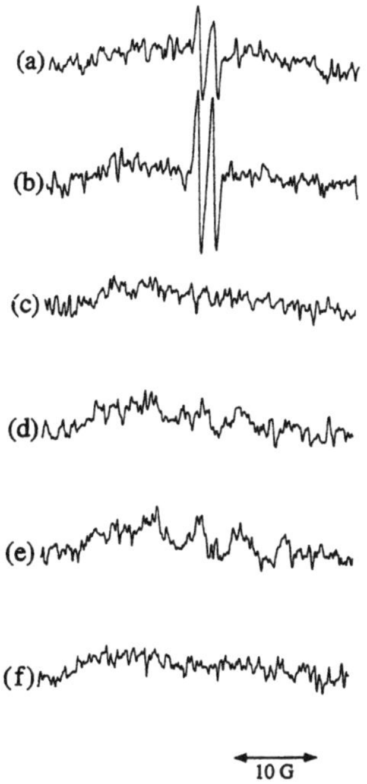

Fig. 4. ESR spectrum of ascorbyl radical signal obtained from UV irradiated mouse skin homogenates in the presence of PMC (10 mM, a-d and f; and 20 mM, e). (a) no UV irradiation; (b) immediately after the start of UV irradiation; (c) UV irradiation for 14 min; (d) for 20 min; (e) for 15 min; (f) for 25 min.

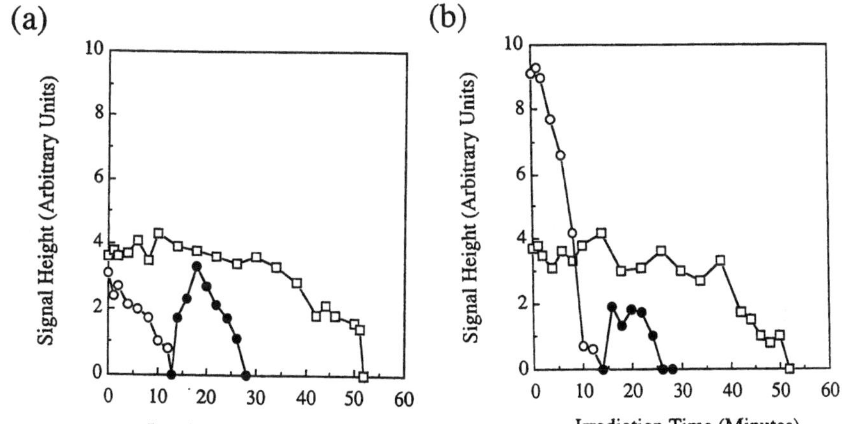

Fig. 5. Typical time course of ESR signals in UV irradiated mouse skin homogenates in the presence of Trolox (10 mM, a) or PMC (10 mM, b). (a) o Ascorbyl radical signal in the absence of Trolox (vehicle) and m ascorbyl radical signal and • Trolox free radical signal in the presence of Trolox; (b) o Ascorbyl radical signal in the absence of PMC (vehicle) and m ascorbyl radical signal and • PMC free radical signal in the presence of PMC.

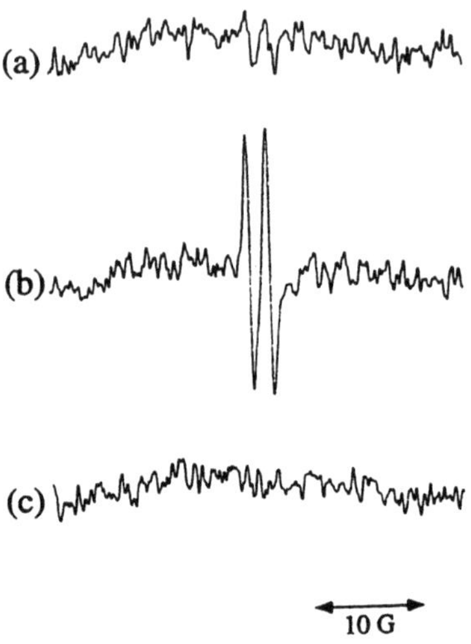

Fig. 6. ESR spectrum of ascorbyl radical signal obtained from UV irradiated mouse skin homogenates in the presence of riboflavin (2 μM). (a) no UV irradiation; (b) immediately after the start of UV irradiation; (c) after 5 min of UV irradiation

These studies are a direct demonstration of the interactions of thiols, ascorbate, and vitamin E, again emphasizing the function of the antioxidant network in skin.

UVA interactions Riboflavin is a chromophore which absorbs in the UVA region. A dramatic increase in ascorbate oxidation was observed following the addition of riboflavin to the skin homogenates (Table II). Riboflavin is excited to its triplet state during UV-induced photochemical reactions as a result of its high photosensitivity . Reaction of this triplet photosensitizer with triplet oxygen immediately yields singlet oxygen, which is much more reactive than triplet oxygen, and can react with proteins, membranes, and DNA (Ogura et al. 1991). Production of superoxide radical also occurs during UV irradiation of riboflavin (Dalle Carbonare and Pathak 1992). This can lead to hydroxyl radical formation through the Fenton reaction. Lipid peroxidation induced by either of these pathways ultimately leads to formation of the ascorbyl radical via operation of the antioxidant network (Fig. 7).

These studies support the model of the antioxidant network in skin. They also highlight the unique nature of skin, which is regularly exposed to UV light. Interactions of UVB with tocopherol and UVA with photosensitizers may both serve to deplete antioxidants, and oxidative damage may then be induced. Hence, skin antioxidant depletion and resultant damage may be produced by both UVA and UVB irradiation.

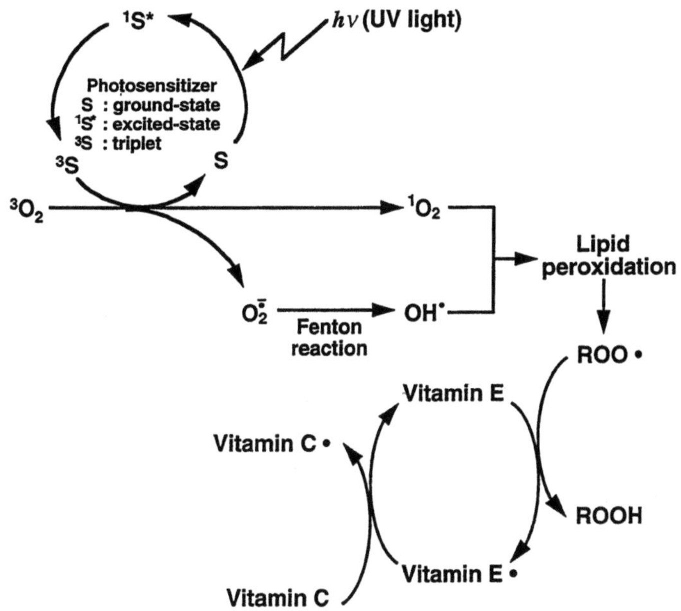

Fig. 7. UVA interactions with photosensitizers leading to ascorbyl radical formation. Singlet oxygen or superoxide can be formed, ultimately leading to lipid peroxidation, production of the vitamin E radical and subsequent formation of vitamin C radical produced when vitamin C quenches the vitamin E radical.

D. In vivo evidence for the antioxidant network

The antioxidant network model predicts that bolstering vitamin C should protect against vitamin E deficiency and bolstering thiols should protect both vitamins C and E. Various studies in animal models have tested this hypothesis.

Accumulating evidence supports the model of an antioxidant network operating in vivo. Increasing levels of vitamin C spares vitamin E in vivo, based on indirect evidence in guinea pigs (Chen and Chang 1978; Hruba et al. 1982; Kunert and Tappel 1983; Bendich et al. 1984; Miyazawa et al. 1986), rats (Chen 1981; Dillard et al. 1984; Chen and Thacker 1986; Chen and Thacker 1987), and humans (Keith et al. 1980; Arad et al. 1985). However, two studies in which dietary levels of vitamins E and C were manipulated and the effects on tissue levels of vitamin E were examined directly produced conflicting results.

Burton and Ingold (Burton et al. 1990) measured tissue vitamin E and its turnover (using deuterated tocopherol) in 9 tissues of guinea pigs over the course of 8 weeks. The animals were fed either low or high vitamin E diets, then each vitamin E group was further divided into groups fed high, normal, or low levels of vitamin C. Dietary vitamin C did not change tissue vitamin E levels or its turnover. The authors concluded that vitamin E regeneration by vitamin C, in the absence of oxidative stress, is a minor pathway in vivo (Burton et al. 1990). They speculated that this might not be the case during periods of oxidative stress, and also that regeneration of vitamin E might well be occurring in vivo via enzymatic pathways rather than mainly via vitamin C. We have shown that vitamin E recycling can be driven by enzymatic and electron-transport mechanisms which occur in mitochondria and microsomes (Maguire et al. 1989; Maguire et al. 1992). Another possibility is that the tissues contained sufficient vitamin C to regenerate vitamin E; it is unfortunate that vitamin C levels were not measured, nor was there a vitamin C deficient group.

Igarashi et al., using a mutant strain of rats (osteogenic disorder Shionogi, or ODS, rats) which cannot synthesize vitamin C, performed a similar experiment, feeding groups of animals vitamin E-deficient and E-sufficient diets that were further subdivided into low-, normal-, and high-vitamin C groups (Igarashi et al. 1991). There was a modest (10-50%) but consistent and statistically significant increase in vitamin E in all tissues of rats on the high vitamin C diet compared to those on the normal vitamin C diet. These researchers speculate that their results differ from those of Burton and Ingold (Burton et al. 1990) because the ODS rats display great uniformity in vitamin E levels, whereas the guinea pigs used by Burton and Ingold had highly variable vitamin E levels which would have masked the relatively moderate increase in vitamin E that Igarashi et al. observed. It should be noted that Igarashi et al. used growing rats, in which oxidative stress may be greater, whereas Burton and Ingold used adult guinea pigs—both vitamin requirements as well as species differences may contribute to the disparate results.

The preponderance of evidence supports the antioxidant network and the model of regeneration of vitamin E by vitamin C in vivo. The next question is whether thiols can regenerate vitamin C in vivo and, indirectly, vitamin E. This question is more difficult to answer, in part because the major thiol antioxidant present in the body, glutathione, is not absorbed in usable form from the diet, and thus cannot be easily supplemented in animals.

An alternative approach is to deplete glutathione (GSH); this has been done through administration to animals of buthionine sulfoximine (BSO), an inhibitor of γ-glutamylcysteine synthetase, which catalyzes the first step in glutathione synthesis. GSH deficiency induced in newborn rats with this compound caused a decrease in tissue ascorbic acid levels and increase in dehydroascorbate-to-ascorbate ratios; administration of ascorbate to these animals decreased mortality (Meister 1992). These data are consistent with a role of glutathione in maintaining cellular ascorbate concentrations through regeneration of ascorbate. If vitamin E is dependent on ascorbate for regeneration, a decrease in its concentration would also be expected. We demonstrated that lenses of newborn rats treated with BSO contained decreased levels of both ascorbate (44%) and α–tocopherol (70%) (Maitra et al. 1995).

Cellular thiols, including glutathione, can also be increased by dietary supplementation of α-lipoic acid. We have shown that α–lipoic acid is absorbed from the diet, taken up by tissues, and reduced to dihydrolipoic acid (Handelman et al. 1994; Podda et al. 1994; Podda et al. 1994; Han et al. 1995; Han et al. 1995). Furthermore, α–lipoic acid administration increases intracellular glutathione concentrations both in cell systems and in vitro (Haba et al. 1990; Busse et al. 1992; Han et al. 1995). Since both glutathione and the reduced form of α–lipoic acid, dihydrolipoic acid, have been shown to regenerate vitamin C and, indirectly, vitamin E, in a number of in vitro systems , it is an ideal compound to study the thiol segment of the antioxidant network in vivo.

Over thirty-five years ago Rosenburg and Culik administered α–lipoic acid to both vitamin E-deficient rats and vitamin C-deficient guinea pigs (Rosenberg and Culik 1959). They found that administration of α–lipoate to vitamin E-deficient rats prevented symptoms of E deficiency, and administration to vitamin C-deficient guinea pigs also prevented vitamin C deficiency symptoms. Although these researchers did not measure tissue levels of vitamins C or E, these results are consistent with the ability of DHLA to regenerate vitamin C directly and, indirectly, to recycle vitamin E, thus sparing both vitamins. Indeed, this seems more probable than α–lipoate replacing the antioxidant functions of both vitamin E and vitamin C in these deficient animals.

In the above study of BSO-treated newborn rats, we found that intraperitoneal injection of the rats with α–lipoic acid in addition to BSO maintained both vitamin C levels (89% of normal) and vitamin E levels (90% of normal) (Maitra et al. 1995), again demonstrating that DHLA regenerates vitamin C, which in turn regenerates vitamin E.

We have also found that dietary administration of α–lipoate can protect vitamin E-deficient hairless mice (Podda et al. 1994) in a similar manner. In the absence of α-lipoic acid, these animals show vitamin E deficiency symptoms within 5 weeks, but they are protected when fed α-lipoic acid (Podda et al. 1994). We have extended these studies to include measurements of tissue antioxidants, as well as resistance to oxidative stress (exposure of skin to UV irradiation).

<u>Materials and Methods</u> Adult hairless mice (6-8 weeks old) were divided into three dietary groups of six or seven animals each: vitamin E deficient diet (-E -LA), vitamin E deficient diet with added LA (1.65 gm/kg diet, -E+LA), or vitamin E sufficient diet (30 mg all-rac-α-tocopheryl acetate/kg diet, +E-LA). Animals were sacrificed at 5 weeks, and heart, liver, kidney, brain and skin were removed. Tissues were immediately frozen in liquid nitrogen and kept at -80 °C until extracted. Tissues were assayed for vitamin E and vitamin C. Tissues were extracted and analyzed for lipid soluble antioxidants using HPLC with electrochemical detection and an in-line UV detector, as previously described by Lang et al. (Lang et al. 1986). Ascorbate in tissue extracts was quantitated using HPLC and electrochemical detection (Motchnik et al. 1994). Statistical significance was assessed by

ANOVA and if significant main effects were observed ($p < 0.05$), then least square mean analyses were performed.

Results Mice in the -E-LA group displayed muscular dystrophy and weight loss, symptoms of vitamin E deficiency. Mice in the +E-LA and -E+LA groups maintained their weights and displayed normal behavior, with no physical symptoms of vitamin E deficiency. Nonetheless, vitamin E levels in all tissues except brain were significantly depleted in the -E-LA and -E+LA groups compared to the mice in the +E-LA group. There were no differences in the tissue vitamin E contents between the -E-LA and -E+LA (Fig 8). Ascorbate levels were also significantly reduced in liver and kidney in the -E-LA group compared with the +E-LA group (Fig 9). Remarkably, dietary α-lipoic acid (-E+LA group) prevented ascorbate depletion (Fig 9); the same trend was seen in skin but the ascorbate depletion did not reach statistical significance.

During oxidative stress, when skin was irradiated with UV light, a loss of tocopherol was observed in skin from both -E-LA and +E-LA mice, . However, in skin from the -E+LA group, tocopherol levels were preserved and not depleted by UV irradiation (Table III).

TABLE III

Effect of UV irradiation on skin vitamin E and vitamin C in rats fed various diets

	vitamin E (nmol/gm wet weight)[a]		vitamin C (μmol/gm wet weight)[a]	
Dietary condition	-UV	+UV	-UV	+UV
+E-LA	7.34 ± 1.07	5.24 ± 0.64	0.642 ± 0.10	0.548 ± 0.06
-E-LA	0.71 ± 0.094	0.29 ± 0.05[b]	0.525 ± 0.07	0.292 ± .07
-E+LA	0.53 ± 0.69	0.48 ± 0.14	0.642 ± 0.10	0.483 ± 0.10

[a] Values are mean ± SE, n = 6 or 7

[b] Different from non-irradiated, $p < 0.05$

Discussion Several predictions based on the antioxidant network are confirmed by this study. In vitamin E-deficient animals ascorbate levels should decrease because ascorbate is depleted when it regenerates vitamin E. This did, indeed, happen in the -E-LA group in liver and kidney (and the same trend was seen in skin). One would also predict that dihydrolipoic acid, which directly regenerates ascorbate from dehydroascorbate, would preserve ascorbate levels in the -E+LA group. This, too, was observed.

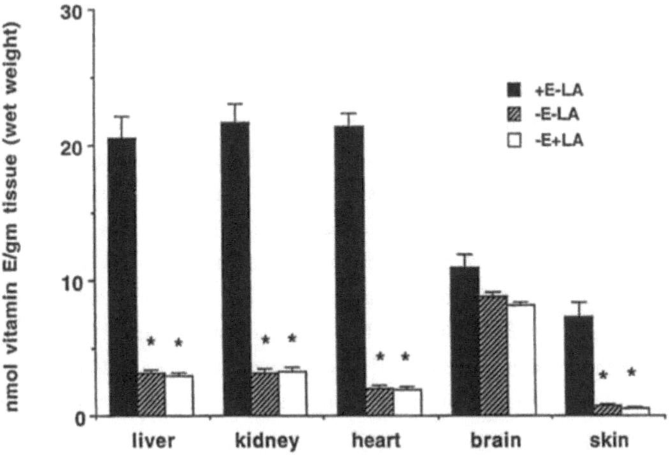

Fig. 8. Vitamin E levels in tissues of hairless mice fed various diets. Adult hairless mice were kept on the indicated diets for 5 weeks, then killed. Tissues were extracted and analyzed as described in Materials and Methods. * Different from +E-LA, $p < 0.01$.

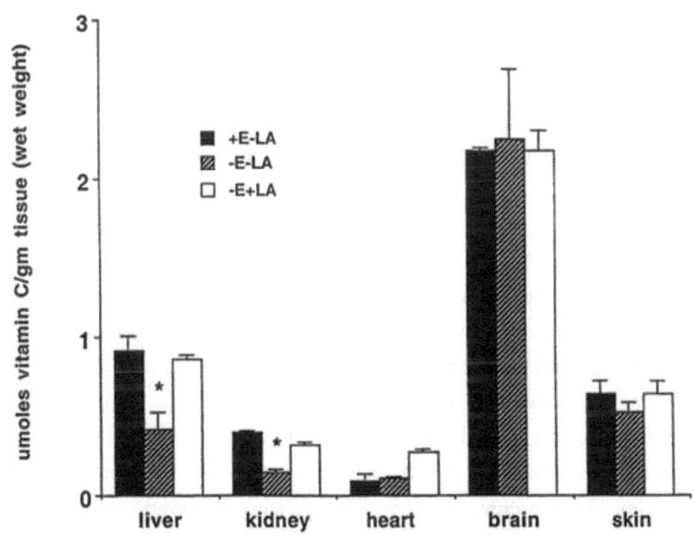

Fig. 9. Vitamin C levels in tissues of hairless mice fed various diets. Adult hairless mice were kept on the indicated diets for 5 weeks, then killed. Tissues were extracted and analyzed as described in Materials and Methods. * Different from +E-LA, $p < 0.05$.

One would further predict that this regeneration of vitamin C would serve to also bolster vitamin E levels in the -E+LA group, due to ascorbate-driven recycling of vitamin E, however this was not observed. It is possible that vitamin E was depleted earlier in the -E-LA group but by five weeks it was also depleted in the -E+LA group. It is also possible that lipoate acts to prevent deficiency symptoms in the mice by substituting for vitamin E in some of its antioxidant functions, rather than by regenerating vitamin E.

The preservation of vitamin E in UV-irradiated skin is another intriguing aspect of this study. Although skin vitamin E levels in the -E+LA group were less than one sixth those in the +E-LA group, there was no significant decrease in the -E+LA group upon UV irradiation, whereas in the -E-LA group vitamin E levels decreased 49% ($p < 0.05$); skin vitamin E levels also decreased upon irradiation in the +E-LA group but these differences were not statistically significant (Table III). These data are consistent with vitamin E recycling by lipoate in skin in the face of an oxidative challenge. Indeed, Burton and Ingold speculated that sparing of vitamin E through its regeneration by vitamin C (and, by extension, indirectly by thiols) is an important pathway only in the face of oxidative stress (Burton et al. 1990) . In our study α–lipoic acid supplementation exhibited a clear protective effect on vitamin E during oxidative stress induced by UV irradiation.

Taken together, these experiments provide support for the hypothesis that the antioxidant network operates in vivo. Further experiments in which the time-course of depletion of vitamin E in E-deficient animals is followed should clarify the contribution of lipoate to vitamin C and vitamin E regeneration.

II. Redox and oxidant regulation of nuclear transcription: NF–κB and the antioxidant network

In response to stresses such as injury and infection, organisms must rapidly marshal a host of responses. Various genes must be activated, and a central, coordinating factor in this activation of rapid response genes is nuclear factor κB (NF–κB). In addition, NF–κB plays a crucial role in modulating gene expression during growth and development. A host of genes have been shown to be modulated by NF–κB. These include genes for cytokines and growth factors, immunoreceptors, adhesion molecules, acute phase proteins, transcription factors and regulators, NO-synthase, and viral genes (Siebenlist et al. 1994).

In pathological conditions such as viral infection, NF–κB is used by the invading pathogen to assist in viral gene expression. Hence, the modulation of NF–κB has far-reaching significance for a variety of pathological conditions in which inflammation, growth, or viral activation occur, such as atherosclerosis or HIV infection (AIDS).

NF–κB is a member of the family of Rel transcription factors which consists of hetero- or homodimeric proteins (p50, p65) in association with an inhibitory protein family, IκB. Phosphorylation and dissociation of IκB from Rel protein dimers, followed by its degradation allows the dimeric DNA-binding NF–κB to enter the nucleus and regulate genes (Brown et al. 1993; Siebenlist et al. 1994).

There are several lines of evidence that reactive oxygen species (ROS) may be a common signal for the various stimuli that activate NF–κB(Brown et al. 1993). First, many of the stimuli which activate NF–κB are also known to cause intracellular increases in ROS. These include TNF-α, IL-1, phorbol myristate acetate (PMA), lipopolysaccharide (LPS), UV light, and gamma irradiation (Schreck et al. 1992; Schreck et al. 1992; Geng et al. 1993; Schieven et al. 1993). Second, administration of hydrogen peroxide can directly stimulate NF–κB activation (Schreck et al. 1991; Schreck et al. 1992). Third, a diverse array of antioxidants can block NF–κB activation. These include N-acetyl-L-cysteine (Meyer et al. 1992), dithiocarbamates (Meyer et al. 1992; Schreck et al. 1992), vitamin E derivatives (Suzuki and Packer 1993; Suzuki and Packer 1994), catechol derivatives (Suzuki and Packer 1994), dithiolthione (Sen et al. 1996) and α–lipoate (Suzuki et al. 1992; Packer and Suzuki 1993; Bessho et al. 1994; Suzuki and Packer 1994; Suzuki et al. 1995). Conversely, depletion of cellular thiol antioxidants induces NF–κB activation (Staal et al. 1990).

A. Modulation of NF–κB activation by vitamin E

We have reported that vitamin E inhibits NF–κB activation. Several compounds which are structurally related to vitamin E—vitamin E acetate, α–tocopheryl succinate, and 2,2,5,7,8-pentamethyl-6-hydroxychromane (PMC)—inhibited TNF-α induced NF–κB activation(Suzuki and Packer 1993). Incubation of Jurkat T cells for 30 min with vitamin E acetate or vitamin E succinate (100 uM - 1 mM) blocked the activation of NF–κB in a dose-dependent manner. The same concentrations of α–tocopherol failed to do so. The lack of effect of α–tocopherol in this system suggests that its uptake and intracellular mobility were so poor that simple addition to the medium did not deliver α–tocopherol in an effective manner. In contrast, both vitamin E acetate and vitamin E succinate are taken up by cells and are de-esterified to the biologically active form, α–tocopherol (Carini et al. 1990). These data suggest that intracellular trafficking of tocopherol may be critical to its function in signal transduction.

The effectiveness of vitamin E in preventing NF-κB activation implies that free radical processes involved in TNF-α induced NF–κB activation occur, at least in part, proximal to membranes, and membrane oxidation may be an integral step in the signal transduction pathway. The ineffectiveness of direct addition of α–tocopherol in inhibiting the NF–κB activation also implies that the membrane oxidation processes required for the cell signaling are localized in the internal compartments of cell architecture such as mitochondria, where vitamin E acetate, but not α–tocopherol, can reach.

PMC was the most potent inhibitor among the vitamin E derivatives examined, as 10 μM significantly blocked the activation. The high mobility of the short-chain homologue suggests that the chromanol portion of the vitamin E molecule is important for inhibition of NF–κB activation.

B. Modulation of NF–κB activation by α-lipoate

We (Suzuki et al. 1992) observed that incubation of Jurkat T cells for 2 hr with lipoate (0.5 to 4mM) prior to the stimulation with TNF-α or phorbol 12-myristate 13-acetate (PMA) inhibited NF–κB activation in a dose-dependent manner; 4 mM α-lipoate was required for complete inhibition. α–Lipoate at these concentrations did not affect either cell viability or oct-1 DNA binding activity. Direct supplementation of DHLA in the incubation medium also inhibited the NF–κB activation induced by TNF-α (Suzuki and Packer 1994).

We also observed this phenomenon in Wurzburg T cells, a subclone of Jurkat T cells in which NF–κB activation is achieved by treatment with hydrogen peroxide. Pretreatment for 18 hr with α–lipoate suppressed NF–κB activation induced by hydrogen peroxide, phorbol ester, or TNF-α. Previous results from our laboratory have also shown that α–lipoate pretreatment of T cells doubles intracellular glutathione (Han et al. 1995). However, using BSO-treated Wurzburg T cells, we observed that the effect of α–lipoate on NF–κB was not the result of its pro-glutathione effect (Sen and Packer 1996).

The inhibition of NF–κB activation by lipoate is consistent with either an antioxidant effect of α–lipoate which directly affects a redox-sensitive step in the process, or recycling of vitamin E which acts as an antioxidant to affect a redox-sensitive step in the process, or both.

C. Synergy of vitamin E and α-lipoate effects on modulation of NF–κB activation

Vitamin E, a peroxyl radical scavenger, and α-lipoic acid, a redox active thiol, most likely function at different sites to cause a decrease in NF-kB activation. Therefore, simultaneous administration could potentiate their actions. Additionally, their interactions through the antioxidant network suggests that cellular metabolism could enhance the ability of cells to respond to oxidative stress.

To test this idea we attempted to modulate cells using conditions as close to physiologic as possible. Therefore we used liposomes for vitamin E delivery. Delivery of vitamin E in liposomes may allow more effective incorporation into both outer and inner membranes, and vitamin E may be delivered to intracellular compartments more efficiently.

HIV has an NF-κB promoter in its long terminal repeat portion. Therefore, we tested the effects of vitamin E and lipoic acid, inhibitors of NF-κB, in OM-10.1 cells. We measured both NF-κB activation and HIV replication in response to TNF-α stimulation.

Materials and Methods The OM-10.1, HIV infected promyelocytic leukemia, human monocyte cell line was used. These cells have minimal constitutive production of HIV-1, and HIV-1 expression was increased by treatment for 24 hours with 5 ng/ml of TNF-α. Cells were or were not preincubated with antioxidants for 24 hours, then treated with TNF-α. Lipoic acid (100 uM) was added in phosphate buffered saline. HIV-1 replication in response to TNF-α was estimated by measuring p24 antigen in cell culture supernatants. NF-kB activation was assessed by electrophoretic mobility shift assay (EMSA) as previously described (Suzuki and Packer 1994).

Results We determined that vitamin E was taken up by cells and cellular concentrations of vitamin E increased approximately 200-fold following 24 hours of incubation with liposomes (medium final concentration 5 μM α–tocopherol). Longer incubations and/or higher concentrations of vitamin E did not increase cellular vitamin E concentrations further, demonstrating that delivery was highly effective.

In previous studies vitamin E esters, but not vitaimin E alone, decreased NF-κB activation (Suzuki and Packer 1993). In this study, vitamin E was delivered to the cells via liposomes, which we predicted would allow it to move to cellular compartments in which it would affect NF-κB activation. EMSA showed that vitamin E did decrease TNF-α activated NF-κB activation in these cells (Fig. 10). Control experiments in which TNF-α activated cells were exposed to liposomes alone showed that the liposomes themselves had a stimulatory effect on NF-κB activation, probably due to oxidized lipids in the liposomes, but that when an equal volume of liposomes + vitamin E was added, there was less activation of NF-κB (Fig. 10).

Addition of 10 μM α–tocopherol alone to the cells 24 hr prior to TNF-α stimulation reduced HIV-1 activation by 21%, while addition of 100 μM lipoic acid alone decreased activation by 9% (Fig 11). Remarkably, addition of both together decreased activation by 68%, clearly a synergistic, rather than an additive effect (Fig 11). This is consistent with a role of lipoic acid in regenerating vitamin E, and thus increasing its antioxidant potency and its effect on NF–κB.

Discussion Our results with vitamin E alone inhibited NF-κB activation when delivered via liposomes (Fig. 11), and previous results from our laboratory have shown that a-lipoate also inhibits NF-κB activation (Suzuki et al. 1992). Hence, the inhibition of HIV-1 replication observed in this study was probably due to the action of these antioxidants on NF-κB. However, there are many other steps in signal transduction besides NF–κB that could also be affected by either or both of these antioxidants, and this study cannot rule them out. For example, vitamin E may reduce membrane fluidity and thereby change receptor or ligand interactions in membranes, thus changing the ability of cells to transduce signals.

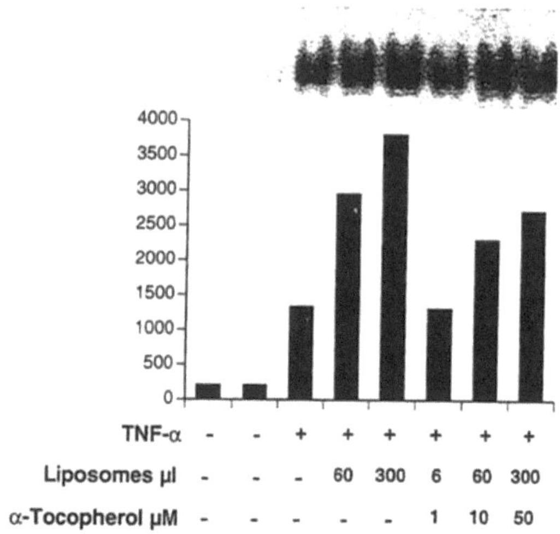

Fig. 10. Inhibition of NF-κB activation in TNF-α stimulated OM-10.1 cells. Top: Electrophoretic mobility shift assay (EMSA) gel; the intensity of the band corresponds to NF-κB acitvation. Bottom: Densitometric scan of the EMSA gel.

However, the synergistic effect emphasizes the importance of the interaction of these antioxidants, presumably in regeneration of oxidized vitamin E, in dramatically decreasing activation of HIV-1 virus, and link the heretofore theoretical studies with an important health problem, AIDS.

III. Conclusions

Vitamin E is the most important antioxidant in the protection of membranes and other lipid environments (e.g., low density lipoproteins). The fact that vitamin E is a metabolic antioxidant that is interacting with other redox antioxidants in the aqueous phase is a key feature of its effectiveness in the protection of biological systems. These interactions minimize the degree to which vitamin E is destroyed in biological systems. In vivo evidence for the antioxidant network from work presented here strengthens this concept. These interactions may be particularly important in skin, a tissue repeatedly exposed to UV light, which can directly and indirectly oxidize vitamin E. In addition, oxidants and antioxidants modulate NF-κB signal transduction and we have found synergistic effects of vitamin E and the thiol antioxidants in modifying these processes. Hence, the operation of the antioxidant network may have clinical relevance both in degenerative diseases (e.g., skin cancer) as well as in infectious diseases (e.g. AIDS).

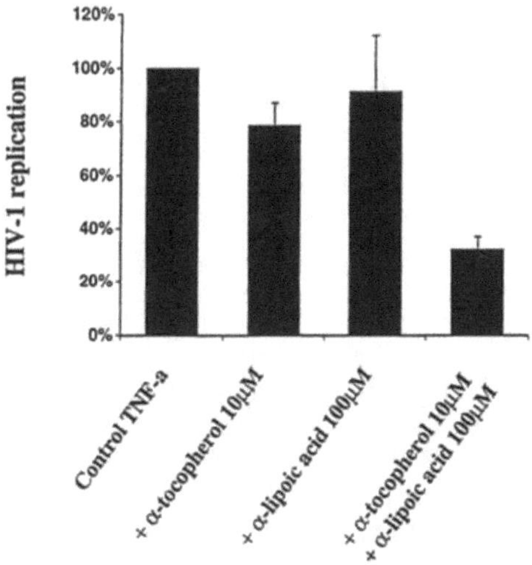

Fig. 11. Effect of α-tocopherol and α–lipoic acid on the inhibition of TNF–α-induced HIV-1 replication. OM-10.1 cells were pretreated for 24 hours as follows: no additions or plus liposomes (control), liposomes containing 10 μM α–tocopherol, 100 μM α–lipoic acid or liposomes containing 10 μM α–tocopherol plus 100 mM α–lipoic acid. Then cells were incubated for 24 hours with 5 ng/ml TNF-α to induce HIV-1 expression. HIV-α replication is given as a percent of control (mean ± SD of three separate experiments).

Acknowledgements

The authors acknowledge the intellectual contributions of Valerian Kagan, Elena Serbinova, and other colleagues who helped to establish in vitro evidence for the redox antioxidant network from studies with membranes and human low density lipoproteins. Research supported by NIH grants CA 47597 and DK50430

References

Arad, I.D., Y. Dgani and F. G. Eyal (1985). Vitamin E and vitamin C plasma levels in premature infants following supplementation of vitamin E. Int. J. Vit. Nutr. Res. 55: 395-397.

Bendich, A,. P. D'Apolito, E. Gabriel and L.K. Machlin (1984). Interaction of dietary vitamin C and vitamin E on guinea pig immune responses to mitogens. J. Nutr. 114(1588-1593).

Bessho, R., K. Matsubara, M. Kubota, K. Kuwakado, H. Hirota, Y. Wakazono, Y.W. Lin, A Okuda, M. Kawai and R. Nishikomori (1994). Pyrrolidine dithiocarbamate, a potent inhibitor of nuclear factor kappa (NF-kappa B) activation, prevents apoptosis in human promyelocytic leukemia HL-60 cells and thymocytes. Biochem. Pharmacol. 48: 1883-1889.

Brown, K., S. Park, T. Kanno, G. Franzoso and U. Siebenlist (1993). Mutual regulation of the transcriptional activator NF-kB and its inhibitor, I kappa B-a. Proc. Natl. Acad. Sci. USA 90: 2532-2536.

Burton, G. W. and K. U. Ingold (1981). Autoxidation of biological molecules. 1. The antioxidant activity of vitamin E and related chain-breaking phenolic antioxidant in vitro. J. Am. Chem. Soc. 103: 6472-6477.
Burton, G. W., U. Wronska, L. Stone, D. O. Foster and K. U. Ingold (1990). Biokinetics of dietary RRR-alpha-tocopherol in the male guinea pig at three dietary levels of vitamin C and two levels of vitamin E. Evidence that vitamin C does not "spare" vitamin E in vivo. Lipids 25: 199-210.
Busse, E., G. Zimmer, B. Schopohl and B. Kornhuber (1992). Influence of alpha-lipoic acid on intracellular glutathione in vitro and in vivo. Arzneimittel-Forschung 42: 829-831.
Carini, R., G. Poli, M. U. Dianzani, S. P. Maddix, T. F. Slater and K. H. Cheeseman (1990). Comparitive evaluation of the antioxidant activity of α-tocopherol, α-tocopherol polyethylene glycol 1000 succinate and α-tocopherol succinate in isolated hepatocytes and liver microsomal suspensions. Biochem. Pharmacol. 39: 1597-1601.
Chen, L. H. (1981). An increase in vitamin E requirement induced by high supplementation of vitamin C in rats. Am. J. Clin. Nutr. 34: 1036-1041.
Chen, L. H. and H. M. Chang (1978). Effect of dietary vitamin E and vitamin C on respiration and swelling of guinea pig liver mitochondria. J. Nutr. 108: 1616-1620.
Chen, L. H. and R. R. Thacker (1986). Effects of dietary vitamin E and high level of ascorbic acid on iron distribution in rat tissues. Int. J. Vit. Nutr. Res. 56: 253-258.
Chen, L. H. and R. R. Thacker (1987). Effect of ascorbic acid and vitamin E in biochemical changes associated with vitamin E deficiency in rats. Int. J. Vit. Nutr. Res. 57: 385-390.
Constantinescu, A., D. Han and L. Packer (1993). Vitamin E recycling in human erythrocyte membranes. J Biol Chem 268: 10906-10913.
Dalle Carbonare, M. and M. A. Pathak (1992). Skin photosensitizing agents and the role of reactive oxygen species in photoaging. J. Photochem. Photobiol. B, Biol. 14: 105-124.
Dillard, C. J., J. E. Downey and A. L. Tappel (1984). Effect of antioxidants on lipid peroxidation in iron-loaded rats. Lipids 19: 127-133.
Doba, T. G., W. Burton and K. U. Ingold (1985). Antioxidant and co-antioxidant activity of vitamin C. The effect of vitamin C, either alone or in the presence of vitamin E or a water-soluble vitamin E analogue, upon the peroxidation of aqueous multilamellar phospholipid liposomes. Biochim. Biophys. Acta 835: 298-303.
Evans, H. M. and K. S. Bishop (1922). On the existence of a hitherto unrecognized dietary factor essential for reproduction. Science 56: 650-651.
Geng, Y., B. Zhang and M. Lotz (1993). Protein tyrosine kinase activation is required for lipopolysaccaride induction of cytokines in human blood monocytes. J. Immunol. 151: 6692-6700.
Golumbic, C. and H. A. Mattill (1941). Antioxidants and the autoxidation of fats. XIII. The antioxygenic action of ascorbic acid in association with tocopherols, hydroquinones and related compounds. J. Am. Chem. Soc. 63: 1279-1280.
Haba, K., K. Ogawa, K. Mizukawa and A. Mori (1990). Time course of changes in lipid peroxidatin, pre- and postsynaptic cholinergic indices, NMDA receptor binding and neuronal death in the gerbil hippocampus following transient ischemia. Brain Res. 40: 116-122.
Han, D., G. J. Handelman and L. Packer (1995). Analysis of reduced and oxidized lipoic acid biological samples by high-performance liquid chromatography. Meth. Enzymol. 251: 315-325.
Han, D., H. J. Tritschler and L. Packer (1995). Alpha-lipoic acid increases intracellular glutathione in a human T-lymphocyte Jurkat cell line. Bioch. Biophys. Res. 207: 258-264.
Handelman, G. J., D. Han, H. Tritschler and L. Packer (1994). Alpha-lipoic acid reduction by mammalian cells to the dithiol form, and release into the culture medium. Biochem. Pharmacol. 47: 1725-1730.
Hopkins, F. G. and J. M. C. Morgan (1936). Some relations between ascorbic acid and glutathione. Biochem. J. 30: 1446-1462.
Hruba, F., V. Novakova and E. Ginter (1982). The effect of chronic marginal vitamin C deficiency on the α-tocopherol content of the organs and plasma of guinea pigs. Experimentia 38: 1454-1455.
Igarashi, O., Y. Yonekawa and Y. Fujiyama-Fujihara (1991). Synergistic action of vitamin E and vitamin C in vivo using a new mutant of Wistar-strain rats, ODS, unable to synthesize vitamin C. J. Nutr. Sci Vitaminol. 37: 359-369.

Kagan, V. E. and L. Packer (1994). Light-induced generation of vitamin E radicals. Assessing vitamin E regeneration. Meth Enzymol. 234: 316-320.

Kagan, V. E. and L. Packer (1994). Light-induced generation of vitamin E radicals: assessing vitamin E regeneration. Meth. Enzymol. 234: 316-320.

Kagan, V. E., E. A. Serbinova, T. Forte, G. Scita and L. Packer (1992). Recycling of vitamin E in human low density lipoproteins. J. Lipid Res. 33: 385-397.

Kagan, V. E., A. Shvedova, E. Serbinova, S. Khan, C. Swanson, R. Powell and L. Packer (1992). Dihydrolipoic acid--A universal antioxidant both in the membrane and in the aqueous phase. Biochem. Pharmacol. 44: 1637-1649.

Keith, R. E., B. M. Chrisley and J. A. Driskell (1980). Dietary vitamin C supplementation and plasma vitamin E levels in humans. Am. J. Clin. Nutr. 33: 2394-2400.

Kunert, K. J. and A. L. Tappel (1983). The effect of vitamin C on in vivo lipid peroxidation in guinea pigs as measured by pentane and ethane production. Lipids 18: 271-274.

Lang, J. K., K. Gohil and L. Packer (1986). Simultaneous determination of tocopherols, ubiquinols, and ubiquinones in blood, plasma, tissue homogenates, and subcellular fraction. Anal. Biochem. 157: 106-116.

Maguire, J. J., V. Kagan, B. A. Ackrell, E. Serbinova and L. Packer (1992). Succinate-ubiquinone reductase linked recycling of alpha-tocopherol in reconstituted systems and mitochondria: requirement for reduced ubiquinone. Arch. Bioch. Biophys. 292: 47-53.

Maguire, J. J., D. S. Wilson and L. Packer (1989). Mitochondrial electron transport-linked tocopheroxyl radical reduction. J. Biol. Chem. 264: 21462-21465.

Maitra, I., E. Serbinova, H. Tritschler and L. Packer (1995). Alpha-lipoic acid prevents buthionine sulfoximine-induced catract formation in newborn rats. Free Rad. Biol. Med. 18: 823-829.

Mehlhorn, R., J. Fuchs, S. Sumida and L. Packer (1990). Preparation of tocopheroxy radicals for detection by electron spin resonance. Meth. Enzymol. 186: 197-205.

Meister, A. (1992). On the antioxidant effects of ascorbic acid and glutathione. Biochem. Pharmacol. 44: 1905-1915.

Meyer, R., W. H. Caselmann, V. Schluter, R. Schreck, H. P.H. and B. P.A. (1992). Hepatitis B virus transactivator MHBst: activation of NF-kappa B, selective inhibition by antioxidants and integral membrane localization. EMBO J. 11: 2992-3001.

Miyazawa, T., T. Ando and T. Kaneda (1986). Effect of dietary vitamin C and vitamin E on tissue lipid peroxidation of guinea pigs fed with oxidized oil. Agric. Biol. Chem. 50: 71-78.

Motchnik, P. A., B. Frei and B. N. Ames (1994). Measurement of antioxidants in human blood plasma. Meth. Enzymol. 234: 269-279.

Niki, E., A. Kawakami, Y. Yamamoto and Y. Kamiya (1982). Synergistic inhibition of oxidation of soybean phosphatidylcholine liposomes in aqueous dispersion by vitamin E and vitamin C. Bull. Chem. Soc. Jpn. 58: 1971-1975.

Niki, E., T. Saito, A. Kawakami and Y. Kamiya (1984). Inhibition of oxidation of methyl linoleate in solution by vitamin E and vitamin C. J. Biol. Chem. 259: 4177-4182.

Ogura, R., M. Sugiyama, J. Nishi and N. Haramaki (1991). Mechanism of lipid radical formation following exposure of epidermal homogenate to ultraviolet light. J. Invest. Dermatol. 97: 1044-1047.

Packer, J. E., T. F. Slater and R. L. Willson (1979). Direct observation of a free radical interaction between vitamin E and vitamin C. Nature 278: 737-738.

Packer, L. and Y. J. Suzuki (1993). Vitamin E and alpha-lipoate: role in antioxidant recycling and activation of the NF-kB transcription factor. Molec. Aspects Med. 14: 229-239.

Podda, M., D. Han, B. Koh, J. Fuchs and L. Packer (1994). Conversion of lipoic acid to dihydrolipoic acid in human keratinocytes. Clin. Res. 42: 41a.

Podda, M., H. J. Tritschler, H. Ulrich and L. Packer (1994). α-Lipoic acid supplementation prevents symptoms of vitamin E deficiency. Biochem. Biophys. Res. Commun. 204(1): 98-104.

Rosenberg, H. R. and R. Culik (1959). Effect of α-lipoic acid on vitamin C and vitamin E deficiencies. Arch. Biochem. Biophys. 80: 86-93.

Schieven, G. L., J. M. Kirihara, D. E. Myers, J. A. Ledbetter and F. M. Uckun (1993). Reactive oxygen intermediates activate NF-kB in a tyrosine-kinase dependent mechanism and in combination with vanadate activate the p56 lck and p59 fyn tyrosine kinase in human lymphocytes. Blood 82: 1212-1220.

Scholich, H., M. E. Murphy and H. Sies (1989). Antioxidant activity of dihydrolipoate against microsomal lipid peroxidation and its dependence on α-tocopherol. Biochim. Biophys. Acta. 1001: 256-261.
Schreck, R., K. Albermann and P. A. Baeuerle (1992). NF-kB: an oxidative stress-responsive transcription factor of eukaryotic cells (a review). Free Radical Res. Commun. 17: 221-237.
Schreck, R., B. Meier, D. N. Maennel, W. Droge and A. Baeuerle (1992). Dithiocarbamates as potent inhibitors of nuclear factor kB activation in intact cells. J. Exp. Med. 175: 1181-1194.
Schreck, R., P. Rieber and P. Baeuerle (1991). Reactive oxygen intermediates are apparently widely used messengers in the activation of NF-kB transcription factor and HIV-1. EMBO J 10: 2247-2258.
Sen, C. K. and L. Packer (1996). Antioxidant and redox regulation of gene transcription. FASEB J. 10: 709-720.
Sen, C. K., K. Traber and L. Packer (1996). Inhibition of NF-kB activation in human T-cell lines by anetholdithiolthione. Biochem. Biophys. Res. Commun. 218: 148-153.
Shibamoto, T. (1994). The role of lipid peroxidation caused by ultraviolet light in skin diseases. J. Toxicol.—Cut. Ocular. Toxicol. 13: 193-202.
Shindo, Y., E. Witt, D. Han and L. Packer (1994). Dose-response effects of acute ultraviolet irradiation on antioxidants and molecular markers of oxidation in murine epidermis and dermis. J. Invest. Dermatol. 102: 470-475.
Shindo, Y., E. Witt and L. Packer (1993). Antioxidant defense mechanisms in murine epidermis and dermis and their responses to ultraviolet light. J. Invest. Dermatol. 100: 260-265.
Siebenlist, U., G. Franzoso and K. Brown (1994). Structure, regulation and function of NF-kB. Ann. Rev. Cell Biol. 10: 405-455.
Staal, F. J. T., M. Roederer, L. A. Herzenberg and L. A. Herzenberg (1990). Intracellular thiols regulate activation of nuclear factor kappa B and transcription of human immunodeficiency virus. Proc. Natl. Acad. Sci. USA 87: 9943-9947.
Suzuki, Y. J., B. B. Aggarwal and L. Packer (1992). Alpha-lipoic acid is a potent inhibitor of NF-kB activation in human T cells. Biochem. Biophys. Res. Commun. 189: 1709-1715.
Suzuki, Y. J., M. Mizuno, H. J. Tritschler and L. Packer (1995). Redox regulation of NF-kB DNA binding activity by dihydrolipoate. Bioch. Mol. Biol. Int. 36: 241-246.
Suzuki, Y. J. and L. Packer (1993). Inhibition of NF-kB activation by vitamin E derivatives. Biochem. Biophys. Res. Comm. 193: 277-283.
Suzuki, Y. J. and L. Packer (1994). Alpha-lipoic acid is a potent inhibitor of NF-kappa B activation in human T cells: does the mechanism involve antioxidant activities. Biological oxidants and antioxidants. L. Packer and E. Cadenas. Stuttgart, Hippokrates Verlag.
Suzuki, Y. J. and L. Packer (1994). Inhibition of NF-kB DNA binding activity by alpha-tocopheryl succinate. Biochem. Mol. Biol. Intl. 31: 693-700.
Suzuki, Y. J. and L. Packer (1994). Inhibition of NF-kB transcription factor by catechol derivatives. Biochem. Mol. Bio. Intl. 32: 299-305.
Tappel, A. L. (1968). Will antioxidant nutrients slow aging processes? Geriatrics 23: 97-105.
Winkler, B. S., S. M. Orselli and T. S. Rex (1994). The redox couple between glutathione and ascorbic acid: a chemical and physiological perspective. Free Rad. Biol. Med. 17: 333-349.

THE MOLECULAR BASIS FOR PLEIOTROPIC DRUG RESISTANCE IN THE YEAST *SACCHAROMYCES CEREVISIAE*: REGULATION OF EXPRESSION, INTRACELLULAR TRAFFICKING AND PROTEOLYTIC TURNOVER OF ATP BINDING CASSETTE (ABC) MULTIDRUG RESISTANCE TRANSPORTERS.

Karl Kuchler, Ralf Egner, Friederike Rosenthal and Yannick Mahé
Department of Molecular Genetics
University and Biocenter Vienna
Dr. Bohr-Gasse 9/2
A-1030 Vienna, Austria
Phone: 43-1-79515-2111
FAX: 43-1-79515-2900
e-mail: **kaku@mol.univie.ac.at**

The major research goal of our laboratory is to understand the function mechanism of certain eukaryotic ABC transporters, including those from yeast as well as selected mammalian ABC proteins of medical importance. ABC transporters comprise the largest membrane transport protein family known to date, as more than 200 different ABC transporters have been identified operating from bacteria to man (Higgins, 1992; Kuchler and Thorner, 1992). The hallmark characteristics of all ABC-proteins include the presence of two highly conserved domains for ATP-binding (ABC), and two membrane domains each containing several predicted membrane-spanning α-helices (TMS). These four domains are normally arranged in an $(TMS\text{-}ABC)_2$ or $(ABC\text{-}TMS)_2$ configuration, although "half-size" transporters with an TMS-ABC or ABC-TMS topology and other topologies are also frequently found (Figure 1).

Despite their highly conserved structural organization (Figure 1), most ABC proteins appear to be of rather limited or dedicated substrate specificity. As a consequence, ABC transporters are implicated in a remarkable variety of transport processes, including the transmembrane transport of ions, heavy metals, carbohydrates, anticancer drugs, amino acids, oligopeptides, phospholipids, steroids, glucocorticoids, bile acids, mycotoxins, xenobiotics, antibiotics, pigments, and even polypeptides. However, the mechanism by which transport of such a substrate and size diversity can be achieved, while each ABC-transporter maintains selectivity for its particular substrate, represents an intriguing and yet unsolved problem (Egner *et al.*, 1995a; Higgins, 1992; Kuchler and Thorner, 1992).

NATO ASI Series, Vol. H 101
Molecular Mechanisms of Signalling
and Membrane Transport
Edited by Karel W. A. Wirtz

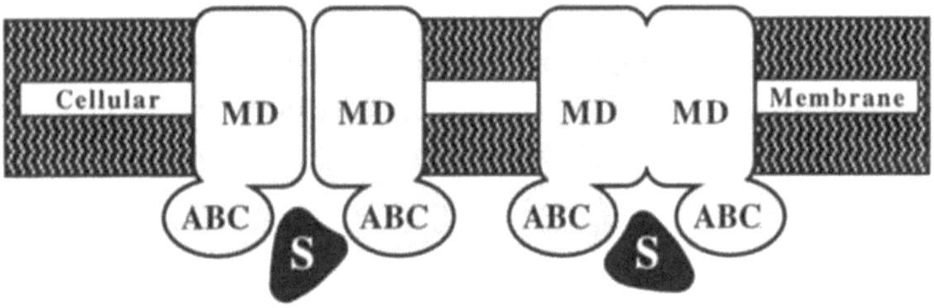

Figure 1: Domain organization of eukaryotic ABC Transporters
The membrane domains (MD) of ABC proteins contain several predicted membrane-spanning helices. Some substrates (S) may also be recognized or bound from within the lipid bilyaer.

Several human ABC transporters are of considerable medical importance and some are implicated in genetic disease. For instance, the cystic fibrosis transmembrane conductance regulator CFTR is associated with cystic fibrosis, a fatal genetic disease occurring most frequently in Caucasians (Harris and Argent, 1993; Riordan and Chang, 1992). Further, overexpression of the mammalian P-glycoprotein (Pgp; the *MDR1* gene product) and MRP ("multidrug resistance related protein") is associated with a multidrug resistance (MDR) phenotype in cultured cells and in many tumors (Borst *et al.*, 1993; Gottesman and Pastan, 1993; Kane, 1996). The hallmark of MDR, which poses a serious obstacle to successful cancer chemotherapy, is resistance to a single agent to which tumors or cultured cells are exposed, and the concomitant development of cross-resistance to a variety of structurally and functionally unrelated drugs (Kane, 1996). Likewise, P-glycoprotein homologues of the malarial parasite *Plasmodium* (Cowman and Karcz, 1993) and from *Leishmania* (Callahan and Beverley, 1991) have been implicated in chloroquine and heavy metal resistance phenomena, respectively. The peroxisomal ALDp transporter implicated in the pathogenesis of adrenoleukodystrophy and Zellweger syndrome (Aubourg *et al.*, 1993), and a novel ABC transporter, SUR, linked to the familial hyperinsulinemic hypoglycemia of infancy (Aguilar-Bryan *et al.*, 1995), represent additional examples of mammalian ABC transporters linked to genetic disease. Finally, antigen presentation of endogenously proteolyzed viral peptide antigens via the MHC class I pathway (Hill and Ploegh, 1995) requires the action of a heterodimeric Tap1/Tap2 ABC transporter of the ER membrane, which translocates antigenic peptides from the cytoplasm into the ER lumen (Hill and Ploegh, 1995).

While only a few years ago, the Ste6 **a**-factor pheromone transporter (Kuchler *et al.*, 1989; McGrath and Varshavsky, 1989) was the only known yeast ABC transporter, work from our and several other laboratories and the completion of the world-wide yeast genome sequencing project revealed the existence of more than 30 different ABC protein genes in the unicellular eukaryote *Saccharomyces cerevisiae* (Egner *et al.*, 1995a). Thus, yeast is the eukaryotic organism with the largest number of distinct ABC proteins existing in the same cell. The functions of yeast ABC proteins range from an essential role in peptide pheromone secretion to mediating multidrug and heavy metal resistance phenomena (Egner *et al.*, 1995a), although in many cases the physiological roles of sequenced genes are unknown.

Export of the yeast a-factor mating pheromone requires a dedicated ABC transporter

Mating or conjugation of haploid yeast cells leads to a diploid cell and requires extracellular peptide hormones. These hormones, known as mating pheromones **a**-factor and α-factor, are released from haploid *MAT***a** and *MAT*α cells, respectively (Sprague and Thorner, 1992). The biologically active pheromone **a**-factor is a 12 amino acid lipopeptide that is modified with a farnesyl moiety attached via a thioether linkage to the C-terminal cysteine residue which is also carboxymethylated (Anderegg *et al.*, 1988). The signal triggering these modifications was identified as the so-called 'CAAX-box' (where C is cysteine, A is an aliphatic residue and X may be any amino acid) found at the extreme C-terminus of the unmodified peptide. Compelling genetic and biochemical evidence suggested that **a**-factor is translocated to the extracellular space by the substrate-specific Ste6 ABC transporter without traversing the membrane of an exocytic secretory compartment (Kuchler *et al.*, 1989). Thus, export of **a**-factor bypasses the classical ER-Golgi secretory pathway, for extracellular pheromone is still being produced when *MAT***a** cells carrying temperature-sensitive secretion-defective (*sec*) mutations are shifted to the restrictive temperature (Kuchler *et al.*, 1989; Kuchler, 1993).

Initially, **a**-factor is synthesized on polysomes as a unglycosylated precursor that lacks a hydrophobic signal peptide but has a hydrophilic N-terminal extension. Immediately after its biosynthesis, pro-**a**-factor is lipid-modified by the heterodimeric Ram1/Ram2 farnesyltransferase (Hrycyna and Clarke, 1992). The CAAX-box farnesylation targets the pheromone to the membrane for further maturation such as proteolytic clipping of CAAX-box, followed Ste14-mediated carboxymethylation (Sterne-Marr *et al.*, 1990). The proteases involved in **a**-factor maturation were only recently identified. Most interestingly, the *AXL1* gene product, which also may play a role in bud site selection (Adames *et al.*, 1995), represents a homologue of the mammalian interleukin-1 converting enzyme (Fujita *et al.*, 1994). Efficient proteolytic CAAX Box clipping, which precedes carboxymethylation, requires two novel genes (J. Rine; personal communication). Hydrophobicity analysis of the *STE14* carboxymethyltransferase (Ashby *et al.*, 1993) suggested that Ste14 is membrane associated, but the exact subcellular localization of Ste14 has not yet been determined (Hrycyna and Clarke, 1992).

The rate-limiting translocation step across the plasma membrane is mediated by Ste6, which consumes ATP during pheromone translocation (Kuchler, 1993; Kuchler *et al.*, 1989). It is not clear why only fully processed **a**-factor can be efficiently secreted from cells, but one may assume that **a**-factor release is somehow coupled to its processing, as very little, if any, mature **a**-factor can be detected intracellularly (Kuchler *et al.*, 1989). Although it seems reasonably clear that Ste6 mediates the actual **a**-factor transport step, accessory proteins that associate with Ste6 and/or Ste14 in an pore-forming translocation complex may also be necessary for pheromone export. It is also not known whether any of these proteins transiently or permanently interact with Ste6 to form a putative **a**-factor transport complex in the plasma membrane or if a functional translocation complex is assembled at another subcellular location. For the moment, the molecular mechanism by which Ste6 extrudes **a**-factor through the plasma membrane, and the structure of a putative membrane pore remains mysterious. Remarkably, mutations in *STE6* were recently identified which affect mating but not pheromone export, implying a novel function for Ste6 in cell fusion (Elia and Marsh, 1996) in addition to its essential role in pheromone secretion (Kuchler, 1993).

Ste6 is a typical member of the ABC-transporter super-family being most closely related both in predicted membrane topology and in primary sequence to the mammalian P-glycoproteins (Kuchler *et al.*, 1989; McGrath and Varshavsky, 1989). In fact, we and others have demonstrated that wild type human Mdr1 (Kuchler *et al.*, 1992), mouse Mdr3 (Raymond *et al.*, 1992), human MRP (Ruetz *et al.*, 1996) and *Plasmodium* pfMdr (Volkman *et al.*, 1995) can partially compensate for the loss of Ste6 function when functionally expressed in yeast. In contrast, however, overexpression of Ste6 in yeast is generally not associated with a multidrug resistance phenotype. We have therefore set out to identify novel yeast genes of the ABC transporter family whose expression is linked to multidrug resistance phenomena.

Biogenesis, turnover and intracellular targeting yeast ABC drug efflux pumps

To isolate novel yeast genes associated with drug resistance phenomena, we have used multicopy-based genetic screen in which we selected yeast transformants able to grow in the presence of high concentrations of the mycotoxin sporidesmin. This approach indeed allowed for the cloning of a sporidesmin resistance gene, *STS1* (for Sporidesmin Toxicity Suppressor; Bissinger and Kuchler, 1994). The *STS1* gene, which turned out to be identical to the *PDR5* (Balzi *et al.*, 1994), *LEM1* (Kralli *et al.*, 1995) and *YDR1* (Hirata *et al.*, 1994), encodes a 1511-residue ABC transporter most closely related to the *SNQ2* gene product originally cloned as a nitroquinoline N-oxide resistance determinant (Servos *et al.*, 1993). Interestingly, the domain organization of both Pdr5 and Snq2 is reversed as compared to the mammalian P-glycoproteins (Kane, 1996), because the first ABC domain is positioned at the extreme N-terminus of the polypeptides. A similar topology is present in *white, brown* and *scarlet* from *Drosophila melanogaster* (Ewart *et al.*, 1994), in a newly identified mammalian homologue of the fly ABC proteins (Chen *et al.*, 1996; Savary *et al.*, 1996) and in a plant *PDR5*-like ABC transporter (Smart and Fleming, 1996).

Expression of *PDR5* and *SNQ2,* both of which represent true functional homologues of human P-glycoprotein, is linked to typical multidrug resistance phenotypes, conferring resistance to a remarkable variety of cytotoxic compounds (Bissinger and Kuchler, 1994). To better understand the mechanisms underlying multidrug resistance development, we have begun to characterize yeast ABC transporters such as Pdr5 and Snq2 at the biochemical and cell biological level. Subcellular localization studies demonstrated that both Pdr5 and Snq2 localize to the plasma membrane (Egner *et al.*, 1995b). Notably, Pdr5 and Snq2 are rather short-lived proteins whose turnover requires a functional vacuole, since no turnover of Pdr5 or Snq2 occurs in *pep4* mutants blocked in vacuolar proteolysis. The Ste6 pheromone transporter is also a very short-lived protein, whose turnover requires cell surface targeting, followed by rapid endocytosis and delivery to the vacuole for proteolytic turnover (Kölling and Hollenberg, 1994; Berkower *et al.*, 1994). Vacuolar delivery of Ste6 from the cell surface may require ubiquitination of Ste6 at the plasma membrane, since Ste6 was shown to accumulate in a ubiquitinated form in mutants blocked in endocytosis (Kölling and Hollenberg, 1994; Egner and Kuchler, 1996). Likewise, vacuolar delivery of Pdr5 and Snq2 from the cell surface requires a functional endocytic pathway (Egner *et al.*, 1995b).

The observation that the Ste6 **a**-factor pheromone transporter accumulates in a ubiquitinated form in *end4* endocytosis mutants, suggests that a covalent ubiquitin modification of Ste6 might be linked to either its endocytic trafficking or degradation (Kölling and Hollenberg, 1994). Therefore, we tested if a covalent ubiquitin modification is also present in Pdr5, and if Pdr5 ubiquitination is necessary for its vacuolar delivery. Using *c-myc* epitope-tagged ubiquitin, we have shown that Pdr5 is a ubiquitinated plasma membrane protein *in vivo* (Egner and Kuchler, 1996). Ubiquitination of Pdr5 and Ste6 was detected in both wild type and conditional *end4* mutants defective in endocytic vesicle formation. By contrast, the plasma membrane ATPase Pma1, a metabolically very stable protein, was found not to be ubiquitinated (Egner and Kuchler, 1996). To date, ubiquitination has been appreciated as the primary determinant for degradation of short-lived proteins by the proteasome, a multi-catalytic protease found both in the cytoplasm and the nucleus (Peters, 1994). However, despite the ubiquitin modification of Pdr5, and Ste6, their turnover does not require a functional proteasome, since the Pdr5 and Ste6 half-life was unaffected in either *pre1-1* or *pre1-1 pre2-1* mutants defective in the multicatalytic cytoplasmic proteasome (Egner and Kuchler, 1996; Egner and Kuchler, unpublished data). Our results suggest a novel function for ubiquitin in protein trafficking and indicate that ubiquitination of certain short-lived plasma membrane proteins may trigger their endocytic delivery to the vacuole for proteolytic turnover (Egner *et al.*, 1995b).

Although the physiological function of Pdr5 and Snq2 remains ill-defined, the pleiotropic drug resistance phenotype of cells overexpressing Pdr5 and Snq2 raises the possibility that these ABC transporters may be components of an endogenous defense or detoxification system for toxic metabolites (Egner *et al.*, 1995a). Indeed, the intracellular trafficking pattern of several yeast ABC pumps is in line with that idea. One could appreciate that a yeast cell would be able to create a very efficient detoxification machinery if there were several different efflux pumps with a largely distinct substrate specificity guarding both the exocytic and endocytic pathway. Thus, detoxification could be accomplished by pumping toxic metabolites into exocytic compartments, across the plasma membrane into the medium, and, finally, into endocytic vesicles destined for vacuolar delivery (Egner *et al.*, 1995b).

The yeast PDR network and the substrate specificity of ABC multidrug transporters

The regulation of expression of ABC drug efflux pumps appears as a key to understanding multidrug resistance development in yeast. Therefore, we have begun to analyze the transcriptional control mechanisms that govern pleiotropic drug resistance (PDR) development. A complex transcriptional regulatory network seems to regulate PDR in yeast, since a compound PDR phenotype can also arise from gain-of-function mutations in genes encoding transcription factors. The genes *PDR1* (Balzi *et al.*, 1987), *PDR3* (Balzi *et al.*, 1987), the allelic *SNQ3/YAP1* (Hertle *et al.*, 1991; Moye-Rowley *et al.*, 1989) and *YAP2* (Bossier *et al.*, 1993) encode transcription factors implicated in drug and heavy metal resistance development. For instance, mutations in *PDR1* (Meyers *et al.*, 1992) and *PDR3* (Delaveau *et al.*, 1994; Katzmann *et al.*, 1994).are associated with PDR, since Pdr1 was shown to control expression of *PDR5* (Balzi *et al.*, 1994; Katzmann *et al.*, 1996) via *cis*-acting elements, the so-called PDRE boxes, found in the promoters of PDR-responsive genes (Delahodde *et al.*, 1995; Katzmann *et al.*, 1994).

Notably, a *pdr1-3* gain-of-function mutant exhibits a PDR phenotype (Meyers *et al.*, 1992), including elevated resistance to the mutagen 4-nitroquinoline-N-oxide, a known substrate for Snq2 but not for Pdr5. Northern analysis and immunoblotting demonstrated that the *SNQ2* gene is ten-fold overexpressed in a *pdr1-3* gain-of-function mutant, whereas Snq2 expression is severely reduced in a Δ*pdr1* deletion strain, and almost abolished in a Δ*pdr1* Δ*pdr3* double mutant when compared to the isogenic *PDR1* strain (Mahé *et al.*, 1996b). Overexpression of Snq2 and Pdr5 in *pdr1-3* mutants also leads to higher levels of the transporters in the plasma membrane (Decottignies *et al.*; 1995; Mahé *et al.*, 1996b). In addition, DNA footprint analysis revealed the presence of three Pdr3 binding sites in the *SNQ2* promoter (Mahé *et al.*, 1996b). Our results identify *SNQ2* as a novel target for both Pdr1 and Pdr3, and demonstrate that the PDR phenotype of a *pdr1-3* mutant results from overexpression of more than one ABC drug efflux pump (Mahé *et al.*, 1996b). Additional components of the PDR network include *YOR1*, a recently identified oligomycin resistance gene (Katzmann *et al.*, 1995), and additional homologues of Pdr5 and Snq2, including Pdr10, Pdr11, Pdr12, and Pdr15 (Egner *et al.*, 1995a; Mahé *et al.*, unpublished data). Interestingly, Ngg1, a transcription factor involved in the regulation of carbon source utilization (Berger *et al.*, 1992; Brandl *et al.*, 1993), was recently shown by cross-linking and co-immunoprecipitation experiments to interact with Pdr1 (Martens *et al.*, 1996), suggesting that additional transcription factors might contribute to the regulation of the PDR network.

Overexpression of Pdr5 and Snq2 exhibits the hallmarks of P-glycoprotein-mediated MDR in animal cells. Yet again, as in animal cells, the physiological function(s) of the yeast multidrug transporters remain an enigma. Results from two studies, however, implied a possible *in vivo* role for yeast Pdr5 in intracellular steroid transport. First, a *pdr1* mutant allele was previously cloned in a genetic screen by its ability to suppress the squelching toxicity mediated by an estradiol-inducible chimeric VP16-human estrogen receptor (VEO) expressed in yeast (Gilbert *et al.*, 1993). One possible mechanism for the Pdr1-mediated suppression of VEO-toxicity would be that Pdr1 represents a limiting transcriptional intermediary factor interacting with VEO. However, the authors proposed that overexpression of estradiol-specific efflux pumps in the isolated *pdr1* mutant would be a more likely explanation for the apparent squelching suppression (Gilbert *et al.*, 1993). Second, *PDR5* was independently recloned in the *LEM1* gene as a dexamethasone-specific transporter in a genetic screen aimed at identifying factors capable of regulating the glucocorticoid response signal transduction pathway expressed in yeast (Kralli *et al.*, 1995). Gene dosage variation of both *PDR5* and *SNQ2*, allowed us to show that both Pdr5 and Snq2 can transport steroids such as estradiol *in vivo* (Mahé *et al.*, 1996a). We could demonstrate that relief of estradiol-toxicity in yeast cells expressing VEO, requires functional *PDR5* and *SNQ2* genes, since a Δ*pdr5* Δ*snq2* double deletion leads to an increased estradiol toxicity.

Furthermore, using *URA3* as an estradiol-inducible reporter gene, we showed that Pdr5 and Snq2, when overexpressed from high-copy plasmids, can reduce the intracellular concentration of estradiol (Mahé *et al.*, 1996a). In contrast, a Δ*pdr5* Δ*snq2* double deletion mutant accumulates almost 30-fold more intracellular estradiol than the isogenic wild type. Indirect immunofluorescence also showed that a *pdr1-3* mutant massively overexpress Pdr5 at the plasma membrane, suggesting that estradiol efflux from the cells occurs via the plasma membrane (Mahé *et al.*, 1996a).

Taken together, these results demonstrate that Pdr5 and Snq2 can transport steroid and glucocorticoid substrates *in vivo*, and suggest that steroids and/or related membrane lipids could represent physiological substrates for certain yeast ABC transporters, which are otherwise involved in the development of pleiotropic drug resistance. Moreover, our results show that suppression of VEO-squelching by a *pdr1* gain-of-function allele (Gilbert *et al.*, 1993) is the consequence of overexpression of estradiol-specific efflux pumps such as Pdr5 and Snq2, both of which are tightly controlled by Pdr1 and Pdr3 (Mahé *et al.*, 1996b). Hence, we propose that a possible *in vivo* role of Pdr5 and Snq2 or closely related ABC pumps may be to function as sensors of membrane sterol or phospholipid composition, thereby maintaining lipid homeostasis of cellular membranes, so as to regulate membrane fluidity and permeability (Lipowsky, 1991). It seems plausible, that a mechanism must exist by which membrane fluidity and rigidity is controlled, since this is a prerequisite for processes such regulated transport across membranes or vesicle formation and budding to occur (Lipowsky, 1991). Intriguingly, analysis of the sterol composition of various yeast membranes demonstrated that the plasma membrane represents the cellular membrane with the highest sterol content (Parks *et al.*, 1995; Zinser *et al.*, 1993). Thus, the expression level of yeast multidrug transporters in the plasma membrane could be responsible for maintaining a desired or required ergosterol composition in the membrane. However, the mechanism by which such regulation would occur is not easy to envisage. The intracellular trafficking pattern of ABC transporters could represent a possible means of removing/transporting ergosterol or other membrane lipids from/to the plasma membrane. This could be achieved by tight binding of lipids to the hydrophobic TMS domains of Pdr5 or Snq2, similar to the proposed vacuum cleaner model for the interaction of hydrophobic drugs and steroids with mammalian Mdr1 (Gottesman and Pastan, 1993; Kane, 1996).

Strikingly, a function for human Mdr1 in intracellular cholesterol transport was recently proposed, although only circumstantial evidence could be provided (Metherall *et al.*, 1996a; Metherall *et al.*, 1996b). Mammalian P-glycoproteins have also been demonstrated to interact with several steroids, and they were shown to mediate transcellular transport of steroids and phospholipids both *in vitro* and *in vivo* (van Helvoort *et al.*, 1996; Ruetz and Gros, 1994; Smit *et al.*; 1994; van Kalken *et al.*, 1993; Ueda *et al.*, 1992). In addition, human Mdr1 and Mdr3 were recently demonstrated to act as phospholipid-specific flippases (van Helvoort *et al.*, 1996), which seems to confirm an earlier hypothesis on the true physiological role of P-glycoproteins (Higgins and Gottesman, 1992). Several yeast PDR-genes are structurally and functionally related to mammalian ABC multidrug transporter such as Pgp and MRP. Since expression of human Pgp and MRP in yeast results in a typical drug resistance phenotype, we are now employing yeast as a heterologous expression system to study the structure/function relationships of medically important ABC transporters from mammalian cells. For instance, we are using yeast as a model system for the isolation and characterization of novel Pgp transport mutants with *distinct and limited substrate* specificity for drugs frequently used in clinical cancer chemotherapy (Kane, 1996). To demonstrate this to be a feasible approach, we have randomly mutagenized the yeast *PDR5* gene in order to isolate ABC transport mutants with altered substrate specificity (R. Egner and K. Kuchler, in preparation). A number of mutants were isolated that were able to confer a dramatically changed drug resistance profile when compared to wild type Pdr5. The molecular analysis of the mutations indicated that mutations at many different locations in *PDR5* can modulate substrate specificity, implying that folding and membrane structure of ABC transporters may be a major way to control substrate specificity (R. Egner and K. Kuchler, in preparation). Thus, by using this approach we are not only able to modulate substrate specificity of endogenous yeast ABC drug efflux pumps, but we were also able to generate "custom-made" drug transporters with desired substrate specificities by phenotypic selection in yeast. This approach can now be applied to the human *MDR1* gene, as well as related drug resistance genes, to generate custom-made Pgp transporters with desired drug substrate specificity. Such Pgp transporter genes would be highly sought-for therapeutic genes to achieve chemoprotection *in vivo* after their transfer into hematopoietic stem cells obtained from patients undergoing high dose cancer chemotherapy (Kane, 1996).

Acknowledgments

I wish to thank my colleagues and collaborators, in particular P. Bissinger, P. Chambon, A. Delahodde, A. Goffeau, C. Jacq, S. Kane, R. Kölling, N. Kralli, Y. Lemoine, R. Losson, P. Piper, J. Subik, D. Sanglard, J. Thorner, and D. Wolf and for their generosity in sharing research materials, yeast strains and reagents, but also for many fruitful discussions and communication of unpublished data. Work in my laboratory is supported by grants from the Austrian Science Foundation (Project P-09537, P-MOB-10123 and FWF-SFB-604), the Austrian National Bank (OENB project 5638), by funds from the "Herzfelder'schen Familienstiftung" and in part by NIH grant #RO1-CA64645-01A1. R. E. is a recipient of a postdoctoral fellowship from the "Deutsche Forschungsgemeinschaft".

References

Adames N, Blundell K, Ashby MN, Boone C (1995) Role of yeast insulin-degrading enzyme homologs in propheromone processing and bud site selection. Science 270: 464-467
Aguilar-Bryan L, Nichols CG, Wechsler SW, Clement JPt, Boyd AEr, Gonzalez G, Herrera Sosa H, Nguy K, Bryan J, Nelson DA (1995) Cloning of the beta cell high-affinity sulfonylurea receptor: a regulator of insulin secretion. Science 268: 423-426
Anderegg RJ, Betz R, Carr SA, Crabb JW, Duntze W (1988) Structure of the *Saccharomyces cerevisiae* mating hormone **a**-factor: identification of S-farnesyl cysteine as a structural component. J Biol Chem 263: 18236-18240
Ashby MN, Errada PR, Boyartchuk VL, Rine J (1993) Isolation and DNA sequence of the *STE14* gene encoding farnesyl cysteine: carboxyl methyltransferase. Yeast 9: 907-913
Aubourg P, Mosser J, Douar AM, Sarde CO, Lopez J, Mandel JL (1993) Adrenoleukodystrophy gene: unexpected homology to a protein involved in peroxisome biogenesis. Biochimie 75: 293-302
Balzi E, Chen W, Ulaszewski S, Capieaux E, Goffeau A (1987) The multidrug resistance gene *PDR1* from *Saccharomyces cerevisiae*. J Biol Chem 262: 16871-16879
Balzi E, Wang M, Leterme S, van Dyck L, Goffeau A (1994) PDR5, a novel yeast multidrug resistance conferring transporter controlled by the transcription regulator *PDR1*. J Biol Chem 269: 2206-2214
Berger SL, Pina B, Silverman N, Marcus GA, Agapite J, Regier JL, Triezenberg SJ, Guarente L (1992) Genetic isolation of *ADA2*: a potential transcriptional adaptor required for function of certain acidic activation domains. Cell 70: 251-265
Berkower C, Loayza D, Michaelis S (1994) Metabolic instability and constitutive endocytosis of *STE6*, the **a**-factor transporter of *Saccharomyces cerevisiae*. Mol. Biol. Cell 3: 633-654
Bissinger PH, Kuchler K (1994) Molecular cloning and expression of the *Saccharomyces cerevisiae STS1* gene product. A yeast ABC transporter conferring mycotoxin resistance. J Biol Chem 269: 4180-4186
Borst P, Schinkel AH, Smit JJ, Wagenaar E, Van Deemter L, Smith AJ, Eijdems EW, Baas F, Zaman GJ (1993) Classical and novel forms of multidrug resistance and the physiological functions of P-glycoproteins in mammals. Pharmacol Ther 60: 289-99.
Bossier P, Fernandes L, Rocha D, Rodrigues-Pousada C (1993) Overexpression of *YAP2*, coding for a new yAP protein, and *YAP1* in *Saccharomyces cerevisiae* alleviates growth inhibition caused by 1,10- phenanthroline. J Biol Chem 268: 23640-23645
Brandl CJ, Furlanetto AM, Martens JA, Hamilton KS (1993) Characterization of *NGG1*, a novel yeast gene required for glucose repression of GAL4p-regulated transcription. EMBO J 12: 5255-5265
Callahan HL, Beverley SM (1991) Heavy metal resistance: a new role for P-glycoproteins in *Leishmania*. J Biol Chem 266: 18427-18430
Chen H, Rossier C, Lalioti MD, Lynn A, Chakravarti A, Perrin G, Antonarakis SE (1996) Cloning of the cDNA for a human homologue of the *Drosophila white* gene and mapping to chromosome 21q22.3. Am J Hum Genet 59: 66-75
Cowman AF, Karcz S (1993) Drug resistance and the P-glycoprotein homologues of *Plasmodium falciparum*. Semin Cell Biol 4: 29-35
Decottignies A, Lambert L, Catty P, Degand H, Epping EA, Moye-Rowley WS, Balzi E, Goffeau A (1995) Identification and characterization of *SNQ2*, a new multidrug ATP binding cassette transporter of the yeast plasma membrane. J Biol Chem 270: 18150-18157
Delahodde A, Delaveau T, Jacq C (1995) Positive autoregulation of the yeast transcription factor Pdr3p, which is involved in control of drug resistance. Mol Cell Biol 15: 4043-4051
Delaveau T, Delahodde A, Carvajal E, Subik J, Jacq C (1994) *PDR3*, a new yeast regulatory gene, is homologous to *PDR1* and controls the multidrug resistance phenomenon. Mol Gen Genet 244: 501-511

Egner R, Kuchler K (1996) The yeast multidrug transporter Pdr5 of the plasma membrane is ubiquitinated prior to endocytosis and degradation in the vacuole. FEBS Lett 378: 177-181

Egner R, Mahé Y, Pandjaitan R, Huter V, Lamprecht A, Kuchler K (1995a) ATP binding cassette transporters in yeast: from mating to multidrug resistance. In, Rothman S (ed) "*Membrane Protein Transport.*" Greenwich JAI Press Inc., pp 57-96

Egner R, Mahé Y, Pandjaitan R, Kuchler K (1995b) Endocytosis and vacuolar degradation of the plasma membrane localized Pdr5 ATP binding cassette multidrug transporter in *Saccharomyces cerevisiae.* Mol Cell Biol 15: 5879-5887

Elia L, Marsh L (1996) Role of the ABC transporter Ste6 in cell fusion during yeast conjugation. J Cell Biol 135: 741-751

Ewart GD, Cannell D, Cox GB, Howells AJ (1994) Mutational analysis of the traffic ATPase (ABC) transporters involved in uptake of eye pigment precursors in *Drosophila melanogaster.* Implications for structure-function relationships. J Biol Chem 269: 10370-10377

Fujita A, Oka C, Arikawa Y, Katagai T, Tonouchi A, Kuhara S, Misumi Y (1994) A yeast gene necessary for bud-site selection encodes a protein similar to insulin-degrading enzymes. Nature 372: 567-570

Gilbert DM, Heery DM, Losson R, Chambon P, Lemoine Y (1993) Estradiol-inducible squelching and cell growth arrest by a chimeric VP16-estrogen receptor expressed in *Saccharomyces cerevisiae*: suppression by an allele of *PDR1*. Mol Cell Biol 13: 462-472

Gottesman M, Pastan I (1993) Biochemistry of multidrug resistance mediated by the multidrug transporter. Ann Rev Biochem 62: 385-427

Harris A, Argent BE (1993) The cystic fibrosis gene and its product CFTR. Semin Cell Biol 4: 37-44

Hertle K, Haase E, Brendel M (1991) The *SNQ3* gene of *Saccharomyces cerevisiae* confers hyper-resistance to several functionally unrelated chemicals. Curr Genet 19: 429-433

Higgins CF (1992) ABC-transporters: from microorganisms to man. Ann Rev Cell Biol 8: 67-113

Higgins CF, Gottesman MM (1992) Is the multidrug transporter a flippase? Trends Biochem Sci 17: 18-21

Hill A, Ploegh H (1995) Getting inside out: The transporter associated with antigen processing (TAP) and the presentation of viral antigen. Proc Natl Acad Sci (USA) 92: 341-343

Hirata D, Yano K, Miyahara K, Miyakawa T (1994) *Saccharomyces cerevisiae YDR1*, which encodes a member of the ATP-binding cassette (ABC) superfamily, is required for multidrug resistance. Curr Genet 26: 285-294

Hrycyna CA, Clarke S (1992) Maturation of isoprenylated proteins in *Saccharomyces cerevisiae*. Multiple activities catalyze the cleavage of the three carboxyl-terminal amino acids from farnesylated substrates *in vitro*. J Biol Chem 267: 10457-10464

Kane SE (1996) Multidrug resistance of cancer cells. In B. Testa a, Meyer U. A. (ed) "*Advances in Drug Research.*" San Diego: Academic Press, pp 181-252

Katzmann D, Hallstrom TC, Mahé Y, Moye-Rowley WS (1996) Multiple Pdr1/Pdr3 binding sites are essential for normal expression of the ATP binding cassette transporter protein encoding gene *PDR5*. J Biol Chem 271: 23049-23054

Katzmann DJ, Burnett PE, Golin J, Mahé Y, Moye-Rowley WS (1994) Transcriptional control of the yeast *PDR5* gene by the *PDR3* gene product. Mol Cell Biol 14: 4653-4661

Katzmann DJ, Hallstrom TC, Voet M, Wysock W, Golin J, Volckaert G, Moye-Rowley, WS (1995) Expression of an ATP-binding cassette transporter-encoding gene (*YOR1*) is required for oligomycin resistance in *Saccharomyces cerevisiae*. Mol Cell Biol 15: 6875-6883

Kölling R, Hollenberg CP (1994) The ABC-transporter Ste6 accumulates in the plasma membrane in a ubiquitinated form in endocytosis mutants. EMBO J 13: 3261-3271

Kralli A, Bohen SP, Yamamoto KR (1995) *LEM1*, an ATP-binding-cassette transporter, selectively modulates the biological potency of steroid hormones. Proc Natl Acad Sci (USA) 92: 4701-4705

Kuchler K (1993) Unusual routes of protein secretion: the easy way out. Trends Cell Biol. 3: 421-426

Kuchler K, Göransson M, Visnawathan M, Thorner J (1992) Dedicated transporters for peptide export and intercompartemental traffic in yeast. CSH Symp Quant Biol 57: 579-592

Kuchler K, Sterne RS, Thorner J (1989) *Saccharomyces cerevisae STE6* gene product: A novel pathway for protein export in eukaryotic cells. EMBO J 8: 3973-3984

Kuchler K, Thorner J (1992) Secretion of peptides and proteins lacking hydrophobic signal sequences: The role of ATP-driven membrane translocators. Endocrine Rev 13: 499-514

Lipowsky R (1991) The conformation of membranes. Nature 349: 475-481

Mahé Y, Lemoine Y, Kuchler K (1996a) The ATP-binding cassette multidrug transporters Pdr5 and Snq2 of *Saccharomyces cerevisiae* can mediate transport of steroids *in vivo.* J Biol Chem 271: 25167-25172

Mahé Y, Parle-McDermott A, Nourani A, Delahodde A, Lamprecht A, Kuchler K (1996b) The ATP-binding cassette multidrug transporter Snq2 of *Saccharomyces cerevisiae*: a novel target for the transcription factors Pdr1 and Pdr3. Mol Microbiol 20: 109-117

Martens JA, Genereaux J, Saleh A, Brandl CJ (1996) Transcriptional activation by yeast *PDR1* is inhibited by its association with *NGG1ADA3*. J Biol Chem 271: 15884-15890

McGrath JP, Varshavsky A (1989) The yeast *STE6* gene encodes a homologue of the mammalian multidrug resistance P-glycoprotein. Nature 340: 400-404

Metherall JE, Li H, Waugh K (1996a) Role of multidrug resistance P-glycoproteins in cholesterol biosynthesis. J Biol Chem 271: 2634-2640

Metherall JE, Waugh K, Li H (1996b) Progesterone inhibits cholesterol biosynthesis in cultured cells. J Biol Chem 271: 2627-2633

Meyers S, Schauer W, Balzi E, Wagner M, Goffeau A, Golin J (1992) Interaction of the yeast pleiotropic drug resistance genes *PDR1* and *PDR5*. Curr Genet 21: 431-436

Moye-Rowley WS, Harshman KD, Parker CS (1989) Yeast *YAP1* encodes a novel form of the *jun* family of transcriptional activator proteins. Genes Dev 3: 283-292

Parks LW, Smith SJ, Crowley JH (1995) Biochemical and physiological effects of sterol alterations in yeast-a review. Lipids 30: 227-30

Peters JM (1994) Proteasomes: protein degradation machines of the cell. Trends Biochem Sci 19: 377-382

Raymond M, Gros P, Whiteway M, Thomas DY (1992) Functional complementation of yeast *ste6* by a mammalian multidrug resistance *mdr* gene. Science 256: 232-234

Riordan JR, Chang XB (1992) CFTR, a channel with the structure of a transporter. Biochim Biophys Acta 1101: 221-222

Ruetz S, Gros P (1994) Phosphatidylcholine translocase: a physiological role for the *mdr2* gene. Cell 77: 1071-1181

Ruetz S, Brault M, Kast C, Hemenway C, Heitman J, Grant CE, Cole SP, Deeley RG, Gros P (1996) Functional expression of the multidrug resistance-associated protein in the yeast *Saccharomyces cerevisiae*. J Biol Chem 271: 4154-4160

Savary S, Denizot F, Luciani M, Mattei M, Chimini G (1996) Molecular cloning of a mammalian ABC transporter homologous to *Drosophila white* gene. Mamm Genome 7: 673-676

Servos J, Haase E, Brendel M (1993) Gene *SNQ2* of *Saccharomyces cerevisiae*, which confers resistance to 4-nitroquinoline-N-oxide and other chemicals, encodes a 169 kDa protein homologous to ATP-dependent permeases. Mol Gen Genet 236: 214-218

Smart CC, Fleming AJ (1996) Hormonal and environmental regulation of a plant *PDR5*-like ABC transporter. J Biol Chem 271: 19351-19357

Smit JJM, Schinkel AH, Elferink RPJ, Groen AK, Wagenaar E, van Deemter L, Mol CAAM, Ottenhoff R, van der Lug NMT, van Room MA, van der Kalk MA, Offerhaus GJA, Berns AJM, Borst P (1993) Homozygous disruption of the murine *mdr2* P-glycoprotein gene leads to a complete absence of phospholipid from bile and to liver disease. Cell 75: 451-462

Sprague GF, Thorner J (1992) Pheromone response and signal transduction during the mating process of *Saccharomyces cerevisiae*: "The Molecular Biology of the Yeast *Saccharomyces cerevisiae*, Second Edition.", pp 657-744

Sterne-Marr RE, Blair LC, Thorner J (1990) *Saccharomyces cerevisiae STE14* gene is required for COOH-terminal methylation of **a**-factor mating pheromone. J Biol Chem 265: 20057-20060

Ueda K, Okamura N, Hirai M, Tanigawara Y, Saeki T, Kioka N, Komano T, Hori R (1992) Human P-glycoprotein transports cortisol, aldosterone, and dexamethasone, but not progesterone. J Biol Chem 267: 24248-24252

van Helvoort A, Smith AJ, Sprong H, Fritzsche I, Schinkel A, Borst P, van Meer G (1996) *MDR1* P-glycoprotein is a lipid translocase of broad specificity. while *MDR3* P-glycoprotein specifically translocates phosphatidylcholine. Cell: 507-517

van Kalken CK, Broxterman HJ, Pinedo HM, Feller N, Dekker H, Lankelma J, Giaccone G (1993) Cortisol is transported by the multidrug resistance gene product P-glycoprotein. Br J Cancer 67: 284-289

Volkman SK, Cowman AF, Wirth DF (1995) Functional complementation of the *ste6* gene of *Saccharomyces cerevisiae* with the *pfmdr1* gene of *Plasmodium falciparum*. Proc Natl Acad Sci (USA) 92: 8921-8925

Zinser E, Paltauf F, Daum G (1993) Sterol composition of yeast organelle membranes and subcellular distribution of enzymes involved in sterol metabolism. J Bacteriol 175: 2853-2858

Regulation of Carbon Metabolism in Bacteria

Marga Gunnewijk, Grietje Sulter, Pieter Postma[1] and Bert Poolman

Groningen Biomolecular Sciences and Biotechnology institute
Department of Microbiology
University of Groningen
Kerklaan 30
9751 NN Haren
The Netherlands
phone: +31 50 3632170
fax: +31 50 3632154
e-mail: B.Poolman@biol.rug.nl

Abstract

Bacteria have different regulatory systems to control *carbon metabolism*, in a number of cases employ the *phospho*enol*pyruvate phosphotransferase system* (PTS) is used, to tune the transport of carbohydrate and/or transcription of genes specifying catabolic enzymes. Although the physiological consequences of this regulation are similar in Gram-negative bacteria and Gram-positive bacteria, the mechanisms are entirely different. In the Gram-negative negative bacteria the state of phosphorylation of the glucose-specific PTS IIA protein plays a central role in *inducer exclusion* and *cAMP synthesis*, whereas regulatory processes are triggered by the phosphorylation state of HPr in Gram-positive bacteria. In both mechanisms the activity of target enzymes is modulated upon allosteric interaction with a particular form of IIA^{glc} or HPr. In some Gram-positive bacteria, however, the regulation takes place via direct PEP-dependent enzyme I/HPr-mediated phosphorylation of a IIA-like domain of the target protein. The functional and structural (dis)similarities of this IIA-like domain and IIA^{glc} are discussed in this paper.

---

[1] E.C. Slater Institute, BioCentrum, University of Amsterdam, Plantage Muidergracht 12, 1018 TV Amsterdam, The Netherlands

NATO ASI Series, Vol. H 101
Molecular Mechanisms of Signalling
and Membrane Transport
Edited by Karel W. A. Wirtz

Introduction

The uptake of carbohydrates by bacteria can be catalysed by *primary* and *secondary transport* mechanisms, as well as *group translocation* systems. In primary sugar transport the uptake is driven by ATP (or related energy-rich compound), whereas in secondary transport the uptake is driven by the electrochemical gradients of the translocated solutes (Poolman and Konings, 1993). Among the secondary transport systems one can distinguish *symporters* (cotransport of two or more solutes), *uniporters* (transport of one molecule) and *antiporters* (countertransport of two or more solutes). Sugar symporters usually couple the uphill movement of the sugar to the downhill movement of a proton (or sodium ion), i.e., the electrochemical proton (or sodium ion) gradient drives the accumulation of sugar. The lactose transport proteins LacY of *Escherichia coli* and LacS of *Streptococcus thermophilus* are examples of galactoside-H^+ symporters. These systems, however, also catalyse sugar exchange, which in case of LacS protein may involve the uptake of lactose ('forward stroke' of the carrier) in exchange for galactose ('back stroke'). The lactose/galactose exchange reaction is approximately two-orders of magnitude faster than lactose-H^+ symport and represents the physiologically relevant translocation reaction as *S. thermophilus* is unable to metabolize the galactose moiety of lactose (Poolman, 1990; Knol *et al.*, 1996). Sugar uptake by group translocation is unique to bacteria and involves phospho*enol*pyruvate:sugar phosphotransferase systems (PTSs) (Postma *et al.*, 1993). The PTS catalyses the uptake of sugar concomitant with its phosphorylation. The phosphoryl group is transferred from phospho*enol*pyruvate (PEP) via the general energy coupling proteins enzyme I and HPr and the sugar-specific phosphoryl transfer proteins/domains IIA and IIB; IIB~P transfers the phosphoryl group to the sugar that is translocated via the sugar-specific IIC protein/domain (Fig 1). It should be stressed that IIA, IIB and IIC can be separate proteins, domains in a single polypeptide (IIABC) or linked as pairs (i.e., IIAB together with IIC or IIA together with IIBC). In case of the glucose-PTS of enteric bacteria (Fig. 1), the IIA^{glc} is a cytoplasmic protein that interacts with the membrane complex $IIBC^{glc}$.

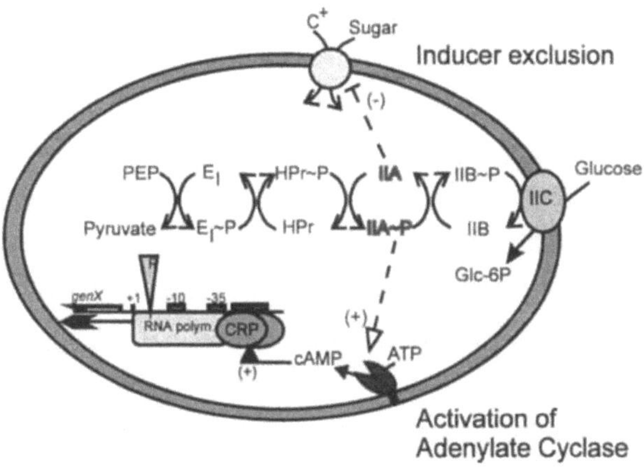

Fig. 1. Schematic representation of the phospho*enol*pyruvate:glucose phosphotransferase system found in enteric bacteria. The central role of IIA^{glc} (~P) in regulating the activity of adenylate cyclase and sugar transport proteins is illustrated. C^+, cation; EI, enzyme I; CRP, cAMP receptor protein; R, repressor.

Role of PTS in the regulation of carbohydrate metabolism

Apart from its function in the uptake and phosphorylation of carbohydrates, the PTS regulates transport and subsequent metabolism of non-PTS carbohydrates, such as glycerol, maltose, melibiose and lactose (Postma *et al.*, 1993). In the Gram-negative enteric bacteria this regulation is mediated by the phosphorylation state of IIA^{glc}, which is determined by the relative rates of phosphorylation by HPr~P and dephosphorylation by $IIBC^{glc}$. The available data suggest that IIA^{glc}~P is involved in the stimulation of adenylate cyclase, whereby the expression of many catabolic enzymes is regulated through changes in cAMP levels (Botsford *et al.*, 1992). Unphosphorylated IIA^{glc}, on the other hand, binds to several enzymes essential in carbohydrate metabolism and thereby inhibits their activities; the mechanism by which IIA^{glc} inhibits has been called *inducer exclusion*.

The interaction of IIA^{glc} with one of its targets, glycerol kinase (GlpK), has been elucidated to a great extent by analyzing the crystal structure of *E. coli* glycerol kinase in complex with *E. coli* IIA^{glc} (Hurley *et al.*, 1993). This study revealed that IIA^{glc} binds to glycerol kinase at a region that is distant from the catalytic site of glycerol kinase, which suggests that long-range conformational changes mediate the inhibition of glycerol kinase by IIA^{glc}. The binding of glycerol kinase and IIA^{glc} involves mainly hydrophobic and electrostatic interactions. One hydrogen bond, involving an uncharged aspartate, and a Zn(II) binding site also appear to play a role in the interaction (Feese *et al.*,1994). The Zn(II) binding site is made up of the two active-site histidines of IIA^{glc} (His75 and His90), Glu478 of glycerol kinase and a H_2O molecule. Phosphorylation of IIA^{glc} detroys the intermolecular Zn(II) binding site and disrupts the interactions between IIA^{glc} and glycerol kinase.

Other known targets of IIA^{glc} are the galactoside transport proteins LacY, MelB and RafB, and the ATP-coupling subunit (MalK) of the maltose transport system. Based on mutations that render LacY, RafB and MalK resistant to inducer exclusion, i.e., to inhibition by PTS sugars, a consensus sequence for IIA^{glc} binding has been proposed (Titgemeyer *et al.*, 1994). This sequence VGANXSLXSX, however, is not present in GlpK and MelB, and it remains to be established whether this region indeed forms part of the contact site with IIA^{glc}.

In contrast to the Gram-negative enteric bacteria, where the glucose-specific phosphocarrier IIA^{glc} has a central role in the regulation of carbohydrate uptake, in Gram-positive bacteria the serine-phosphorylated form of HPr [HPr(Ser-P)] seems to control carbohydrate metabolism both at the protein and the gene level, i.e., transport activities (*inducer exclusion* of both PTS and non-PTS sugars and/or *inducer expulsion*; see below) and transcription. In Gram-positive bacteria HPr cannot only be phosphorylated by PEP/EI on a histidine (His15~P), but, in addition, a metabolite-activated ATP-dependent protein kinase can phosphorylate a serine residue at position 46 in HPr (Deutscher and Saier 1983) . The activity of this HPr(Ser) kinase is dependent on divalent cations as well as on one of several intermediary metabolites, e.g., fructose 1,6-bisphosphate (FBP) or gluconate 6-phosphate (Gnt 6P) (Reizer *et al.*, 1984; Deutscher and Engelmann, 1984). Thus, the activity of HPr(Ser) kinase is directly linked to the glycolysis and/or the pentose phosphate pathway, thereby forming a intricate network to control sugar metabolism.

Several, but not all, low-GC Gram-positive bacteria exhibit the *inducer expulsion* phenomenon. In this process, accumulated sugar(-P) is released from the cell following hydrolysis by a cytoplasmic membrane-associated sugar(-P) phosphatase (PaseII), which is activated upon binding of HPr(Ser-P) (Reizer *et al,* 1983, 1985; Sutrina *et al,* 1988; Ye *et al.*, 1995c). The transporter responsible for this efflux has not yet been identified but its properties mimic that of a uniporter which allows efflux of free sugar down the concentration gradient. *Lactococcus brevis* and other heterofermentative lactobacilli exhibit metabolite-activated sugar expulsion, that is controlled by a mechanism that does not depend on sugar-P hydrolysis. Transport experiments suggest that the H^+/glucose symporter and H^+/lactose symporters are mechanistically converted into uniport systems upon binding of HPr(Ser-P) to the cytoplasmic surface of the carrier protein (Romano *et al,* 1987; Ye *et al,* 1994). Binding of HPr(Ser-P) only occurs in the presence of substrate (Ye and Saier, 1995a,b).

HPr(Ser-P) also controls the transcription of target genes (operons) in Gram-positive bacteria by interacting with the transcription factor CcpA (Hueck and Hillen, 1995). CcpA is a repressor that binds to DNA and thereby retards or blocks transcription initiation. There is evidence that HPr(Ser-P) interacts with and activates CcpA; i.e., increases the affinity of CcpA for its target site (catabolite responsive element) on the DNA.

For both Gram-positive and Gram-negative bacteria it is clear that components of the PTS are major factors (signal transducers) in the regulation of carbohydrate metabolism. In Gram-negative bacteria external sugars are sensed, whereas in Gram-positive bacteria intracellular metabolites are sensed as a primary signal. A second signal is then transmitted via the enzymes IIB/HPr by (de)phosphorylation of IIA^{glc} in Gram-negative bacteria, while in Gram-positive bacteria a phosphorylating signal is transmitted to HPr via the HPr kinase to effect allosteric regulation of the target proteins.

Role of IIA domain of LacS in sugar transport

The involvement of IIA^{glc} or IIA-like proteins in PTS-mediated regulation in Gram-positive bacteria is unknown. However, several non-PTS sugar transporters have a C-terminal domain that is homologous to IIA^{glc} of *E. coli.* To this family of transporters belong the lactose transport protein LacS of *Streptococcus thermophilus, Lactobacillus bulgaricus* and

Leuconostoc lactis and the raffinose transport protein RafP of *Pediococcus pentosaceus*. Interestingly, these proteins are, amongst others, homologous to the melibiose transport proteins of *Salmonella typhimurium* and *E. coli,* which lack a IIA-like domain, but are regulated by IIA^{glc} as discussed above (Poolman *et al.*, 1996). The LacS protein of *S. thermophilus* catalyses the uptake of galactosides in symport with a proton or exchanges lactose for intracellularly formed galactose (Poolman, 1990; Foucaud *et al.*, 1992). The amino-terminal (carrier) domain of LacS is typical for a polytopic membrane protein and is composed of 12 α-helical transmembrane segments. The carboxyl-terminal IIA domain is located in the cytoplasm, approximately 160 amino acids in size and has 34% identical residues with IIA^{glc} (Poolman *et al.*, 1989).

To asses the role of the IIA domain in LacS-mediated lactose transport, several carboxy-terminal truncation mutants were constructed and expressed in *E. coli* and their properties were analyzed (Poolman *et al.*, 1995). Remarkably, the entire IIA domain (160 amino acids) could be deleted without significant effect on lactose-H^+ symport and galactoside equilibrium exchange. To study the properties of the IIA domain of LacS in more detail, the carboxy-terminal domain was overexpressed in *E. coli* followed by isolation and purification of the protein. Purified IIA can be phosphorylated by PEP and the general PTS energy coupling proteins EI and HPr of *E. coli* and *Bacillus subtilis*. Phosphorylation most likely occurs at His-552 (Poolman *et al.*, 1992;1995). This histidine residue corresponds with His-90 of IIA^{glc} in *E. coli*, which has been shown to be the phosphoryl accepting site (Dörschug *et al.*, 1984; Presper *et al.*, 1989) (Fig. 2).
PEP-dependent enzyme I/HPr-mediated phosphorylation of the LacS IIA domain inhibits the transport activity of LacS. This was studied in membrane vesicles fused to cytochrome c oxidase containing liposomes (proton motive force-generating mechanism). When appropriate, PEP plus purified enzyme I and HPr were incorporated into the hybrid membranes. Generation of a proton motive force (Δp) in the hybrid membranes resulted in LacS-dependent accumulation of lactose. With PEP and the energy coupling proteins enzyme I and HPr present on the inside, the Δp-driven lactose uptake by wildtype LacS was inhibited (Fig. 3). This inhibition was not observed with LacS(Δ160) and LacS(H552R), indicating that PEP-dependent enzyme I/HPr-mediated phosphorylation of the IIA domain (possibly the conserved His-552 residue) modulates lactose-H^+ symport activity (Poolman *et al.*, 1995).

Conserved Residues in Active-site of IIA proteins

Non-PTS IIA protein(s) domains

```
                     * ***: * :  ::**  : :     *:* : :    *:  :**:*:**:
LacSSt      DEHFASGSMGKGFAIKPTDGAVFAPISGTIRQILPTRHAVGIESEDGVIVLIHVGIGTVK
RafPPa      DPTFAAGTLGDGFAIKPSDGRILAPFDATVRQVFTTRHAVGLVGDNGIVLLIHIGLGTVK
LacSLb      DPVFADKKLGDGFALVPADGKVYAPFAGTVRQLAKTRHSIVLENEHGVLVLIHLGLGTVK
LacSLl      NEVDGNTLTGIGFAIDPEEGNLFAPFDGKVDFTFSTKHVLGVVSNNGLKAIIHVGIGTIN
```

PTS IIA protein(s) domains

```
            s hh s hh                        h   Z            h Z   @ h
            *  *     * *:*: *    :  *  : :     : * :      * * ::*:*:**:
IIASty      DVVFAEKIVGDGIAIKPTGNKMVAPVDGTIGKIFETNHAFSIESDSGIELFVHFGIDTVE
IIAEc       DVVFAEKIVGDGIAIKPTGNKMVAPVDGTIGKIFETNHAFSIESDSGVELFVHFGIDTVE
PtsGBs      DQVFSGKMMGDGFAILPSEGIVVSPVRGKILNVFPTKHAIGLQSDGGREILIHFGIDTVS
BglFEc      DTTFASGLLGKGIAILPSVGEVRSPVAGRIASLFATLHAIGIESDDGVEILIHVGIDTVK
NagEKp      DEAFASKAVGDGIAVKPTDNIVVAPAAGTVVKIFNTNHAFCLETNNGAEIVVHMGIDTVA
PtsGBl      DPIFAAGKLGPGIAIEPTGNTVVAPADATVILVQKSGHAVALRLESGVELLIHIGLDTVQ
PtsMACg     DPIFAAGKLGPGIAIQPTGNTVVAPADATVILVQKSGHAVALRLDSGVEILVHVGLDTVQ
NagEEc      DEAFASKAVGDGVAVKPTDKIVVSPAAGTIVKIFNTNHAFCLETEKGAEIVVHMGIDTVA
```

Fig. 2. Alignment of active site of PTS IIA protein(s) and non-PTS IIA protein(s) domains. *, conserved residue; :, similar residue; s, salt-bridge with residue in Glycerol Kinase; h, hydrophobic interaction with glycerol kinase; z, coordination of Zn(II); @, charged group near phosphorylation-site.

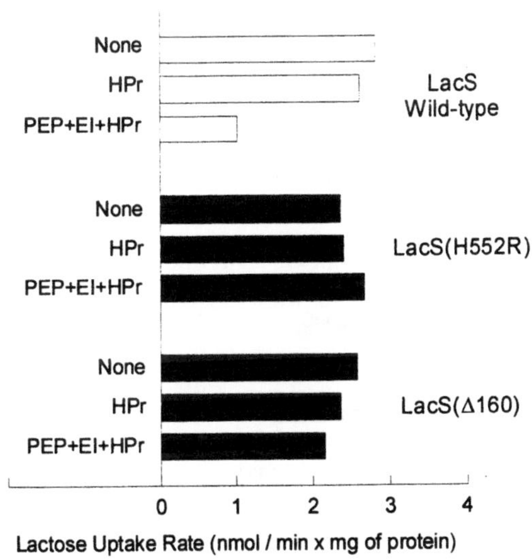

Fig. 3. Effect of PEP-dependent enzyme I/ HPr- mediated phosphorylation on lactose transport by wild-type LacS, LacS(H552R), and LacS(Δ160). Membrane vesicles of *S. thermophilus* ST11/(ΔlacS)/pGKGS8 (wild type), ST11 (ΔlacS)/pGKGS8(lacS H552R) and ST11 (ΔlacS)/pGKGS8 (lacS Δ160) were fused with cytochrome c oxidase-containing liposomes in the absence or presence of PEP and/or PTS enzymes as indicated; the molar ratio of HPr~P over LacS was < 0.1.

The involvement of the LacS IIA domain in the regulation of the transport activity was also observed *in vivo* in *S. thermophilus*. HPLC analysis of sugar utilization by *S. thermophilus* expressing wildtype or mutant LacS protein is shown in Fig. 4. In *S. thermophilus* sucrose is taken up by a specific PTS, whereas lactose is transported in exchange for galactose by the secondary transport protein LacS. The experiments indicate that (i) lactose and sucrose are cometabolized; (ii) for every lactose molecule metabolized a galactose is excreted into the medium; (iii) mutation of the putative phosphorylation site (His-552) in the IIA domain or deletion of the entire IIA domain increases the rate of lactose utilization relative to that of the wildtype; (iv) loss of feedback control of lactose utilization results in a slowing down of sucrose metabolism.

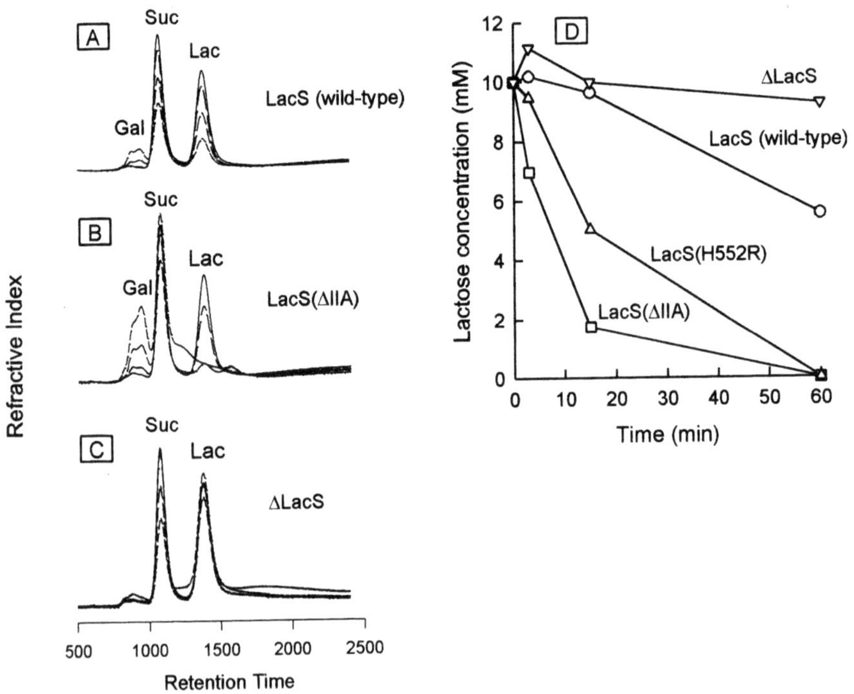

Fig. 4. Sugar consumption of *S. thermophilus* ST11(ΔlacS) expressing wild-type LacS (A), LacS(ΔIIA) (B), or no LacS protein (C). *S. thermophilus* cells were grown in Belliker broth in the presence of lactose (10 mM) plus sucrose (10 mM), the cells were washed and resuspended in 50 mM potassium phosphate, pH 7.0, plus 5 mM $MgSO_4$, and sucrose plus lactose were added to 10 mM. The utilization of sugars (and production of galactose) was analyzed by HPLC (Biorad column) for 0, 3, 15 and 60 min. The utilization of lactose by *S. thermophilus* ST11(ΔlacS) expressing wild-type LacS, LacS(H552R), LacS(ΔIIA) or no LacS protein, plotted as a function of time is shown in panel D.

These data confirm the *in vitro* data and suggest that lactose uptake is negatively regulated by phosphorylation of the IIA domain. Since PEP is formed not only from the metabolism of a PTS sugar but also from lactose and, consequently, HPr(His~P) will be present in both cases (EI and HPr are constitutively expressed), we propose that regulation of lactose transport by HPr(His~P) serves to prevent unbridled uptake of lactose (control of glycolysis) rather than to affect the hierarchy of sugar utilization as is the consequence of inducer exclusion. Such a mechanism allows lactose and sucrose to be cometabolized, which is indeed observed (Fig. 4). The decreased rate of sucrose utilization in the LacS(ΔIIA) [and LacS(H552R), not shown] mutants is most likely due to the increased utilization of lactose, which uses the same (glycolytic) pathway and therefore competes with sucrose utilization at the level of intermediary metabolism.

We also tested whether lactose transport is affected by HPr(Ser-P) by including HPr(S46D), mutant of HPr, which mimics HPr(Ser-P) (Reizer *et al.*, 1992), in the hybrid membrane system as described in Fig. 3. Up to a molar ratio of HPr(S46D) over LacS of 2.5, we observed no specific inhibition of lactose transport (Poolman and Reizer, unpublished).

Regulation of glycerol kinase

The phosphoryl transfer activity of LacS IIA domain was studied in *E. coli* LM1 *(crr/manA)* strains; these cells are defective in glucose transport due to an inactive IIA^{glc} and the lack of the mannose PTS. Expression of the IIA domain did not restore glucose uptake in this strain. Apparently, the IIA domain of LacS cannot transfer the phosphoryl group to the $IIBC^{glc}$ component of the glucose PTS. In another set of experiments we tested whether the IIA domain could regulate glycerol utilization by inhibiting glycerol kinase (Fig. 5). The IIA domain of LacS was expressed to a high level in *S. typhimurium* PP2178 (*crr*::Tn10 *nagE*). Upon addition of 2-deoxyglucose the uptake (utilization) of glycerol was inhibited, most likely because IIA^{LacS}~P is dephosphorylated through the redirection of phosphoryl groups to the PTS, i.e. HPr, for the uptake of 2-deoxyglucose, resulting in the interaction of IIA^{LacS} with glycerol kinase. The extent of inhibition of glycerol uptake was dependent on the actual uptake rate [amount of GlpK (Fig. 5D)], which is expected when IIA^{LacS} inhibits glycerol kinase by forming a (stoichiometric) complex as observed for IIA^{glc} and glycerol kinase (Hurley *et al.*, 1993; van der Vlag *et al.*, 1994).

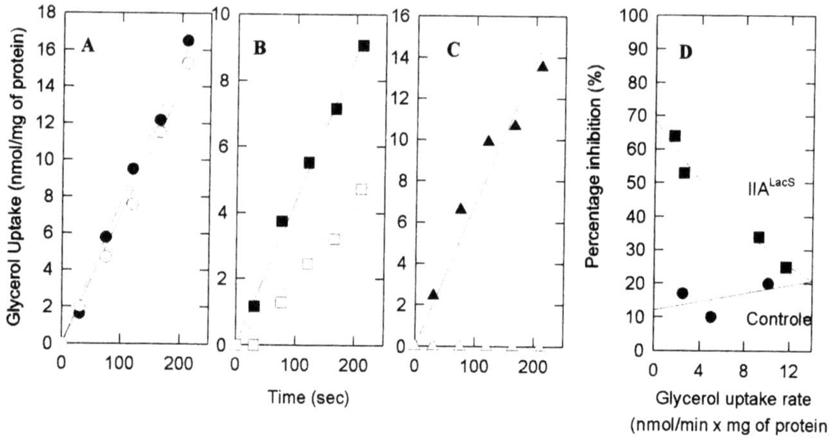

Fig. 5 Inhibition of glycerol kinase activity by IIA domain of LacS. *Salmonella typhimurium PP2178 (crr*::Tn10 *nag*E*)* was complemented *in trans* with plasmids containing genes specifying the IIA domain of LacS (B), IIA^{glc} (C) or no IIA^{glc}-like protein (A). The uptake of 0.5 mM [^{14}C]glycerol was measured in the presence (open symbols) and absence (filled symbols) of 10 mM 2-deoxyglucose.The inhibition of GlpK by IIA^{LacS} as a function of glycerol kinase activity in four independent experiments is shown in panel D.

In a previous study (van der Vlag *et al.*, 1995), we have analysed the activity of glycerol kinase in the presence of LacS, i.e., the wild type protein in which the IIA domain is fused to the membrane-bound carrier domain. No inhibition of glycerol kinase activity was observed, which could be due to the lower expression of the LacS protein in comparison to the IIA^{LacS} or to the inability of "membrane-bound" IIA^{LacS} to interact functionally with GlpK. It should be noted that membrane-bound IIA^{Nag}, as part of the $IIABC^{Nag}$ complex, is able to inhibit GlpK (van der Vlag *et al.*, 1995), indicating that inducer exclusion is not exclusively mediated by cytosolic IIA^{glc}.

Hurley *et al (*1993*)* identified the interactions between glycerol kinase and E. coli IIA^{glc}, which mainly involve hydrophobic and electrostatic interactions and the Zn(II) binding site (Fig. 2). Except for the histidines that coordinate the Zn-atom, the other positions are poorly conserved in the non-PTS IIA protein(s) domain(s) (Fig. 2). Since the LacS IIA domain of *S.*

thermophilus can inhibit glycerol kinase, we suggest that the two active-site histidines are important for the interaction.

Conclusion

In the present study, we show that the IIA domain of LacS is involved in the regulation of lactose transport. Inhibition of lactose/H^+ symport is PEP-dependent and enzyme I/HPr-mediated and serves to down regulate lactose metabolism (feedback control) under conditions of excess of carbon and energy sources. (Fig. 6, top). The mechanism is different from other PTS-mediated regulatory mechanisms in Gram-positive and Gram-negative bacteria, which, although mechanistically dissimilar, serve to reduce the cytoplasmic inducer levels under conditions that a rapidly metabolizable PTS sugar is present (inducer exclusion).

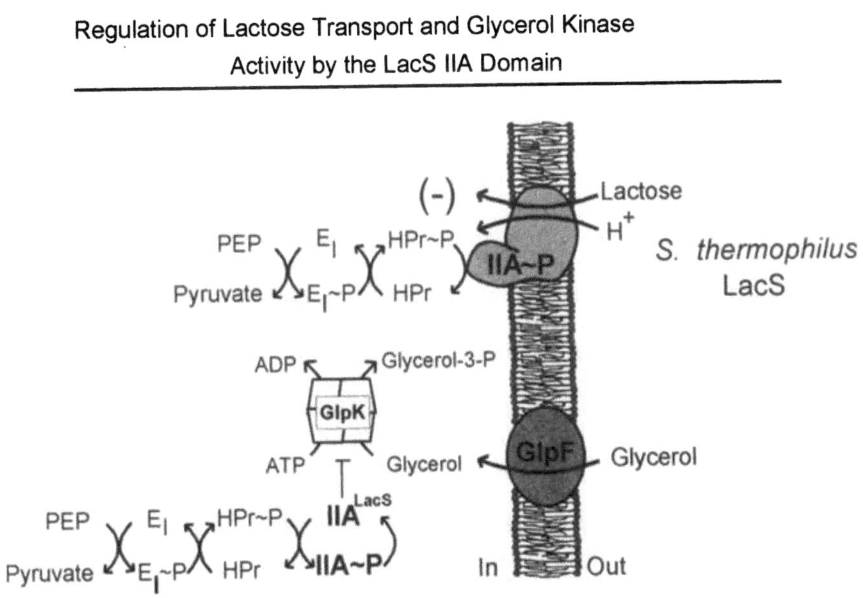

Fig. 6. Schematic representation of regulation of carbon utilization by PEP-dependent enzyme I/HPr-mediated phosphorylation of the LacS IIA domain. IIA-mediated regulation is indicated by the inhibition of lactose/H^+ symport of LacS (-)and glycerol phosphorylation of glycerol kinase (GlpK) (⊥).

References

Botsford JL and Harman JG (1992) Microbiol. Rev. 56: 100-122
Dörschug M, Frank R, Kalbitzer H R, Hengstenberg W and Deutscher J (1984) Eur. J. Biochem. 144: 113-119
Deutscher J and Engelmann R (1984) FEMS Microbiol Lett 23: 157-162
Deutscher J and Saier MH Jr (1983) Proc Natl Acad Sci USA 80, 6790-6794
Feese M D W, Pettigrew ND, Meadow S, Roseman S and Remington S J (1994) Proc. Natl. Acad. Sci. USA 91: 3544-3548
Foucaud C and Poolman B (1992) J. Biol. Chem. 267: 22087-22094
Hueck CJ and Hillen W (1995) Mol. Microbiol. 15: 395-401
Hurley JH, Faber HR, Wortylake D, Meadow ND, Roseman S, Pettigrew W and Remington S (1993) Science 259: 673-677
Knol J, Veenhoff L, Liang WJ, Henderson PJF, Leblanc G and Poolman B (1996) J.Biol. Chem. 271: 15358-15366
Poolman B, Royer TJ, Mainzer SE and Schmidt BF (1989) J. Bacteriol. 171: 244-253
Poolman B, (1990) Mol. Microbiol. 4: 1629-1636
Poolman B and Konings WN (1993) Biochim. Biophys. Acta 1183: 5-39
Poolman B, Knol J, Mollet B, Nieuwenhuis B and Sulter G (1995) Proc. Natl. Acad. Sci. USA 92: 778-782
Poolman B, Knol J, van der Does C, Henderson PJF, Liang W-J, Leblanc G, Pourcher T and Mus-Veteau I (1996) Mol. Microbiol. 19: 911-922
Postma PW, Lengeler J and Jacobsen GR (1993) Microbiol. Rev. 57: 543-594
Presper KA, Wong C-Y, Liu L, Meadow ND and Roseman S (1989) Proc. Natl. Acad. Sci. USA 86: 4052-4055.
Reizer J, Novotny MJ, Panos C and Saier MH Jr (1983) J. Bacteriol. 156: 354-361
Reizer J, Novotny MJ, Hengstenberg W and Saier MH, Jr (1984) J. Bacteriol 160: 333-340
Reizer J, Deutscher J, Sutrina S, Thompson J and Saier MH, Jr. (1985) Trends Biochem. Sci. 1: 32-35
Reizer J, Sutrina SL, Wu L.-F, Deutscher J, Reddy P and Saier MH, Jr (1992) J. Biol. Chem. 267: 9158-9169
Romano AH, Brino G, Peterkofsky A and Reizer J (1987) J. Bacteriol. 169: 5589-5596
Sutrina S, Reizer J and Saier MH, Jr (1988) J. Bacteriol. 170: 1874-1877
Titgemeyer F, Walkenhorst J, Cui X, Reizer J and Saier MH, Jr (1994) Res Microbiol. 145: 89-92
Van der Vlag J, Postma PW and van Dam K (1994) J. Bacteriol. 176: 3518-3526
Van der Vlag J and Postma PW (1995) Mol. Gen. Genet. 248: 236-241
Ye J-J, Reizer J, Cui X and Saier MH, Jr (1994) Proc Natl .Acad. Sci. USA 91: 3102-3106.
Ye J-J and Saier MJ, Jr (1995a) Proc Natl Acad Sci USA 92: 417-421
Ye J-J and Saier MJ, Jr (1995b) J. Bacteriol. 177:1900-1902
Ye J-J and Saier MJ, Jr (1995c) J. Biol. Chem. 270:15740-16744

INDEX

NATO ASI Series H

Vol. 1: **Biology and Molecular Biology of Plant-Pathogen Interactions.** Edited by J.A. Bailey. 415 pages. 1986.

Vol. 2: **Glial-Neuronal Communication in Development and Regeneration.** Edited by H.H. Althaus and W. Seifert. 865 pages. 1987.

Vol. 3: **Nicotinic Acetylcholine Receptor: Structure and Function.** Edited by A. Maelicke. 489 pages. 1986.

Vol. 4: **Recognition in Microbe-Plant Symbiotic and Pathogenic Interactions.** Edited by B. Lugtenberg. 449 pages. 1986.

Vol. 5: **Mesenchymal-Epithelial Interactions in Neural Development.** Edited by J. R. Wolff, J. Sievers, and M. Berry. 428 pages. 1987.

Vol. 6: **Molecular Mechanisms of Desensitization to Signal Molecules.** Edited by T M. Konijn, P J. M. Van Haastert, H. Van der Starre, H. Van der Wel, and M.D. Houslay. 336 pages. 1987.

Vol. 7: **Gangliosides and Modulation of Neuronal Functions.** Edited by H. Rahmann. 647 pages. 1987.

Vol. 8: **Molecular and Cellular Aspects of Erythropoietin and Erythropoiesis.** Edited by I.N. Rich. 460 pages. 1987.

Vol. 9: **Modification of Cell to Cell Signals During Normal and Pathological Aging.** Edited by S. Govoni and F. Battaini. 297 pages. 1987.

Vol. 10: **Plant Hormone Receptors.** Edited by D. Klämbt. 319 pages. 1987.

Vol. 11: **Host-Parasite Cellular and Molecular Interactions in Protozoal Infections.** Edited by K.-P. Chang and D. Snary. 425 pages. 1987.

Vol. 12: **The Cell Surface in Signal Transduction.** Edited by E. Wagner, H. Greppin, and B. Millet. 243 pages. 1987.

Vol. 13: **Toxicology of Pesticides: Experimental, Clinical and Regulatory Perspectives.** Edited by L.G. Costa, C.L. Galli, and S.D. Murphy. 320 pages. 1987.

Vol. 14: **Genetics of Translation. New Approaches.** Edited by M.F. Tuite, M. Picard, and M. Bolotin-Fukuhara. 524 pages. 1988.

Vol. 15: **Photosensitisation. Molecular, Cellular and Medical Aspects.** Edited by G. Moreno, R. H. Pottier, and T. G. Truscott. 521 pages. 1988.

Vol. 16: **Membrane Biogenesis.** Edited by J.A.F Op den Kamp. 477 pages. 1988.

Vol. 17: **Cell to Cell Signals in Plant, Animal and Microbial Symbiosis.** Edited by S. Scannerini, D. Smith, P. Bonfante-Fasolo, and V. Gianinazzi-Pearson. 414 pages. 1988.

Vol. 18: **Plant Cell Biotechnology.** Edited by M.S.S. Pais, F. Mavituna, and J. M. Novais. 500 pages. 1988.

NATO ASI Series H

Vol. 19: **Modulation of Synaptic Transmission and Plasticity in Nervous Sys tems.**
Edited by G. Hertting and H.-C. Spatz. 457 pages. 1988.

Vol. 20: **Amino Acid Availability and Brain Function in Health and Disease.**
Edited by G. Huether. 487 pages. 1988.

Vol. 21: **Cellular and Molecular Basis of Synaptic Transmission.**
Edited by H. Zimmermann. 547 pages. 1988.

Vol. 22: **Neural Development and Regeneration. Cellular and Molecular Aspects.**
Edited by A. Gorio, J. R. Perez-Polo, J. de Vellis, and B. Haber. 711 pages. 1988.

Vol. 23: **The Semiotics of Cellular Communication in the Immune System.**
Edited by E.E. Sercarz, F. Celada, N.A. Mitchison, and T. Tada. 326 pages. 1988.

Vol. 24: **Bacteria, Complement and the Phagocytic Cell.**
Edited by F. C. Cabello und C. Pruzzo. 372 pages. 1988.

Vol. 25: **Nicotinic Acetylcholine Receptors in the Nervous System.**
Edited by F. Clementi, C. Gotti, and E. Sher. 424 pages. 1988.

Vol. 26: **Cell to Cell Signals in Mammalian Development.**
Edited by S.W. de Laat, J.G. Bluemink, and C.L. Mummery. 322 pages. 1989.

Vol. 27: **Phytotoxins and Plant Pathogenesis.**
Edited by A. Graniti, R. D. Durbin, and A. Ballio. 508 pages. 1989.

Vol. 28: **Vascular Wilt Diseases of Plants. Basic Studies and Control.**
Edited by E. C. Tjamos and C. H. Beckman. 590 pages. 1989.

Vol. 29: **Receptors, Membrane Transport and Signal Transduction.**
Edited by A. E. Evangelopoulos, J. P. Changeux, L. Packer, T. G. Sotiroudis, and K.W.A. Wirtz. 387 pages. 1989.

Vol. 30: **Effects of Mineral Dusts on Cells.**
Edited by B.T. Mossman and R.O. Begin. 470 pages. 1989.

Vol. 31: **Neurobiology of the Inner Retina.**
Edited by R. Weiler and N.N. Osborne. 529 pages. 1989.

Vol. 32: **Molecular Biology of Neuroreceptors and Ion Channels.**
Edited by A. Maelicke. 675 pages. 1989.

Vol. 33: **Regulatory Mechanisms of Neuron to Vessel Communication in Brain.**
Edited by F. Battaini, S. Govoni, M.S. Magnoni, and M. Trabucchi. 416 pages. 1989.

Vol. 34: **Vectors asTools for the Study of Normal and Abnormal Growth and Differentiation.**
Edited by H. Lother, R. Dernick, and W. Ostertag. 477 pages. 1989.

Vol. 35: **Cell Separation in Plants: Physiology, Biochemistry and Molecular Biology.** Edited by D. J. Osborne and M. B. Jackson. 449 pages. 1989.

NATO ASI Series H

Vol. 36: **Signal Molecules in Plants and Plant-Microbe Interactions.** Edited by B.J.J. Lugtenberg. 425 pages. 1989.

Vol. 37: **Tin-Based Antitumour Drugs.** Edited by M. Gielen. 226 pages. 1990.

Vol. 38: **The Molecular Biology of Autoimmune Disease.** Edited by A.G. Demaine, J-P. Banga, and A.M. McGregor. 404 pages. 1990.

Vol. 39: **Chemosensory Information Processing.** Edited by D. Schild. 403 pages. 1990.

Vol. 40: **Dynamics and Biogenesis of Membranes.** Edited by J. A. F. Op den Kamp. 367 pages. 1990.

Vol. 41: **Recognition and Response in Plant-Virus Interactions.** Edited by R. S. S. Fraser. 467 pages. 1990.

Vol. 42: **Biomechanics of Active Movement and Deformation of Cells.** Edited by N. Akkas. 524 pages. 1990.

Vol. 43: **Cellular and Molecular Biology of Myelination.** Edited by G. Jeserich, H. H. Althaus, and T. V. Waehneldt. 565 pages. 1990.

Vol. 44: **Activation and Desensitization of Transducing Pathways.** Edited by T. M. Konijn, M. D. Houslay, and P. J. M. Van Haastert. 336 pages. 1990.

Vol. 45: **Mechanism of Fertilization: Plants to Humans.** Edited by B. Dale. 710 pages. 1990.

Vol .46: **Parallels in Cell to Cell Junctions in Plants and Animals.** Edited by A. W Robards, W. J . Lucas, J . D. Pitts, H . J . Jongsma, and D. C. Spray. 296 pages. 1990.

Vol. 47: **Signal Perception and Transduction in Higher Plants.** Edited by R. Ranjeva and A. M. Boudet. 357 pages. 1990.

Vol. 48: **Calcium Transport and Intracellular Calcium Homeostasis.** Edited by D. Pansu and F. Bronner. 456 pages. 1990.

Vol. 49: **Post-Transcriptional Control of Gene Expression.** Edited by J. E. G. McCarthy and M. F. Tuite. 671 pages. 1990.

Vol. 50: **Phytochrome Properties and Biological Action.** Edited by B. Thomas and C. B. Johnson. 337 pages. 1991.

Vol. 51: **Cell to Cell Signals in Plants and Animals.** Edited by V. Neuhoff and J. Friend. 404 pages. 1991.

Vol. 52: **Biological Signal Transduction.** Edited by E. M . Ross and K . W. A. Wirtz. 560 pages. 1991.

Vol. 53: **Fungal Cell Wall and Immune Response.** Edited by J. P. Latge and D. Boucias. 472 pages. 1991.

NATO ASI Series H

Vol. 54: **The Early Effects of Radiation on DNA.**
Edited by E. M. Fielden and P. O'Neill. 448 pages. 1991.

Vol. 55: **The Translational Apparatus of Photosynthetic Organelles.**
Edited by R. Mache, E. Stutz, and A. R. Subramanian. 260 pages. 1991.

Vol. 56: **Cellular Regulation by Protein Phosphorylation.**
Edited by L. M. G. Heilmeyer, Jr. 520 pages. 1991.

Vol. 57: **Molecular Techniques in Taxonomy.**
Edited by G . M . Hewitt, A. W. B. Johnston, and J. P. W. Young .
420 pages. 1991.

Vol. 58: **Neurocytochemical Methods.**
Edited by A. Calas and D. Eugene. 352 pages. 1991.

Vol. 59: **Molecular Evolution of the Major Histocompatibility Complex.**
Edited by J. Klein and D. Klein. 522 pages. 1991.

Vol. 60: **Intracellular Regulation of Ion Channels.**
Edited by M. Morad and Z. Agus. 261 pages. 1992.

Vol. 61: **Prader-Willi Syndrome and Other Chromosome 15q Deletion Disorders.**
Edited by S. B. Cassidy. 277 pages. 1992.

Vol. 62: **Endocytosis. From Cell Biology to Health, Disease and Therapie.**
Edited by P. J. Courtoy. 547 pages. 1992.

Vol. 63: **Dynamics of Membrane Assembly.**
Edited by J. A. F. Op den Kamp. 402 pages. 1992.

Vol. 64: **Mechanics of Swelling. From Clays to Living Cells and Tissues.**
Edited by T. K. Karalis. 802 pages. 1992.

Vol. 65: **Bacteriocins, Microcins and Lantibiotics.**
Edited by R. James, C. Lazdunski, and F. Pattus. 530 pages. 1992.

Vol. 66: **Theoretical and Experimental Insights into Immunology.**
Edited by A. S. Perelson and G. Weisbuch. 497 pages. 1992.

Vol. 67: **Flow Cytometry. New Developments.**
Edited by A. Jacquemin-Sablon. 1993.

Vol. 68: **Biomarkers. Research and Application in the Assessment of Environmental Health.** Edited by D. B. Peakall and L. R. Shugart.
138 pages. 1993.

Vol. 69: **Molecular Biology and its Application to Medical Mycology.**
Edited by B. Maresca, G. S. Kobayashi, and H. Yamaguchi. 271 pages.
1993.

Vol. 70: **Phospholipids and Signal Transmission.**
Edited by R. Massarelli, L. A. Horrocks, J. N. Kanfer, and K. Löffelholz.
448 pages. 1993.

NATO ASI Series H

Vol. 71: **Protein Synthesis and Targeting in Yeast.**
Edited by A. J. P. Brown, M. F. Tuite, and J. E. G. McCarthy. 425 pages. 1993.

Vol. 72: **Chromosome Segregation and Aneuploidy.**
Edited by B. K. Vig. 425 pages. 1993.

Vol. 73: **Human Apolipoprotein Mutants III. In Diagnosis and Treatment.**
Edited by C. R. Sirtori, G. Franceschini, B. H. Brewer Jr. 302 pages. 1993.

Vol. 74: **Molecular Mechanisms of Membrane Traffic.**
Edited by D. J. Morré, K. E. Howell, and J. J. M. Bergeron. 429 pages. 1993.

Vol. 75: **Cancer Therapy. Differentiation, Immunomodulation and Angiogenesis.**
Edited by N. D'Alessandro, E. Mihich, L. Rausa, H. Tapiero, and T. R. Tritton. 299 pages. 1993.

Vol. 76: **Tyrosine Phosphorylation/Dephosphorylation and Downstream Signalling.**
Edited by L. M. G. Heilmeyer Jr. 388 pages. 1993.

Vol. 77: **Ataxia-Telangiectasia.** Edited by R. A. Gatti, R. B. Painter. 306 pages. 1993.

Vol. 78: **Toxoplasmosis.** Edited by J. E. Smith. 272 pages. 1993.

Vol. 79: **Cellular Mechanisms of Sensory Processing. The Somatosensory System.**
Edited by L. Urban. 514 pages. 1994.

Vol. 80: **Autoimmunity: Experimental Aspects.**
Edited by M. Zouali. 318 pages. 1994.

Vol. 81: **Plant Molecular Biology. Molecular Genetic Analysis of Plant Development and Metabolism.**
Edited by G. Coruzzi, P. Puigdomènech. 579 pages. 1994.

Vol. 82: **Biological Membranes: Structure, Biogenesis and Dynamics.**
Edited by Jos A. F. Op den Kamp. 367 pages. 1994.

Vol. 83: **Molecular Biology of Mitochondrial Transport Systems.**
Edited by M. Forte, M. Colombini. 420 pages. 1994.

Vol. 84: **Biomechanics of Active Movement and Division of Cells.**
Edited by N. Akkaş. 587 pages. 1994.

Vol. 85: **Cellular and Molecular Effects of Mineral and Synthetic Dusts and Fibres.**
Edited by J. M. G. Davis, M.-C. Jaurand. 448 pages. 1994.

Vol. 86: **Biochemical and Cellular Mechanisms of Stress Tolerance in Plants.**
Edited by J. H. Cherry. 616 pages. 1994.

NATO ASI Series H

Springer and the environment

At Springer we firmly believe that an international science publisher has a special obligation to the environment, and our corporate policies consistently reflect this conviction.
We also expect our business partners – paper mills, printers, packaging manufacturers, etc. – to commit themselves to using materials and production processes that do not harm the environment. The paper in this book is made from low- or no-chlorine pulp and is acid free, in conformance with international standards for paper permanency.

MIX
Papier aus verantwortungsvollen Quellen
Paper from responsible sources
FSC® C105338

If you have any concerns about our products,
you can contact us on
ProductSafety@springernature.com

In case Publisher is established outside the EU,
the EU authorized representative is:
Springer Nature Customer Service Center GmbH
Europaplatz 3, 69115 Heidelberg, Germany

Printed by Libri Plureos GmbH
in Hamburg, Germany